S

An Interactive Environment for Data Analysis and Graphics

The Wadsworth Statistics/Probability Series

Series Editors

Peter J. Bickel, *University of California*
William S. Cleveland, *Bell Laboratories*
Richard M. Dudley, *Massachusetts Institute of Technology*

Becker, R.A.; and Chambers, J.M., *S: An Interactive Environment for Data Analysis and Graphics*

Bickel, P.; Doksum, K.; and Hodges, J. L., Jr., *Festschrift for Erich L. Lehmann*

Breiman, L.; Friedman, J. H.; Olshen, R. A.; and Stone, C. J., *Classification and Regression Trees*

Chambers, J.; Cleveland, W.; Kleiner, B.; and Tukey, P., *Graphical Methods for Data Analysis*

Graybill, F., *Matrices with Applications in Statistics,* Second Edition

S

An Interactive Environment for Data Analysis and Graphics

Richard A. Becker
John M. Chambers

WADSWORTH ADVANCED BOOK PROGRAM
Belmont, California
A Division of Wadsworth, Inc.

ISBN 0-534-03313-X

This book was typeset in Palatino by the authors, using a photo-typesetter driven by a VAX-11/750 running the 8th Edition of the UNIX operating system.

UNIX is a trademark of Bell Laboratories. VAX is a trademark of Digital Equipment Corporation.

Printed in the United States of America
    4 5 6 7 8 9 10—88 87 86 85

Library of Congress Cataloging in Publication Data

Becker, Richard A.
    S : an interactive environment for data analysis and graphics.

    (The Wadsworth statistics/probability series)
    Includes bibliographical references and index.
    1. S (Computer system) 2. Interactive computer systems. 3. Statistics--Data processing. 4. Mathematical statistics--Data processing. I. Chambers, John M. II. Title. III. Title: Interactive environment for data analysis and graphics. IV. Series.
QA276.4.B43   1984   519.5'0285–425   84–3654
ISBN 0-534-03313-X

To Eloise and Bea

Preface

S is an interactive environment for data analysis and graphics, with two primary components: a language and a support system. The S *language* is a very high-level language for specifying computations. S is also a *system* in that it provides a total environment for the user, including data management, documentation, and graphics.

The primary goal of the S environment is *GOOD DATA ANALYSIS*. The facilities in S are directed toward this goal. S encourages the iterative, interactive style of data analysis which leads to understanding. In this way, S is quite unlike most statistical "packages".

S provides the user with interactive computation, both simple and complex, graphical displays on a wide variety of graphics devices, data management and structuring.

S is easy to use, yet it is also powerful. A wide range of people are presently using S in diverse areas—financial analysis, statistics research, management, academia—for analytical computing, graphics, and data analysis. S was created, on the other hand, in a statistics research organization; as a result, it contains many recently developed statistical routines, along with facilities that allow users to extend the system to new methods and new application areas.

How to Read This Book

This book describes the S language and system for interactive data analysis. Because of the diversity of S use, readers may want to follow different paths through the book. We will provide some guidelines in this preface. Whatever path you choose, we recommend combining reading with doing. Have a computer terminal handy, preferably with graphics, and start S running. Read a little, then try out some of the examples or problems, and experiment with variations or with your own data. A good way to review or supplement the discussion of a particular S function, say **plot**, in the book, is to look at the on-line documentation by typing

  help("plot")

In particular, whenever the book suggests looking up something in Appendix 1, you can use the on-line documentation instead.

The book begins with three introductory chapters that should be helpful to all users: a case study that shows S without trying to explain it; a tutorial chapter covering material in an informal way; and a more formal treatment of the basic techniques in the language. Chapters 4 through 8 deal with more advanced or specialized material. Readers may choose to read these according to their needs. Chapter 4 gives a detailed account of graphics within S, ranging from high-level plotting commands to detailed control of plots. The emphasis is on analytical displays: those that are valuable during data analysis. Chapter 5 provides advanced material and Chapter 6 describes the S Macro Processor. These chapters will be essential for anyone building specialized software using S. Chapter 7 covers computations in S for statistical methods and studies; it will be a good place for statisticians to see how S can be used in their research and data analysis. Chapter 8 shows how S can be used for presenting results in the form of reports and graphs. This material involves traditional graphical techniques such as pie charts and bar graphs, as well as publication quality reports.

This ends the narrative material, but is just mid-way through the book. The remainder is devoted to detailed documentation of all of the S functions, macros, and datasets. Documentation for S functions and macros is presented in Appendix 1. Appendix 2 describes the datasets that are available in the system data directory. The material in Appendices 1 and 2 is also available on-line for the S user. Appendix 3 provides one-line descriptions of S functions and macros, broken down by various topics. It can help you to find functions when you can't remember their names. An extensive index at the end of the book should be used for quick access to the places in the text where specific topics are discussed.

Hardware and Software

S is a software system that runs under Bell Laboratories UNIX† operating system on a variety of hardware configurations. S is

---

† UNIX is a trademark of Bell Laboratories.

designed to take advantage of other UNIX tools and languages; see, for example, the interaction with word-processing facilities in Chapter 8. As this book is being written, S runs primarily on VAX computers made by the Digital Equipment Corporation, and on 3B computers manufactured by AT&T. Work is under way to implement S on new hardware, including several varieties of personal workstations running the UNIX system.

For information on obtaining S, in the United States, contact:

AT&T Technologies Software Sales
P.O. Box 25000
Greensboro, North Carolina 27420
(800) 828-UNIX

and, elsewhere:

AT&T International
PO Box 7000B
Basking Ridge, NJ 07920
(201) 953-7581

You will need to specify the model of computer on which S is to be run and the particular UNIX operating system running there.

Statistical Methods

Much of S can be used to analyze data without detailed knowledge of statistical theory. However, a number of the functions are based on formal methods, both classical and relatively recent. In this book and in the S on-line documentation, we try to provide references to papers or books describing such methods. (Including a full statistical presentation would expand the book to enormous proportions, of course, even if we felt competent to describe all the methods.)

At the same time, in spite of the large number of S functions, many statistical techniques are *not* explicitly provided. Because S is designed to be general and extensible, many of these techniques can be implemented simply as S expressions. For the statistically trained user, studying such expressions is a good way to get a feel for the style of the language. In Chapter 7 we present a number of examples of statistical techniques implemented in S.

Bibliography

The following books provide material that is suitable as a background to using S. For each reference, we give a brief description of some of the material covered and the level of sophistication. In addition, specific references are given in Appendix 1 for certain S functions. For users who are new to the UNIX operating system, we particularly recommend the introductory material in the last two books.

DATA ANALYSIS

Paul F. Velleman, and David C. Hoaglin, *Applications, Basics, and Computing of Exploratory Data Analysis*, Duxbury Press, Boston, Massachusetts, 1981.

>   Stem-and-Leaf Displays; Boxplots; Scatter Plots; Resistant Line Fitting; Smoothing; Median Polish.

>   A good introduction. Includes BASIC and FORTRAN programs which are not needed by the S user.

John W. Tukey, *Exploratory Data Analysis*, Addison-Wesley, Reading, Massachusetts, 1977.

>   Stem-and-Leaf Displays; Reexpression; Boxplots; Comparisons; Scatter Plots; Fitting Straight Lines; Smoothing; Median Polish; Two-way Fits; Counts; Shapes of Distributions.

>   The "Bible" of EDA; probably the best place to read about the philosophy of data analysis. Numerous examples.

David C. Hoaglin, Frederick Mosteller, and John W. Tukey, *Understanding Robust and Exploratory Data Analysis*, John Wiley & Sons, New York, 1983.

>   Stem-and-Leaf Displays; Boxplots; Transformations; Resistant Lines; Median Polish; Examining Residuals; Robust Estimators of Location.

>   An advanced book providing the theory underlying EDA. A good reference for those familiar with classical statistical methods.

REGRESSION

Samprit Chatterjee and Bertram Price, *Regression Analysis by Example,* John Wiley & Sons, New York, 1977.

> Simple Linear Regression; Violation of Model Assumptions; Multiple Regression; Correlated Errors; Collinear Data; Biased Estimation; Selection of Variables.

> A basic introduction to classical regression analysis.

Frederick Mosteller and John W. Tukey, *Data Analysis and Regression, a Second Course in Statistics,* Addison-Wesley, Reading, Massachusetts, 1977.

> Approaching Data Analysis; Displays and Summaries for Batches; Reexpression; Jackknife and Cross Validation; Two Way Tables; Robust and Resistant Measures; Regression for Fitting; Examining Regression Residuals.

> An intermediate level text about data analysis and regression. Much emphasis is on attitudes and approaches to situations involving real data. The emphasis is on understanding the value and limitations of techniques.

Norman Draper and Harry Smith, *Applied Regression Analysis,* Second Edition, John Wiley & Sons, New York, 1981.

> Least Squares; Matrix Approach to Linear Regression; Examination of Residuals; Selecting the Best Regression Equation; Multiple Regression Applied to Analysis of Variance Problems; Nonlinear Estimation.

> Intermediate text emphasizing applications, describing many useful techniques.

Cuthbert Daniel and Fred S. Wood, *Fitting Equations to Data,* Second Edition, John Wiley & Sons, New York, 1980.

> Assumptions and Fitting Methods; One Independent Variable; Two or More Independent Variables; Selection of Independent Variables; Nonlinear Least Squares.

> Good examples of thorough applied regression analysis at an intermediate level, although presented with old-fashioned graphical and computing techniques.

David A. Belsley, Edwin Kuh, and Roy E. Welsch, *Regression Diagnostics: Identifying Influential Data and Sources of Collinearity*, John Wiley & Sons, New York, 1980.

> Detecting Influential Observations and Outliers; Detecting and Assessing Collinearity; Applications and Remedies.

> Advanced look at ways of detecting and understanding the problems caused by influential data, outliers, and collinearity.

STATISTICAL COMPUTING

John M. Chambers, *Computational Methods for Data Analysis*, John Wiley & Sons, New York, 1977.

> Programming; Data Management and Manipulation; Numerical Computations; Linear Models; Nonlinear Models; Simulation; Computational Graphics.

> Intermediate level book which describes modern methods of statistical computing. Concise descriptions of many of the algorithms used in S.

MULTIVARIATE ANALYSIS

R. Gnanadesikan, *Methods for Statistical Data Analysis of Multivariate Observations*, John Wiley & Sons, New York, 1977.

> Reduction of Dimensionality; Multivariate Dependencies; Classification and Clustering; Summarization and Exposure.

> An intermediate text on multivariate data analysis, with emphasis on methodology; includes graphical techniques.

K. V. Mardia, J. Kent, and J. M. Bibby, *Multivariate Analysis*, Academic, 1979.

> Multivariate Regression; Econometrics; Principle Component Analysis; Factor Analysis; Canonical Correlation Analysis; Discriminant Analysis; Multivariate Analysis of Variance; Cluster Analysis; Multidimensional Scaling; Directional Data.

> Advanced combination of theoretical and data-analytic methods.

GRAPHICS

John M. Chambers, William S. Cleveland, Beat Kleiner, and Paul A. Tukey, *Graphical Methods for Data Analysis*, Wadsworth International Group, Belmont, California, 1983.

> Distribution of a Set of Data; Comparing Distributions; Two-Dimensional Data; Plots of Multivariate Data; Assessing Distributional Assumptions; Assessing Regression Models.

> Introductory text describing statistical graphics with emphasis on simple, practical methods; concentrates on graphical displays for analysis rather than presentation.

THE UNIX OPERATING SYSTEM

Henry McGilton and Rachel Morgan *Introducing the UNIX System*, McGraw-Hill, 1983.

> Getting Started on the UNIX System; Directories and Files; Commands and Standard Files; Text Manipulation; The Ed and Sed Editors; The EX and VI Editors; Programming the UNIX shell.

> Introductory text describing the UNIX operating system. Particularly useful introduction to text editing and good hints about commonly encountered problems.

B. W. Kernighan and Rob Pike, *The UNIX Programming Environment*, Prentice-Hall, 1984.

> UNIX for Beginners; The File System; Using the Shell; Filters; Shell Programming; Programming with Standard I/O; UNIX System Calls; Program Development.

> A more advanced text, providing important information and "tricks" for those who would like to create new packages or systems. Recommended to anyone who wants to build on S for local use.

Acknowledgments

In writing this book, and generally in our research on S, we owe pehaps our largest debt to our users, who have provided hundreds of comments, questions and complaints. Comments on the draft of this book from a number of intrepid readers have been very helpful: we particularly thank W. W. Everett, T. R. Hamilton, W. J. Hery,

Ross Ihaka, Colin Mallows, Daryl Pregibon, Judy Schilling, W. J. Shugard, Don Swartwout, Paul Tukey, and Allan Wilks. The subject index in Appendix 3 drew on work done by John Kitchin and Gary Perlman. Marylyn McGill's excellent work on the figures contributed greatly to the book.

Contents

S

An Interactive Environment for Data Analysis and Graphics

1

How to Beat the Lottery

One of the best ways of getting acquainted with statistical software like S is to use it to analyze some data. Let's look at a situation where you might be motivated to perform data analysis.

1.1 Data Analysis Using S

The lottery is a common feature of modern life. Lotteries range from the Irish Sweepstakes, with its yearly large drawings and enormous payoffs, to daily numbers games run by state governments (as well as illegal games run by bookies).

You might wonder what lotteries have to do with data analysis. There are several answers. First, there is the traditional association between probability theory and gambling—the foundations of statistics go back to games of chance. Lotteries raise many interesting questions. In fact, data analysis may be the only practical way of answering questions such as "Is the lottery fair?" A second reason is that the ubiquity of gambling and lotteries has acquainted most everyone with the basic concepts involved. A third reason is that a scientific look at lottery data may provide some of you with answers to an important question: "Should I play, and if so, how should I play?"

1.2 New Jersey Pick-It Lottery Data

The specific data we will look at concerns the New Jersey Pick-It Lottery, a daily numbers game run by the state of New Jersey to aid education and institutions. Our data is for 254 drawings just after the lottery was started, from May, 1975 to March, 1976. Pick-It is a parimutuel game, meaning that the winners share a fraction of the money taken in for the particular drawing. Each ticket costs fifty cents; the player picks a three-digit number ranging from 000 to 999. Half of the money bet during the day is placed in a prize pool (the state takes the other half) and anyone who picked the winning number shares equally in the pool.

　　The data available from the NJ Lottery Commission gives for each drawing the winning number and the payoff for a winning ticket. The winning numbers are:*

> **lottery.number　# print the winning numbers**

	810	156	140	542	507	972	431
[8]	981	865	499	20	123	356	15
[15]	11	160	507	779	286	268	698
[22]	640	136	854	69	199	413	192
[29]	602	987	112	245	174	913	828
[36]	539	434	357	178	198	406	79
[43]	34	89	257	662	524	809	527
[50]	257	8	446	440	781	615	231
[57]	580	987	391	267	808	258	479
[64]	516	964	742	537	275	112	230
[71]	310	335	238	294	854	309	26
[78]	960	200	604	841	659	735	105
[85]	254	117	751	781	937	20	348
[92]	653	410	468	77	921	314	683
[99]	0	963	122	18	827	661	918
[106]	110	767	761	305	485	8	808
[113]	648	508	684	879	67	282	928
[120]	733	518	441	661	219	310	771
[127]	906	235	396	223	695	499	42
[134]	230	623	300	380	646	553	182
[141]	158	744	894	689	978	314	337
[148]	226	106	299	947	896	863	239
[155]	180	764	849	87	975	92	701
[162]	402	1	884	750	236	395	999
[169]	744	714	253	711	863	496	214
[176]	430	107	781	954	941	416	243
[183]	480	111	47	691	616	253	477
[190]	11	114	133	293	812	197	358
[197]	7	996	842	255	374	693	383
[204]	99	474	333	467	515	357	694

---

* At this point, don't worry about understanding the boldface S expression **lottery.number**. The text following the # is a comment.

```
[211]      919     424     274     913     919     245     964
[218]      472     935     434     170     300     476     528
[225]      403     677     559     187     652     319     582
[232]      541      16     981     158     945      72     167
[239]       77     185     209     893     346     515     555
[246]      858     434     541     411     109     761     767
[253]      597     479
```

The corresponding payoffs are:

> **lottery.payoff   # print the payoffs**

```
            190.0   120.5   285.5   184.0   384.5   324.5
[   7]     114.0   506.5   290.0   869.5   668.5    83.0
[  13]     188.0   449.0   289.5   212.0   466.0   548.5
[  19]     260.0   300.5   556.5   371.5   112.5   254.5
[  25]     368.0   510.0   102.0   206.5   261.5   361.0
[  31]     167.5   187.0   146.5   205.0   348.5   283.5
[  37]     447.0   102.5   219.0   292.5   343.0   332.5
[  43]     532.5   445.5   127.0   557.5   203.5   373.5
[  49]     142.0   230.5   482.5   512.5   330.0   273.0
[  55]     171.0   178.0   463.5   476.0   290.0   176.0
[  61]     195.0   159.5   296.0   177.5   406.0   182.0
[  67]     164.5   137.0   191.0   298.0   110.0   353.0
[  73]     192.5   308.5   287.0   203.5   377.5   211.5
[  79]     342.0   259.0   231.0   348.0   159.0   130.5
[  85]     176.0   128.5   159.0   290.0   335.0   514.0
[  91]     191.0   304.5   167.0   257.0   640.0   142.0
[  97]     146.0   356.0    96.0   295.0   237.0   312.5
[ 103]     215.0   442.5   127.0   127.0   756.0   228.5
[ 109]     132.0   256.0   374.5   262.5   286.5   264.0
[ 115]     380.5   357.5   478.5   511.5   218.0   353.0
[ 121]     162.5   184.0   548.0   166.5   147.5   240.0
[ 127]     386.0   130.5   287.5   230.0   480.5   247.5
[ 133]     380.0   238.5   237.5   214.5   394.5   416.5
[ 139]     392.5   244.5   202.0   371.5   553.0   293.5
[ 145]     295.0   178.0   334.5   226.0   194.0   388.5
[ 151]     353.0   404.0   348.0   163.5   216.5   283.0
[ 157]     388.5   567.5   250.5   478.0   267.5   326.5
[ 163]     369.0   512.5   341.0   188.5   386.0   239.0
[ 169]     480.5   105.0   227.0   130.5   384.5   294.5
[ 175]     154.0   324.0   116.0   229.0   301.5   334.0
[ 181]     143.5   212.0   448.0   126.5   417.5   276.5
[ 187]     303.0   211.0   373.0   209.5   207.5   195.0
[ 193]     317.0   170.5   230.0   143.0   361.0   452.0
[ 199]     260.5   308.5   206.0   256.5   291.0   421.5
[ 205]     295.5   119.5   268.5   221.0   151.5   314.5
[ 211]     313.5   323.5   204.0   241.0   637.0   214.0
```

[217]	348.0	191.5	384.0	220.0	285.5	335.0
[223]	251.5	131.5	328.0	392.0	509.0	235.5
[229]	249.5	129.5	303.0	201.5	365.0	346.5
[235]	210.5	334.0	376.5	215.5	312.0	239.5
[241]	221.0	388.0	154.5	268.5	127.0	537.5
[247]	427.5	272.0	197.0	167.5	292.0	170.0
[253]	486.5	262.0				

Thus, for the first drawing, the winning number was 810 and it paid $190.00. Streams of numbers like this are both difficult to use and boring. One of the best ways to understand the data is to look at it graphically. Before doing any plots, however, we should think of the questions we might want to ask of the data. For example, there have been notorious cases of fraud in lotteries (see Figure 1).

6 Named in Rigged $1 Million Pennsylvania Lottery

HARRISBURG, Pa., Sept. 19 — The television announcer who conducts Pennsylvania's daily lottery, a lottery official and four other persons conspired to rig a lottery drawing and fraudulently won more than $1 million, a grand jury said today.

The grand jury said the six would have won even more money through the illegal numbers game in Pittsburgh, which uses the same three-digit number as the legal game, had the bookmakers who accepted their bets not refused to pay off.

The bookmakers refused because the heavy betting on the group's number, 666, led them to suspect that the April 24 drawing had been rigged. It was the bookmakers' suspicions that led to the investigation.

The grand jury recommended the indictment of the announcer, Nick Perry; four of his friends and Edward Plevel, the Bureau of State Lotteries official in charge of security that day, on charges of criminal solicitation, criminal conspiracy, theft by deception and rigging a publicly exhibited contest. It recommended that Mr. Perry and Mr. Plevel also be charged with perjury.

Under Pennsylvania law, the Attorney General makes specific, formal charges based on grand jury presentments Attorney General Harvey Bartle 3d said today that the charges would be filed after the jury's report was received. Those named in the presentment could not be reached for comment immediately.

At a news conference today at which he announced the grand jury action, Gov. Dick Thornburgh also announced the resignation, effective Nov. 30, of State Revenue Secretary Howard Cohen. Mr.

Continued on Page 6, Column 3

Associated Press
In the April 24 lottery drawing, the 6 ball popped up three times

Figure 1.1. A case of lottery fraud in September, 1980. (© 1980 by The New York Times Company. Reprinted by permission. Wide World Photos.)

Although a single rigged drawing is something that we could not detect with our data, we may be able to detect long-term irregularities.

Let's look at the winning numbers to see if they appear to be chosen at random. How about this:

> **hist(lottery.number)   # Figure 2**

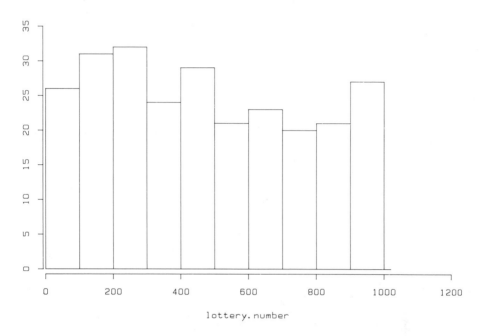

lottery.number

Figure 1.2. Histogram of winning numbers from 254 lottery drawings. Since there are 10 bars, the count should be approximately 25 in each bar, if the winning numbers are drawn at random.

The histogram looks fairly flat—no need to inform a grand jury.

Of course, most of our attention will probably be directed at the payoffs. Elementary probabilistic reasoning tells us that, unless we can predict the future or have rigged the lottery, a single number that we pick has a 1 in 1000 chance of winning. If we play many times, we expect about 1 winning number per 1000 plays. Since we pay fifty cents to play and we win after 1000 plays, we must win $500 to earn back all of the money we put in. (Everyone does want to win, right?)

Lets make a histogram of the payoffs.

> **hist(lottery.payoff)   # Figure 3**

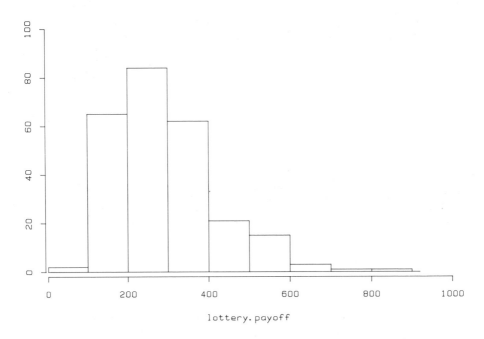

lottery.payoff

Figure 1.3. A histogram of the lottery payoffs shows that payoffs range from less than $100 to more than $800, although the bulk of the payoffs are between $100 and $400.

In our set of data there were a number of payoffs larger than $500— perhaps we have a chance. The widely varying payoffs are primarily due to the parimutuel betting in the lottery; if you win when few others win, you will get a large payoff. If you are *unlucky* enough to win along with lots of others, the payoff may be relatively small. Let's see what the largest and smallest payoffs and corresponding winning numbers were:

```
> prefix( "lottery." )
> max( payoff )   # the largest payoff
     869.5

> number[ payoff==max(payoff) ]
     499

> min( payoff )   # the smallest payoff
     83

> number[ payoff==min(payoff) ]
     123
```

Winners who bet on "123" must have been disappointed; $83 is not a very large payoff. On the other hand, $869.50 is very nice.

Since the winning numbers and the payoffs come in pairs, a number and a payoff for each drawing, we can produce a scatter plot of the data to see if there is any relationship between the payoff and the winning number.

```
> plot( number, payoff )   # Figure 4
```

What do you see in the picture? Does the payoff seem to depend on the position of the winning number? Perhaps it would help to add a "middle" line that follows the overall pattern of the data:

```
> lines( lowess( number, payoff, f=.2 ) )   # Figure 5
```

Can you see the interesting characteristics now in Figure 5? There are substantially higher payoffs for numbers with a leading zero, meaning fewer people bet on these numbers. Perhaps that reflects people's reluctance to think of numbers with leading zeros. After all, no one writes $010 on a ten dollar check! Also note that, except for the numbers with leading zeroes, payoffs seem to increase as the winning number increases.

It would be interesting to see exactly what numbers

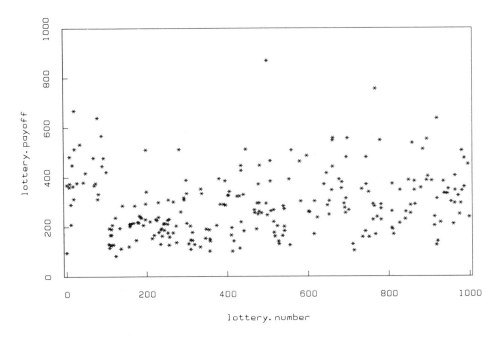

Figure 1.4. Scatter plot of winning number and payoff for the 254 different lottery drawings.

corresponded to the large payoffs. Fortunately, with an interactive graphic input device, we can do that by simply pointing at the "outliers":

> **> identify(number, payoff, number)   # Figure 6**

Can you see the pattern in the numbers with very high payoffs? Spend some time thinking before looking at the footnote which contains the explanation.* Did you find the pattern? If so, you have

---

* Most of the numbers with high payoffs have duplicate digits. The lottery has a mode of betting, called "combination bets" where players win if the digits in their number appear in any order. Ticket 123 would win on 321, 231, etc. The combination bet provides six ways to win and pays a one sixth share. However, combination bettors must pick numbers with three different digits. Payoffs for the numbers with duplicate digits are not shared with combination bettors, and thus are higher.

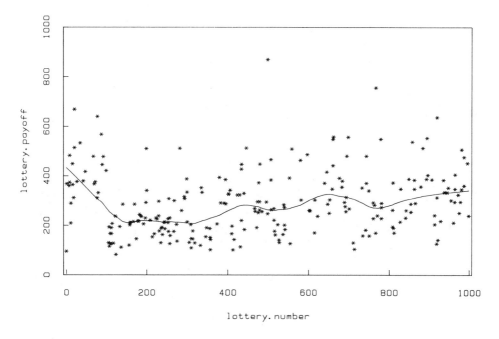

Figure 1.5. A smooth curve is superimposed on the winning number and payoff scatter plot.

accomplished something very important—you learned something new by looking at the data, and afterwards found that it could be explained by the rules of the game.

If you are now ready to put down this book and place your bet on 088, wait just a moment. By looking at similar data from two later periods in the Pick-It lottery, beginning November, 1976 and December, 1980, we can see if the payoffs changed over time.

```
> prefix()   # Figure 7
> boxplot( lottery.payoff, lottery2.payoff, lottery3.payoff )
> abline( h=500 )     # horizontal line at 500
```

The box in a boxplot contains the middle half of the data; the whiskers extending from the box reach to the most extreme non-outlier; outlying points are plotted individually. Clearly, we are not the only people to have seen the patterns in payoffs. Over the period of years for which we have data, the payoffs settle down and it is now quite

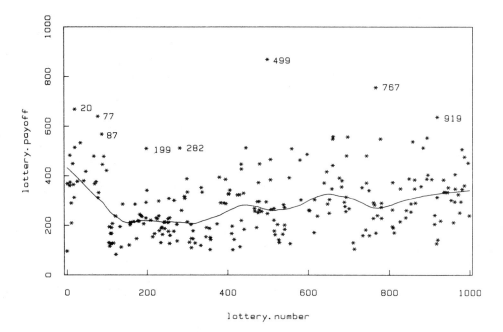

Figure 1.6. Outliers on the scatter plot are labelled with the actual winning number.

rare for a payoff to ever exceed $500.

1.3 A Comment

Throughout this chapter, we have presented, without explanation, a number of S expressions and the resulting output. The reason for this was to concentrate on the data analysis as well as demonstrate that S is an unobtrusive tool; it doesn't force you to think like a programmer when you want to think about your data. You should have seen that the S expressions needed to analyze the lottery data were simple (you may want to go back and look at them). You should also have noticed the power of interactive data analysis combined with well-chosen graphical displays.

In the next section, we will again analyze some data. However, besides looking at the data, we will also turn more attention to explaining a number of basic features of S.

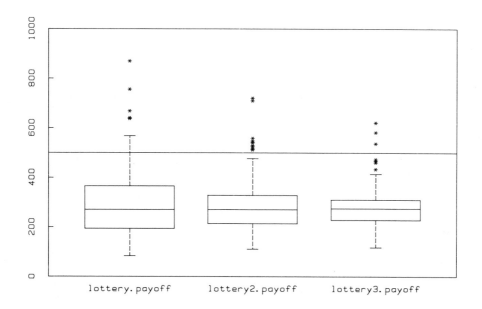

Figure 1.7. Boxplots show lottery payoffs for three different time periods: 1975, 1976, and 1980.

You now know that the title of this chapter is somewhat misleading; perhaps it should be "How to Lose Less in the Lottery". Place your bet if you enjoy gambling. Do not expect to win.

1.4 Related Reading.

Irwin W. Kabak and Jeffrey S. Simonoff, "A Look at Daily Lotteries", *The American Statistician*, Vol. 37, No. 1, pp. 49-52, 1983.

Herman Chernoff, "An Analysis of the Massachusetts Numbers Game", *Statistics and Related Topics*, M. Csorgo, D. A. Dawson, J. N. K. Rao, A. K. Md. E. Saleh (eds.), North-Holland, 1981.

Herman Chernoff, "How to Beat the Massachusetts Numbers Game. An Application of Some Basic Ideas in Probability and Statistics" *The*

Mathematical Intelligencer, Vol. 3, No. 4, pp. 166-172, 1981.

Jagdish S. Rustagi, "Probabilistic Structures of Modern Lottery Games", Technical Report No. 236, Department of Statistics, Ohio State University, May 1981.

2
Tutorial Introduction to S

The use of S requires some basic familiarity with interactive computing: how to login to the UNIX system, what keys to use on the terminal to delete characters and lines and to send interrupts, etc. This elementary training is generally available through the introductory UNIX system manuals and courses. Make sure that you have this training before you go on. (See, for example, the book *Introducing the UNIX System*, listed in the Preface).

2.1 Invoking S

Once you are able to login to the computer and have received the UNIX prompt (**$**), type the capital letter

    $ S

to start running S. When finished with an S session, type **q** to quit.

2.2 Expressions and Data

A good way to start working with S is to think of it as a *very powerful* calculator. In an interactive session with S, you type *expressions*, and S does computations and prints the results.

    > 3*(11.5 + 2.3)
        41.4

(In this and later examples, the user's typing is shown in boldface following the S prompt character (>), and the results printed by S follow in typewriter font). Unlike a calculator however, S has several hundred functions (and allows the user to define many more), and it can generate reports and graphical displays as well as print numbers. S operates on whole vectors and other data structures at once. Also, results can always be kept around for further use by picking a name and *assigning* that name to the result of an expression:

> **pi** ← **3.14159**

keeps a dataset called **pi** containing the single value, 3.14159. The sequence of characters "pi ←" is normally read as "pi gets". (The left-arrow is typed as the two characters "<" and "−".) Once the assignment takes place, **pi** can appear in another S expression, to stand for the data currently assigned to **pi**. In particular, just typing the name of a dataset by itself causes the data to be printed on the terminal:

> **pi**
```
3.14159
```

Expressions are made up by combining constants and dataset names with *operators* and *functions*. Operators include the usual arithmetic operations, comparisons of values and a few special operators. Thus,

> **pi + 1**
```
4.14159
```

is an expression whose value is 1 more than the data value in **pi**. This expression *does not* change the value of **pi**, since no assignment was done.

Most of the statistical, graphical and other analyses in S are done by functions, which are invoked by giving their name followed by their *arguments* in parentheses; e.g., the **c** function

> **c(15.1, 11.3, 7.8)**
```
15.1   11.3   7.8
```

collects the data in all its arguments to form a vector (here with 3 data values). This is a good way to create small datasets:

> **murder** ← **c(15.1, 11.3, 7.8, 10.1, 10.3,**
+ **6.8, 3.1, 6.2, 10.7, 13.9)**

Notice we used a longer name here to help us remember what data was assigned. Generally, names can contain letters, digits, and periods ("."), and must start with a letter. A dataset name need not have anything to do with the contents of the dataset; there are no rules that force datasets containing character data to have names starting with "c", for example. It is good practice to use mnemonic dataset names; however, other than that, users can re-use names arbitrarily. As the example shows, expressions can be spread over several lines, with blanks and tabs inserted between parts of the expression to make them more readable, if you like. (By the way, the data are the murder rates per 100,000 population for the first 10 states alphabetically—Alabama to Georgia—during 1977.)

There are several hundred functions in S: to see a list of those supplied in the standard version of S, look up *function* in the index at the back of the book. Functions compute statistical summaries (**mean(x)**, **rank(x)**, **regress(x,y)**, ...), mathematical transformations (**sin(x)**, **log(x)**, **sqrt(x)**, ...), data manipulations (**sort(x)**, **range(x)**, ...) and many specialized analyses. Some of them are useful for their *side-effects* instead of their value; for example, **stem(x)** computes and prints a summary of a dataset that is essentially a histogram on its side:*

```
> stem( murder )

N = 10    Median = 10.2
Quartiles = 6.8, 11.3

Decimal point is 1 place to the right of the colon

    0 : 3
    0 : 678
    1 : 00114
    1 : 5
```

The function **c(x1,x2,...)** provides a good way to generate small datasets. For somewhat larger datasets, it is usually best to put the data values into a UNIX file, say with a text editor, and then use the S **read** function. We can escape from S to a UNIX editor, say *ed*, and create a file "illitdata" containing the data:

```
> !ed
a
2.1   1.5   1.8   1.9   1.1
0.7   1.1   0.9   1.3   2
w illitdata
38
q
>
```

---

* The preface contains references to statistical texts that describe graphical and statistical techniques such as the stem-and-leaf display.

(The "!" in front of **ed** tells S that we want to type the command to the UNIX system. The lines beginning **a**, **w**, and **q**, are commands to *ed*; 38 is the number of characters that *ed* wrote to the file.) We can then read these numbers (they are 1977 percent illiteracy for the same 10 states as the murder rates) and put them into a dataset with the S expression:

```
> illit ← read( "illitdata" )
Read 10 items
```

S expressions can have character string values as well as numerical or logical ones. For example, if we have created a file, "statecodes":

```
> !cat statecodes
"AL"    "AK"    "AZ"    "AR"    "CA"
"CO"    "CT"    "DE"    "FL"    "GA"
```

we can read it in just the same way:

```
> states ← read( "statecodes" )
Read 10 items
```

(The quotes in the file of character data are not needed if it is obvious to S that the data are not numbers; however, they never hurt, and are necessary if strings contain blanks.) We will use these three datasets, **murder**, **illit**, and **states**, to show how to do some analysis with S.

2.3 Arithmetic

S has the usual arithmetic operators:

$$+ \quad - \quad * \quad / \quad \hat{\ }$$

(the last one raises to a power). Looking at how these operators work will help give the flavor of computations in S. Here are some examples:

```
> murder / illit   # murders per thousand illiterates
         7.190477   7.533333   4.333333   5.315790
[ 5]     9.363636   9.714286   2.818182   6.888889
[ 9]     8.230770   6.950000
> murder − mean( murder )    # residual from mean
         5.570000     1.770000    -1.729999
[ 4]     0.5700006    0.7700005   -2.729999
[ 7]    -6.429999    -3.329999     1.170000
[10]     4.369999
> 2^c(5,8,3)  # some powers of two
         32     256      8
```

(The text beginning with "#" and going to the end of the line is a comment: we can type comments like this anywhere in S expressions.

Such comments are a good way to record what has been found during an analysis.) The first expression divided two vectors of the same length, element-by-element (7.190477 is 15.1 divided by 2.1, and so on). The two vectors don't need to be of the same length; in the second example, **mean(murder)** is a single number. In the subtraction, the single number is in effect replicated 10 times, to match the vector **murder**. In general, the shorter vector is replicated until it is as long as the longer vector.

Try the following problems yourself. They introduce some more S functions.

PROBLEMS

2.1.   Create files "illitdata" and "statecodes" using a UNIX text editor. Invoke S and create the datasets **murder, illit,** and **states** as shown above. Be careful to check that your data agrees with ours.

2.2.   Compute the mean and the residuals from the mean for the illiteracy data.

2.3.   S also has the function **median** as an alternative to **mean**. Compute the medians of **murder** and **illit**. Are the medians different than the means?

2.4.   The function **sort(x)** returns all the values in **x**, sorted smallest to largest. Look at sorted values for **murder** and compare those to **median(murder)**.

2.5.   Make up and enter, either using **c** or **read**, a vector with an odd number of values (say, 9). Compute the median of this data and compare it with the sorted values. Can you guess now, if you don't already know, the definition of **median**?

2.6.   In data analysis, you often want to summarize the size or magnitude of the numbers in a vector. One of many reasonable measures is the median of the absolute values. The S function **abs(x)** returns absolute values of the numbers in **x**; the same values as in **x**, but with negative values made positive. Use the median of the absolute values to summarize the size of the residuals of **murder** from its mean and from its median.

2.4 Logical Operators

Just as S has *arithmetic* operators for addition, subtraction, etc., S also has *logical* operators to compare data values: these compare two expressions and return logical values, T for true and F for false. (If you want to give logical values as input to S, you can use TRUE and FALSE as well as T and F.) For example, x > y returns a vector with T wherever an element of **x** was larger than the corresponding element of **y** and F everywhere else.

> **murder** > 12
>
>   T   F   F   F   F   F   F   F   F   T

The other comparison operators are available too: "<", ">=", "<=". Equality and inequality tests are "==" and "!=". (Note the "==" is two characters long; we have a different use for "=" that you will see shortly.)

S has a few special operators in addition to the standard arithmetic and logical operators; for example, ":" creates the sequence between any two numbers in steps of 1 or −1:

> **0:10**
>
>   0 1 2 3 4 5 6 7 8 9 10
>
> **3:−8**
>
>   3 2 1 0 −1 −2 −3 −4 −5 −6 −7 −8

2.5 Subsets

A subset of the values in a dataset can be extracted by using an expression in square brackets following the name of the dataset. To get the first three values from the dataset **murder**:

> **murder[1:3]**
>
>   15.1 11.3 7.8

As in the example, if the expression in square brackets produces positive integers, the result is the corresponding elements of the dataset (counting from 1 for the first element). Another frequently useful expression extracts all the data corresponding to TRUE values of some logical condition. In S this is done by enclosing in square brackets any logical expression of the same length as the dataset; e.g.,

> **murder[murder > 12]**
>   15.1   13.9

The logical expression often involves the dataset itself, as in this example, but doesn't need to. To find which states had high murder rates, for example:

> **states[murder > 12]**
>   "AL"   "GA"

Or, to find the illiteracy rate for California:

> **illit[states == "CA"]**
>   1.1

Finally, we can choose some elements to *exclude* from the subset and get all the rest of the data by giving minus the indices of the elements to exclude; for example,

> **murder[−1]**
>   11.3   7.8   10.1   10.3   6.8   3.1   6.2
> [8]   10.7   13.9

gives all but the first element. When analyzing data, this gives an easy way to exclude outliers or special cases.

Subsets can be applied to any expression, not just to datasets. Such an expression, other than a dataset name or function call, should be enclosed in parentheses. To find which indices in **murder** correspond to high rates, we can type:

> **which ← (1:10)[murder > 12]**
> **which**
>   1   10

This sort of expression is nice if we want to select a few elements from a large dataset, since **which** will only be as large as the subset, not the whole dataset. In this example, we knew that there were ten numbers in **murder**. To avoid counting by hand the length of the dataset, we can use either of two functions: **len(x)** returns the length of (the number of elements in) its argument, so **1:len(x)** gives the sequence of indices corresponding to **x**. A still simpler way of writing the same thing uses the function **seq** (the functional form of the colon operator): **seq(x)** returns a sequence from 1 to the length of **x**.

> > **seq(murder)[murder > 12]**
> >   1   10

Here are some more problems. Again, these will introduce some useful new functions.

PROBLEMS

2.7.   Summarize the three different kinds of expressions in brackets that can extract a subset.

2.8.   Print the elements of **illit** in reverse order.

2.9.   Print the elements of **illit** that are greater than the median illiteracy. Print the elements of **murder** for which illiteracy is greater than the median.

2.10.   The mean and median are measures of location for a dataset. Other summaries measure the range or spread of the data. For example, **var(x)** computes the sample variance of its argument (approximately, this is the mean of the squared residuals from the mean of **x**). Compute the standard deviation (the square root of the variance) for murder and illiteracy.

2.11.* (This problem is a little harder than the others, which is why it is marked with a star.)

Another measure of spread is what is called the *interquartile range*, the difference between the third quartile and the first quartile. (Quartiles are defined so that one-quarter of the data is smaller than the first quartile, and three-quarters of the data is smaller than the third quartile.)

Find the interquartile range of **murder** and of **illiteracy**. There are several ways to compute this in S. Try to follow both of the following hints to get two different ways:

(i)   Use the elements of the sorted data that correspond to the first and third quartile;

(ii) Use the idea that the first quartile is essentially the median of the smaller half of the data, and that this in turn is the part of the data less than the overall median.

2.6 Graphics

Up to this point, we obtained information from S by printing out numerical values or by the specialized printout of the **stem** function. Printed output is useful, but graphics, such as scatter plots, provide much more information in an easy-to-comprehend form. S has extensive graphical capabilities which can be used with a variety of plotting devices. We will next look at a small fraction of these capabilities.

The first step in doing graphics in S is to tell the system what graphical device you intend to use. (See **devices** in Appendix 1 for the full list.) The device only needs to be specified once per session. Even if your terminal has no true graphic capabilities,

> **printer**

causes plots to be generated using ordinary characters.

Once the device is declared, we can produce a scatter plot; for example,

> **plot(illit, murder)   # Figure 1**

One advantage of plots is that we can look at more data than would be possible with printed output, so let's expand our datasets to all 50 states. The data comes from datasets in the S system data directory (data that is automatically available to all users: **state** in Appendix 2 describes this data). Murder and illiteracy are the fifth and third columns of the matrix **state.x77**, respectively, and the state abbrevia-

---

* Generally, dataset names in S can be made up from letters, numbers, and ".", and must start with a letter.

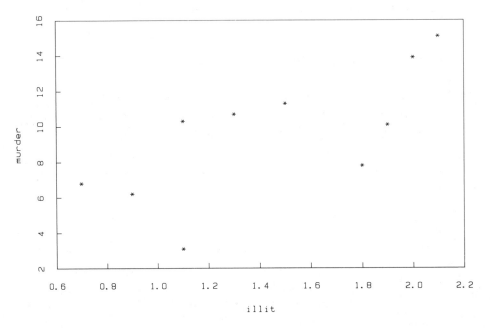

Figure 2.1. Scatter plot of murder versus illiteracy: 10 states.

tions are named **state.abb.*** We execute the following expressions to get the complete data (the expression **state.x77[,5]** extracts the 5th column of the matrix):

> **murder** ← **state.x77[, 5]**
> **illit** ← **state.x77[, 3]**
> **states** ← **state.abb**

These assignments have redefined the three datasets **murder, illit,** and **states**. Their previous definition doesn't restrict the current assignment. Now we can redo the plot of Figure 1.

> **plot(illit, murder)   # Figure 2**

The full 50-state data produces a more interesting picture. Looking at this plot suggests some new questions about the data. (Indeed, the great power of interactive graphics in data analysis is its ability to show unexpected features in the data. We will explore some of the statistical features in the next problem set.) The points appear to fall in a loose pattern, with murder rate rising with illiteracy. Interesting points are those that have particularly large values of either variable

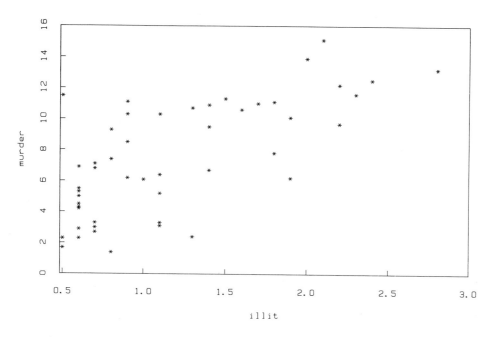

Figure 2.2. Scatter plot of murder versus illiteracy: 50 states.

or that lie on the edge of the overall pattern. The S function **identify** can be used to pick off interesting points interactively (look up your graphic device in Appendix 1 to see how to use its graphic-input capability):

> **identify(illit, murder, states) # Figure 3**
>   18   10   1

was used to pick off three states with large values on both variables. The first two arguments to **identify** should be the same data as were used to make the plot. The third argument says what to plot at each identified point; if you omit it, the indices of the identified points are used. (The term *indices* refers to the integer positions of the points in the **x**, **y** vectors.) These indices are returned in any case as the value of **identify**.

There are several other ways to add information to a scatter plot (or to other plots in S). The function **points(x,y)** plots an asterisk

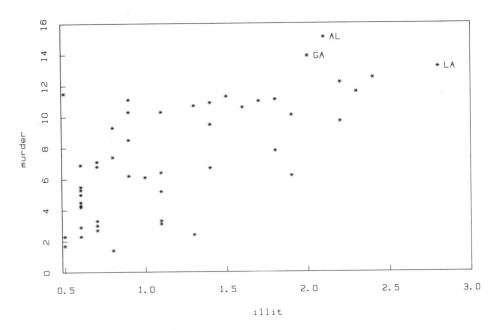

Figure 2.3. Scatter plot of illiteracy and murder rate with several extreme points identified.

at the points defined by **x** and **y**. Suppose we want to identify the points for which murder rates are high, say greater than 10:

> **nasty ← murder>10**

defines the states we want. Since **points** overlays the previous plot, we need to distinguish the new points somehow. For this purpose, we can specify one or more *graphical parameters* in the call to **points**. For example, **pch="O"** changes the plotting character from the default "*" to "O" for this call.

> **points(illit[nasty], murder[nasty], pch="O")   # Figure 4**

Depending on what a particular graphics device can do, we could also change character size (**cex=1.5**) or color (**col=2**). (See section 4.3.1 for a further discussion of graphical parameters.)

The function **text(x,y,label)** plots labels (either character strings

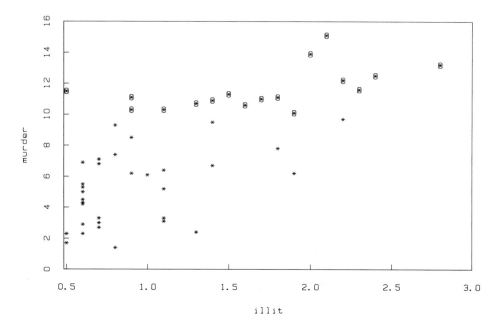

Figure 2.4. Scatter plot of illiteracy and murder rate. Points where the murder rate is larger than 10 are emphasized.

or numerical values) at the specified points. Instead of overstriking the previous points, we can plot the state abbreviations, as in Figure 5:

> **text(illit[nasty], murder[nasty], states[nasty], adj=0)**

(We left-justified the text by using a graphical parameter **adj=0** so that the labels would not overwrite the points. One of the problems in the next set requires you to use **text** *instead* of a scatter of points.)

Scatter plots are particularly useful because they can be applied to so many different analyses. Other graphical displays, however, may be helpful because they represent some special features of the data. S contains a number of such specialized graphical displays, and others can often be created from the basic graphical functions, like **points**, **lines**, and **text**. One kind of specialized plot is a map; for example

> **usa(states=FALSE)**

plots a map of the United States. By default, it also plots the state boundaries: we left these off (by **states=FALSE**) to have more room to add other information to the map. The map becomes a useful tool in data analysis because we can relate the data to the geographical co-

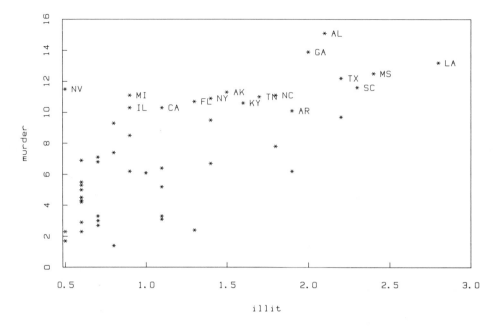

Figure 2.5. Scatter plot of illiteracy and murder rate. Points with murder rate larger than 10 are labelled with the state abbreviation.

ordinates of the states. If you look up **state** in Appendix 2, you may notice that **state.center** is a structure with components **x** and **y** giving the co-ordinates for the state centers. These components can be extracted as follows:

> **state.x** ← **state.center$x**
> **state.y** ← **state.center$y**

(The **s$x** operation returns the component named **x** from the dataset s.) Now we can indicate the states with high murder rate suitably on the map:

> **text(state.x[nasty], state.y[nasty], "HIGH") # Figure 6**

Time for some more problems.

PROBLEMS

2.12.   Use the value returned by **identify** to produce a new scatter plot of illiteracy and murder rates with the outliers omitted.

Figure 2.6. A map of the United States with the word "HIGH" plotted at the center of each state with a high murder rate.

2.13.   The **plot** function has an optional argument, **type**, that controls what type of plot is made (points, lines, etc.). In particular, **type="n"** plots nothing, but sets up the plot and draws axes and axis labels. Use this to generate a scatter plot of illiteracy and murder, but with the *state abbreviations* labelling the points, instead of the default plotting character.

2.14.   Per-capita income is available for the states, also:

> **income ← state.x77[, 2]**

Make a plot that shows states with (1) high and (2) low per-capita income, on a scatter plot of illiteracy and murder. Play around with different versions of the techniques used so far, until you get a plot you like.

2.15.   The function **symbols** is a useful function for drawing circles, rectangles and other symbols on plots. Look up the documentation for **symbols** and use this function to plot a scatter as in the previous problem, but with circles at each of the points. The radius of the circles should be proportional to income.

What would you do to make the *area* proportional to income? Would it be better to use **income** − **3000**?

2.16. Here is a plot to show the geographic distribution of income. Draw the USA map as above. Choose four text strings for very low, low, high and very high values of income. Plot the first string at the co-ordinates of states with less than the first quartile of income, the second at states above the first quartile but below the second (i.e., the median), and so on.

2.17.* The text strings in the previous problem can all be plotted in one call to **text**. How? (Hint: when subscripting, an element from the data can appear in the subset as often as desired.)

2.7 Macros

In solving the preceding problems, you probably created some fairly lengthy S expressions. In order to avoid having to remember (and type) such expressions repeatedly, you can define them as a *macro*, and then refer to them by giving just the macro name. This has the added advantage that, since macros can have arguments in the same way as S functions, we can apply the same expression to a variety of data.

The following expressions make a map and plot the variable **income** on the map, as circles with radius proportional to the amount by which income exceeds $3000.

> **usa(state=F)   # Figure 7**
> **symbols(state.center, circle=income−3000, add=TRUE)**

(The dataset **state.center** is a *plot structure* which contains both the x and y coordinates for a set of points.) We can make this into a general macro to plot any such variable, by first creating a UNIX file named, for example, "my.macro" containing the macro definition,

```
> !cat my.macro
MACRO usacircle(x)
usa( state=F )
symbols( state.center, circle=x, add=TRUE )
END
```

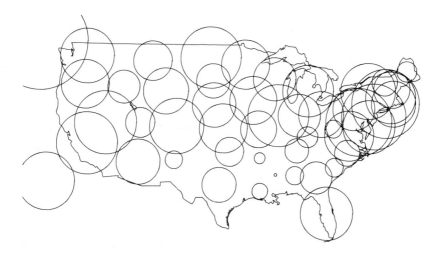

Figure 2.7. Map of the United States with per-capita income level coded by size of circle.

The S function **define** reads the definition and stores it in a dataset named **mac.usacircle**:

> **define("my.macro")**
> mac.usacircle

When macros are defined from a file, the **define** function automatically prints their names. The line beginning with **MACRO** gives the name for the macro itself and its arguments, here one argument, which we chose to call **x**. The next two lines give the definition of the macro. The **END** line signals the end of the macro definition.

Once a macro has been defined, it can be invoked like an S function, but with the name preceded by **?**:

> **?usa.circle(income−3000)**

The argument, **income−3000**, is substituted for every occurrence of the corresponding argument name, **x**, in the macro definition and the resulting text is executed as if it had been typed by the user.

Macros can have more than one argument. For example, suppose we want to plot the values of a variable **z** on a scatter plot of **x** and **y**:

> **MACRO xyzplot(x,y,z)**
> **plot(x, y, type="n")**
> **text(x, y, z)**
> **END**

As with S functions, the user of the macro can give the arguments either by position or by name. Thus we can execute:

> **> ?xyzplot(z=states, income, murder)**

to produce the plot shown in Figure 8.

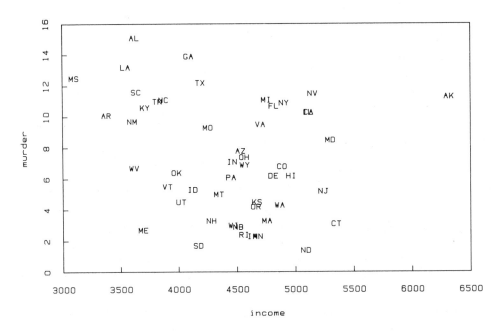

Figure 2.8. Scatter plot of income level and murder rate using state abbreviations as the plotting symbols.

2.8 Wrapup

At this point, you may well feel that this tutorial introduction to S has gone on long enough. The antidote for this feeling is to actually practice some of the things mentioned here. In particular, do at least some of the problems and then try to use S on some (simple) data of your own. Even though later chapters will go into much more detail and introduce many more functions, you should be able to analyze some data with what you now know about S.

3
Basic Use of S

This chapter introduces some key concepts and the basic techniques needed in order to use S. Sections 3.1 and 3.2 introduce some simple and general concepts; grasping them from the start will lead more quickly to good use of S. *You are encouraged to read through this material thoroughly.* Later sections cover the S language, data, data analysis, and graphics. Finally, the chapter ends with a large example.

3.1 Basic Concepts: Expressions and Data

S is an *expression* language. The user types expressions that are interpreted and evaluated by the S system. The expressions include, but are not limited to, conventional algebraic and functional expressions:

```
1 + y
log( 1 + yold * weight )
p > .95
x − mean(x)
```

Expressions are composed from operators and other special characters (plus, minus, parentheses, etc.), from functions (**log** and **mean**, e.g.), and from data (constants and named datasets).

Data in S is organized into named, self-describing *datasets*. These are kept by S in special UNIX directories and retrieved automatically when named in an expression. The fundamental data structure is a *vector*; that is, a sequence of data values (numbers, logical values or character strings). An unlimited variety of other data structures can be built up from vectors. Some structures are recognized by many functions in S; for example, *matrices, arrays* and *time-series.*

Fundamental to the philosophy of S is the uniformity of expressions and data structures. The value of any expression, however complicated, is a data structure. This includes anything from simple arithmetic to high-level statistical functions. So far as the language is concerned, any expression can appear as an argument to an operator or function. Of course, a particular function may generate an error if it cannot interpret the argument sensibly. As we proceed, examples will show how this uniformity leads to flexible, general data analysis.

S is a *case sensitive* language: corresponding upper- and lower-case letters are distinct in the names of functions and datasets. Thus the name **ABC** is not the same as the name **abc.** By convention, functions and system datasets have lower case names. Special built-in values such as TRUE, FALSE, and NA are upper case.

One final concept: it's good for a user to be adventurous. If it isn't clear what S will do in response to a particular expression, try it. You are encouraged to try for yourself the S expressions we give as examples.

3.2 Invoking S

S is invoked from the UNIX shell by typing the capital letter "S":

```
$ S
>  ... do your S computing here ...
>  q
$
```

(Here "$" is the shell's prompt.) S will respond with its prompt character ">". At this point, you type expressions to S; it evaluates them and prints results. When you are finished using S, type **q** to quit.

3.3 The S Language

S provides a complete environment for data analysis. The environment provides data management, device-independent graphics, and many other services to the user. The S language is an important part of this environment, since it is the set of high-level expressions that users type to communicate with the S system.

3.3.1 *Expressions; Operators; Functions*

The S user types expressions to be interpreted and executed. The result of an expression may be to compute a vector or other data structure, to assign data a name for further retrieval, to generate printed or graphical output, or to control the operation of S.

Operators in S include the usual arithmetic and logical operators. These operate on all the data values in the vectors or other operands. Thus,

y + 1

computes its result by adding 1 to each data value in **y**. The arithmetic operators are:

+ $\quad$ — $\quad$ * $\quad$ / $\quad$ ^

The operator "^" raises to a power, e.g. **2^3** is 8. Two less common operations are:

% / $\quad$ %%

The operator "% /" is integer divide; i.e., it divides and then truncates the quotient to an integer. The operator "%%" is a remainder or modulo operator: it gives the remainder when its first operand is divided by its second. There are logical operators for comparison of data values, which produce the values TRUE or FALSE:

< $\quad$ > $\quad$ <= $\quad$ >= $\quad$ == $\quad$!=

for less-than, greater-than, less-than-or-equal-to, greater-than-or-equal-to, equal-to, and not-equal-to. There are also boolean operators

& $\quad$ | $\quad$!

that are used for the operations *and*, *or*, and *not*.

The S operator ":" creates the sequence between any two numbers in steps of ±1:

> **9:23**

```
        9  10  11  12  13  14  15  16  17  18
[  11]  19  20  21  22  23
```

In this and later examples, the user's typing is shown in boldface following the S prompt character ">", and is followed by the response from S. The result of an expression, if not assigned, is printed on the terminal.

The operation of giving a name to a data structure is called *assignment*. Assignment is performed by an operator, which appears as an arrow, "←", in this book. The arrow may be typed as an underline character "_" or as the two characters "<−" (less-than, minus).† On the left of the arrow is a name, on the right any expression. Names may contain letters, digits, and periods, and must start with a letter. Users can choose names for their convenience and can re-use names arbitrarily. The value of the expression is stored with the given name; e.g.,

> x ← 9:23

creates a dataset named **x**. The sequence of characters "x←" is normally read as "x gets". Notice that the result of the assigned expression is not printed.

Subsets of data can be extracted by:

x[expression]

The expression defining the subset can have three forms.
1. a vector of positive integers; the corresponding elements of **x** will appear in the result; for example, **x[1]** to extract the first value from **x**, or **x[1:5]** for the first 5 values.
2. a logical expression of the same length as **x**; the elements of **x** for which the logical expression is true will be selected; for example, **x[x>0]** or **name[height>72 & age<31]**.
3. a vector of negative integers; all elements except those specified will be selected; for example, **x[−1]** gives all but the first element of **x**.

Subset expressions can also be assigned to, with the effect of replacing the corresponding elements of the vector:

x[x<0] ← 0

replaces all negative elements with 0.

Most of the statistical, graphical and other analyses in S are done by *functions*; e.g.,

```
mean(x)        # arithmetic mean
plot(x,y)      # scatter plot of y versus x
c(1,−2.1,3.2,5.78)    # combine data values to make a vector
rnorm(5)       # sample of size 5 from standard normal distribution
```

In general, function calls have the form

---

† The two characters must be adjacent—no space is allowed between the "<" and the "−". Thus, the expression typed "x <− 3" is different from the expression typed "x < −3"; the first is an assignment, the other a comparison.

name(arg1, arg2, ...)

where **name** is the name of the function, and **arg1**, etc. are any expressions, to be taken as arguments to the function. For example **mean(x)** calls the function **mean** to compute the mean value of the data named **x**. Typically, functions will have a few basic arguments (supplying the data on which the function operates, for example). These must always be supplied. There may also be any number of optional arguments, which usually give special features or options to control the function in more detail. These may be omitted, with the function taking some default action.

Usually, optional arguments are most conveniently supplied in the form **name=value**, where **name** is the name of the argument (listed in Appendix 1 under the function name) and **value** is any expression. The user need only supply the options of interest, in any order. To make it easier to use named arguments, the name can be abbreviated to its first few characters, as long as it can be distinguished from other argument names to the same function. As an example of the use of named arguments, one of many optional arguments to **plot** is **ylab**, a label for the y-axis of the plot:

plot(x, log(y), ylab="Logged response values")

Note the distinction between naming arguments with "=", testing equality with "==", and assignment of data with "←".

S also has functions which take an arbitrary number of arguments. For example, the combine function, **c(arg1, ..., argn)** combines all the data in an arbitrary number of arguments to form one vector.

There are a large number of functions in S, all of which are described in Appendix 1. All of them are used with the form of expression illustrated in this section.

PROBLEMS

3.1.   For the numbers

7.3, 6.8, 9, 12, 2.4, 18.9

a)   Find their mean.
b)   Subtract the mean from each number.
c)   Print the square roots of the numbers.
d)   Print the square roots of the numbers rounded to 2 decimal places. *Hint:* there is a function named **round**.
e)   How much do the squares of the rounded roots differ from the

original numbers?

3.2.   One dollar invested at an annual rate of **r** percent compounded
monthly for **m** months is worth

$$(1 + r/1200)^{\wedge} m$$

How much will one dollar be worth at the end of each of ten
years at 6%? What about 17%?

3.3.2 *Interacting with S: Errors and Interrupts*

In general, S users should not be afraid to take a chance with a
new kind of expression. The language and the individual functions
are designed to allow as much flexibility as possible. The effect of
errors when they do occur is not usually serious. In interactive
analysis, errors tend to be detected quickly, and the correction can be
made directly.

Errors may be less likely, particularly when first using S, if
complicated expressions are evaluated in a sequence of simpler steps,
where the intermediate results are assigned names and reused. This
has the added advantage that the intermediate results are available for
examination, to see that they were computed as intended. As you
become more accustomed to S, complicated expressions will grow
more natural. Also, quite complicated expressions can be useful when
turned into source files or macros (section 5.1.1) so that only a simple
macro call is actually typed. To improve readability, you can use
white space (blanks and tabs) freely to separate names, numbers, and
other symbols. Long expressions can be continued over any number
of lines, so long as it is clear that the end of the expression can not
have occurred yet. This can be ensured by ending the line with an
operator or by not supplying enough right parentheses to close the
expression; e.g.,

```
> z ← x+3.14159*(y−      # this line continued
+ mean(y,trim=.1))
```

The first line was doubly sure of being continued since it ended with
the operator "−", and also had an unmatched left parenthesis. As
shown above, S prompts with "+" for continuation lines, and comments can be inserted in expressions, from the character "#" to the
end of the line.

If you have short expressions, you can put any number of expressions on one line by separating them with semicolons (;).

> x ← sqrt(y); z ← log(y)

You may encounter two different kinds of errors in using S: those caused by "expressions" that S cannot understand and those detected by one of the functions during evaluation. The first type of error (a *syntax* error, so named because the input does not follow the syntactic rules for forming expressions), will be reported, by typing back the offending line, with the apparent location of the error marked:

```
> x+7,5        # I meant 7.5
Syntax error
  x+7,5
      ^
```

The second type of error (an *execution* error), is reported by the function which could not compute the desired result:

```
> 1+"abc"
Error: Non-numeric data in arithmetic
Error in +
```

In either case, a message will be printed to describe the error, execution of the current expression will terminate, and S will prompt for further expressions. If an error occurs in a long expression, the expression is written to a file, and the system types "Dumped" on the terminal. The expression can be retyped directly, or the user can type

> **edit**

which invokes a text editor† to allow changes to the expression. The user writes the corrected expression and quits from the editor, after which the expression is automatically re-executed (see **edit** in Appendix 1). The expression can be re-executed without editing by typing **again** after the "Dumped" message.

The user may interrupt execution of any expression by hitting the "del" key or "break" key. For example, you may wish to interrupt long printouts or the execution of an expression which was incorrect. If the expression was long enough, the message "Dumped" will be

---

† Interactive computing inevitably requires some editing of files. If you are not familiar with a text editor, you should learn to use one; see, for example, Chapters 7 and 8 of *Introduction to the UNIX System*. If the default "ed" editor is not your favorite, you can set and export a shell variable named EDITOR to inform S of your choice.

printed and **edit** can be used to correct any mistakes.

3.3.3 *On-Line Documentation; Help*

S comes with quite a large number of functions, macros, and datasets, and it is likely that you will generate many of your own. S provides an on-line **help** function which can provide information to you about how to use specific functions or macros, and describes the contents of datasets. The expression

> > **help("plot")**

for example, causes the documentation (as in Appendix 1) for the function **plot** to be printed on the terminal. Invoking **help** without arguments produces documentation for **help** itself.

A short form of the on-line documentation can be obtained through the **call** function:

> > **call("plot")**

will print just the one-line description of **plot** and a full list of its arguments.

The **help** function can also print documentation for macros or datasets:

> > **help(macro="row")**
> ... *the documentation for macro* **row** *is printed here* ...
> > **help(dataset="state")**
> ... *the documentation for datasets named* **state** *is printed here* ...

As you start creating your own datasets and macros, you can also create your own documentation for them; see section 5.4.5.

Since the S system will continue to evolve, the printed documentation in Appendix 1 can, at best, be correct at the time of publication. However, the on-line documentation is updated whenever S functions change. Thus, the information given by the **help** function should be regarded as the true documentation for the system. You can use the UNIX command

> $ **S NEWS**

to help you find out where the on-line documentation differs from what is in the book. The news printed by this command should specify, in reverse chronological order, what has changed in S since the publication of this book.

The S function documentation is also available from UNIX shell level through the **S HELP** command.

> $ **S HELP plot**
> *... documentation for* **plot** *is printed here ...*

In addition, the UNIX command **S MAN** will print some programmer-oriented documentation.

3.3.4 *UNIX System Interaction*

Problems that persist after you read the documentation and experiment may be reported by sending **mail** to a special user (normally named "s") at each installation. The S function **mail** prompts the user for text, terminated by a line consisting solely of a period. Mail can also be used to report suggestions or other comments about S. You should receive a reply through the UNIX system's **mail** command. See **mail(1)** in the UNIX manual for instructions on how to read mail sent to you.

The S user can communicate easily with the operating system. Any line that begins with the character "!" is passed, untouched, to the operating system for execution. This facility can be used to list files, to enter the text editor, etc. UNIX system commands (e.g., text editors) that interact with the user return to the S environment when that interaction is complete.

> > **!ed mydata**
> *... edit file mydata with text editor ...*
> q
> > # **ready for next S expression**

3.4 Data in S

The previous section described expressions and functions, the language you use to communicate with the S system. The S language allows you to operate on data. This section describes data; how to get data into S, how it is stored, and how it is structured.

3.4.1 *Data Values*

Data values are numbers (real or integer), logical values (TRUE or FALSE), or character strings. Explicit numerical values may be written in the usual integer, decimal fraction or exponent notation:

> 3   −5   6.272   3.4e10

Internal blanks are *not* allowed in numbers.

Character strings may be specified by any string of characters, enclosed in matching quotes (") or apostrophes ('):

> **"Now is the time ...";   ' + − * / '**

The enclosing quotes or apostrophes are only delimiters, not part of the string. A back-slash character may be used to enter a character which would not otherwise be legal, typically a quote or apostrophe:

> **"You are reading the \"S user's guide\""   # or equivalently**
> **'You are reading the "S user\'s guide"'**

Logical data may be entered as T or TRUE or as F or FALSE.

The special data value, NA, stands for **N**ot **A**vailable. It may appear wherever a numerical or logical value is expected. The best way to think of this is as a condition holding whenever one wants to indicate missing data. Not all functions in S know how to handle missing values; if data with NAs is given to these functions, an error will result. In most functions that deal with a single data value at a time, for example arithmetic, any operation upon an NA will generate an NA. Thus:

> **> 1 + NA**
> NA
> **> 1 < NA**
> NA

The function **NA(x)** returns logical TRUE values wherever **x** has missing values, so

> **x[!NA(x)]**

will return the data in **x** with missing values removed. You should consider, however, whether throwing out NAs is sensible for the particular data analysis you are doing. Also, the data value **NA** cannot be used in logical comparisons; use **NA(x)**, *never* **x==NA**.

PROBLEMS

3.3.   How would you expect missing values will be handled in these expressions?

> **> x ← c(1, NA, 7)**
> **> x == NA**
> **> NA == NA**

Does this give you some idea why the function **NA** exists?

3.4.   The "correct" handling of NAs is a difficult problem. With the definition of **x** from the previous problem, what would you expect for the result of the expressions:

>   **> 1 + NA + 7**
>   **> len(x)**
>   **> sum(x)**
>   **> mean(x)**
>   **> median(x)**

Is it bothersome that **sum(x)/len(x)** is not the same as **mean(x)**?

3.4.2 *Vectors*

S supports an unlimited variety of data structures. The simplest are *vectors* containing numbers, logical values, or character strings. Some examples are:

```
> 1:3
    1    2    3
> Weekdays
      "Monday"       "Tuesday"       "Wednesday"       "Thursday"
[5]   "Friday"
> Month.length > 30
      T    F    T    F    T    F    T    T    F    T    F    T
```

Vectors are created either as the result of expressions or when data is read into S (section 3.4.3). Vectors, like all S data structures, are self-describing so that the user need not supply the length or other attributes to S functions that use the vector. There are, however, functions in S to *extract* attributes of vectors and other data structures. For example, the function **len(x)** returns the number of data values in its argument.

3.4.3 *Data Input*

Data values can be read from a file or from the user's terminal to form a vector, by the function **read**:

>   **> x ← read("myfile")**

reads values from the file named "myfile". It is important to keep clear the distinction between a *file* in the UNIX system and a *dataset* in S, as generated by an assignment. S datasets *are* files, created in a directory named "swork" as the result of an assignment. However, for efficiency, the files contain binary data and cannot, for example, be printed or edited by ordinary UNIX commands.

The file ("myfile") used in the expression above might contain, for example,

```
3.5   2.1
7.893   1.3e10   0   0   0
1
-37.892   80000
```

The vector assigned would be of length 10, with the values shown.

The **read** function assumes that the file consists of data items separated by *white space* (that is, by blanks, tabs, or newlines). If the file argument is not given, **read** will read items from the terminal. In this case the user is prompted with the index of the next item to be read:

```
> x← read()
1:  1    2.5    3.14159
4:
3 items read
```

An empty line ends terminal input (but not input from a file). Reading directly from the terminal is reasonable for small amounts of data. However, for larger datasets, most users will find it preferable to have the data items on a file, so that a text editor can be used to edit out typing errors before input.

If the user wants to control the number of items read from the file, the optional argument **len** should be given:

read("myfile",len=6)

reads only the first six items of the file shown.

The input data may also be character strings. By default, **read** examines the first item; if it is numeric, all following fields are assumed to be numeric. Otherwise, the items are all assumed to be character strings. In addition one can force the input to be treated as character strings by using the optional argument, **mode**:

read("file2",mode=CHAR)

Because **read** looks for white space to separate items, character strings which contain blanks or tabs must be enclosed in quotes:

Saturday night and Sunday morning

reads as five character string items, while

"Saturday night and Sunday morning"

reads as one.

An item appearing as the two characters "NA" on input will generate a corresponding missing value in a numeric vector. NAs are not allowed in character vectors, although a character string can be empty, i.e., "".

As shown in the examples here, the best way to generate data for input to S, in most cases, is to create a file of ordinary text, divided into fields by white space. Any convenient UNIX tools can be used to help (*awk*, *sed*, *ed*, ...). Techniques for dealing with different types of files and other forms of data transfer will be discussed in section 5.4.

3.4.4 *Data Directories*

Data structures which are assigned names in S expressions become part of a *data directory*; that is, a collection of datasets that are retrieved automatically when referred to by name in an S expression. Three data directories are available automatically in each S session. The first is the user's *working* directory. Datasets created by assignments are stored here. The second data directory is the user's *save* directory. This is intended for data of permanent value. Placing datasets here is a useful safeguard against accidentally assigning some other dataset the same name. The expression

save(arg1, ...)

puts its arguments onto the save directory. If the argument is of the form **name=value** then **name** is used as the name of the saved dataset. Otherwise, **arg**i should be a dataset (presumably not currently on the save directory) and the saved data will have the same name.

The working directory and save directory are local to the current user. (They are actually UNIX directories named "swork" and "sdata" within the user's current UNIX directory.) A third data directory is the S *system* directory, containing datasets of frequent value to users in general, including some classic statistical examples and standard values (e.g., the lottery data, the names of the states, months, etc.)

When data has been moved to the save directory, or is no longer needed, it may be removed from the working directory by

rm(x)

To obtain a list of the names of all datasets in the working data use the function **ls**.

The available data directories are kept in an ordered list maintained during each S session. (The function **attach** can be used to add new directories to this list; see section 5.4.2.) When a dataset name appears in an expression, the directories are searched in order for the corresponding name until the dataset is found or the search list is exhausted. Notice, in particular, that a dataset will not be retrieved from the save directory or the system directory if a set of the same name appears on the working data.

Functions such as **ls**, **save**, and **rm** can be made to refer to any of the directories currently available by use of the optional argument **pos**. This specifies the position of the directory in the search list; by default, positions 1, 2, and 3 correspond to working, save and system directories. For example:

```
> rm(x,pos=2)   # remove x from save directory
> ls(pos=2)     # what's on the save directory?
  "city" "label" "median" "quartile" "xy"
> ls   # working directory
  "i" "tx" "ty" "tz" "xl" "yl"
```

3.4.5 *Matrices*

General data structures in S are built up from vectors, and usually result from functions that return more than a simple vector. Some data structures are so widely used that they are essentially built into the language. Two examples are *matrices* and *time-series*.

Matrices contain, in addition to a vector of data values, information defining the number of rows and number of columns. Although matrices are actually a special case of general multiway arrays, they are so common that we will introduce them here, and defer to section 5.1 the discussion of general arrays. Subsets of matrices may be specified by two-subscript expressions, **x[e1, e2]**, where **e1** and **e2** define subsets of rows and of columns. If **e1** is omitted, all rows of **x** are included; if **e2** is omitted, all columns are included. For example,

```
x[ , c(1,3) ]    # the first & third cols
x[ res>.5, ]     # all rows i where res[i]>.5
x[1:3,2]     # vector consisting of x[1,2], x[2,2], x[3,2]
```

Logical and negative subscripts have the same meaning for the rows and columns of matrices that they had for elements of vectors.

Matrices are produced as the result of many functions in S. They may also be generated from a vector of data, using the function **matrix**. Besides the data to be used, **matrix** takes the number of rows and columns desired in the matrix. For example

> **matrix(1:12, 3, 4)**

```
Array:
3 by 4
        [,1]   [,2]   [,3]   [,4]
[1,]      1      4      7     10
[2,]      2      5      8     11
[3,]      3      6      9     12
```

When a matrix is printed, the label for a row or column resembles the subset expression to extract that row or column.

By default, **matrix** assumes the data values are given one column at a time. The standard idiom for reading data from a file with the values laid out in rows, uses the optional logical argument **byrow**:

> **x ← matrix(read("matfile"),ncol=5,byrow=TRUE)**

Notice that only one of the number of rows or number of columns need be specified, if the other can be inferred from the number of data values.

A number of S functions create matrices, manipulate them, and give their attributes. The number of rows and the number of columns are attributes obtainable from the functions **nrow(x)** and **ncol(x)**. It is possible, as we saw, to create matrices by extracting rows and columns from an existing matrix. In the other direction, vectors and matrices may be bound together as either the rows or columns of a new matrix. For example, suppose **xold** is a matrix and **update** is a vector:

> **xnew ← cbind(1, xold, update)**

creates a new matrix with a first column of all ones, the next columns the same as the columns of **xold** and the last column taken as **update**. In general, **cbind** can have any number of arguments, any of which may be vectors or matrices. All matrices must have the same number of rows. Notice that the single value, 1, was expanded to a vector of length equal to the number of rows of **xold**. The result has the same number of rows as the matrix arguments (or the length of the longest vector, if there were no matrix arguments). The function **rbind** works similarly with rows:

> **xnew ← rbind(xold, newobs)   # add new observation**

binds the data in the arguments to **rbind** together as rows of the result.

Functions which expect matrices can be given vectors as

arguments. Vectors will be interpreted as matrices with one column. To force a vector to be interpreted as a row matrix, use **matrix(x,nrow=1)**.

Functions which create new matrices or vectors from existing matrices are **t(x)** (the transpose of **x**), **diag(x)** (the vector of the diagonal elements of **x**), **row(x)** and **col(x)** (matrices of the same shape as **x**, but filled with the row number and the column number). For example:

```
> x←matrix(1:12,3,4)
> t(x)
Array:
4 by 3

        [,1]   [,2]   [,3]
[1,]       1      2      3
[2,]       4      5      6
[3,]       7      8      9
[4,]      10     11     12

> diag(x)
         1    5    9

> row(x)

Array:
3 by 4

        [,1]   [,2]   [,3]   [,4]
[1,]       1      1      1      1
[2,]       2      2      2      2
[3,]       3      3      3      3

> col(x)

Array:
3 by 4

        [,1]   [,2]   [,3]   [,4]
[1,]       1      2      3      4
[2,]       1      2      3      4
[3,]       1      2      3      4
```

The functions **row** and **col** are useful for selecting special subsets of a matrix:

```
> x[ row(x) >= col(x) ] <- 0
Array:
3 by 4

          [,1]   [,2]   [,3]   [,4]
[1,]         0      4      7     10
[2,]         0      0      8     11
[3,]         0      0      0     12
```

zeroes the lower triangle of **x**. The function **diag** can also create matrices when given a vector of the desired diagonal elements. If **values** is a vector, **diag(values)** is a square matrix, the number of rows and columns equal to **len(values)**. The diagonal is filled in from **values**, and all other elements 0. Also, if an integer is given to **diag**, it creates an identity matrix of this order; e.g., **diag(10)** is a 10 by 10 matrix with 1 on the diagonal and 0 elsewhere.

Given vectors of character strings which label the rows or columns of a matrix, a specialized printout can be generated from explicit use of the **print** function. (**Print** is implicitly used whenever S automatically prints a result.)

> **print(votes.repub, rowlab=state.abb,**
+   **collab=encode(votes.year))**

The special arguments **rowlab** and **collab** to **print** give character vectors which are used in place of the standard subscript-like row and column labels. Note that the labels may not be numeric: see section 5.5.2 and the detailed documentation for the **encode** function for general ways to construct informative labels.

3.4.6 *Time-Series*

Time-series are defined by a vector of data values, considered to be taken at equally spaced times. Three time-series parameters are used to define the corresponding time domain implicitly: **start, end** and **nper** for the times at which the observations start and end and the number of observations per unit time period. Conventionally, time-series data are used most often relative to units of one year (typically annual, quarterly or monthly). For example, a monthly series starting in June of 1960 could be specified from data on a file "co2data" through the function **ts** as follows:

> **co2 <- ts(read("co2data"), start=c(1960,6), nper=12)**

Notice that, as with **matrix**, we need only two time parameters (**start** and **nper**) since the other (**end**) can be computed from the number of data values. Also note that the vector **c(1960,6)** represents June, 1960.

Monthly and quarterly series are printed with appropriate labels:

> **co2**

```
Time-series:
start: 1960 6     end: 1963 12     Monthly

           Jan      Feb      Mar      Apr      May      Jun
1960                                                   319.74
1961    316.97   317.74   318.63   319.43   320.47   319.71
1962    318.06   318.59   319.74   320.63   321.21   320.83
1963    318.80   319.08   320.15   321.49   322.25   321.50

           Jul      Aug      Sep      Oct      Nov      Dec
1960    318.15   316.00   314.23   314.07   315.04   316.19
1961    318.78   316.84   315.16   315.56   316.14   317.13
1962    319.55   317.75   316.27   315.62   316.84   317.70
1963    319.67   317.61   316.25   316.17   317.01   318.36
```

Many functions in S know how to operate on time-series; in particular, arithmetic operators operate on two time-series to produce a result defined on the intersection of the two time domains.

> **co2 + ts(100,start=1962,end=c(1962,12),nper=12)**

```
Time-series:
start: 1962 1     end: 1962 12     Monthly

           Jan      Feb      Mar      Apr      May      Jun
1962    418.06   418.59   419.74   420.63   421.21   420.83

           Jul      Aug      Sep      Oct      Nov      Dec
1962    419.55   417.75   416.27   415.62   416.84   417.70
```

Special S functions create, manipulate and obtain attributes of time-series. The functions **start(x)**, **end(x)**, **nper(x)** give the starting time, the end time and the number of observations per unit time of **x**. The function **time(x)** returns a time-series on the same domain as **x**, with values equal to the corresponding time. Similarly, the function **cycle(x)** returns a time-series whose values are the period (within the unit of time) corresponding to each observation. For example suppose **x** is a monthly series starting in February 1979 and ending in May 1980. The values in **cycle(x)** are

$$2, 3, 4, \ldots, 12, 1, 2, 3, 4, 5.$$

For some applications (regression, for example) it is helpful to bind together time-series as columns of a matrix. The function **cbind** is not suitable if the series have different time domains, since it ignores the time-series nature of the arguments. The function

tsmatrix(x_1, x_2, ...)

binds its arguments as columns, but computes the intersection of all the time domains for time-series arguments.

Matrices, time-series and other data structures with vectors of data as components are treated as ordinary vectors by certain functions. For example, the function **sort** returns a vector of sorted data values. Even if the argument is a matrix or time-series, the result is no longer a similar structure. Similarly, subsets of matrices and time-series may be selected just as if they were vectors, in which case the result is a vector. For example,

co2[co2>mean(co2)]

produces the vector of all values in **co2** which are larger than the mean value. The function

window(x, start, end)

can be used to produce sub-intervals of time-series.

Functions which expect time-series can be given vectors as arguments, which will be interpreted as yearly time-series starting at time 1.

PROBLEMS

3.5.   Why is the result of sorting a matrix or time-series a simple vector and not a matrix or time-series?

3.4.7 *General Structures*

General data structures arise as the result of statistical and other functions. A general data structure consists of any number of components, each component being either a vector or another general data structure. Each component has a *component name.* For example the function **reg** fits a linear least-squares model. The **reg** function returns a structure with a number of components, including a component named **coef** for the vector of coefficients and another named **resid** for the residuals from the fit. Structure in the input data is preserved in the residuals: a regression which fits a model to a time-series produces residuals which are also a time-series.

Components of a structure are extracted by the operator "$". For example, if one assigns the result of a regression, by **z ← reg(x,y)**, then the vector of coefficients is obtained as:

> **z$coef**

The expression on the left of "$" is usually a dataset name, but it may also be a function call or parenthesized expression (if the rest of the structure is not needed), for example:

> **reg(x,y)$coef**

does the regression, extracts the coefficients, and throws the rest of the regression structure away. The documentation in Appendix 1 describes, for each function, the components of the returned data structure. As in the case of named arguments to functions, it is only necessary to give enough characters of a component name to identify the component uniquely; for example, **reg(x,y)$res** and **reg(x,y)$resid** are identical in effect.

3.5 Data Analysis

Sections 3.3 and 3.4 described the S language and the data it operates upon. With this background, we are ready to deal with our main goal, performing data analysis. This section provides an introduction to data analysis in S. A more comprehensive description is in Chapter 7.

3.5.1 *Statistical Functions*

A few of the more common statistical functions are described here. Elementary estimates of location are:

> **mean(x); median(x)**

for arithmetic mean and median. The mean can be extended to trim a fraction of the largest and smallest values by the optional argument **trim**. Sample variances, covariances and correlations are given by:

> **var(x); var(x,y); cor(x,y)**

If the arguments are matrices, the results are covariance and correlation matrices, treating the columns of **x** and **y** as variables.

A model-fitting function is

> r ← reg(x,y)

which does a linear least-squares regression and returns a *regression structure* containing coefficients (**r$coef**) and residuals (**r$resid**) as well as numerical summaries. The function **regprt** operates on the result of **reg** to print out a summary including significance-test statistics and the covariance matrix of the coefficients. Optional arguments to **reg** allow fitting without an intercept term or fitting by weighted least-squares.

Related functions **l1fit** and **rreg** also fit linear models, but not by least squares. The first fits a general linear model by minimizing the sum of absolute residuals. The other fits a linear model by a robust criterion, essentially trying to down-weight the effect on the fit of observations with large residuals. In some circumstances, either of these methods is preferable to least squares, if there may be a few deviant observations or if the assumption of normal distribution of errors is not appropriate. All the linear fitting functions return components **coef** and **resid** as discussed above.

The function **twoway(x)** fits an additive model to a two-way array; i.e., it decomposes $x_{i,j}$ into

$$\textbf{grand} + \textbf{row}_i + \textbf{col}_j + \textbf{resid}_{i,j}$$

The fit is usually iterative, using medians or trimmed means. The structure returned by **twoway** has components **grand**, **row**, **col**, and **resid** as suggested above. Optional arguments provide control over the method and the number of iterations.

It is often useful to generate pseudorandom values from a specified statistical distribution; e.g., as tests for some analysis or for simulations. S contains random number generators for a number of distributions. They all have names consisting of the letter "r" followed by a code for the distribution, such as "norm" for the normal, "unif" for the uniform and "t" for Students t-distribution. The first argument is the sample size desired. Other arguments, if any, are the parameters of the distribution. Thus,

> xn ← rnorm(100)
> xt ← rt(75,2)

stores in **xn** a sample of 100 from the standard normal and in **xt** a sample of 75 from the t-distribution with parameter (degrees of freedom) equal to 2. The codes for the distributions are:

Code	Distribution	Parameters	Default Value
beta	beta	shape1, shape2	—, —
cauchy	Cauchy	loc, scale	0, 1
chisq	chi-square	df	—
f	F	df1, df2	—, —
gamma	gamma	shape	—
lnorm	log-normal	mean, sd (of log)	0, 1
logis	logistic	loc, scale	0, 1
norm	normal	mean, sd	0, 1
stab	stable	index, skew	—, 0
t	Student's t	df	—
unif	uniform	min, max	0, 1

For information about the parameters to the distributions (including default values), see Appendix 1 under the corresponding distribution code. See also section 7.9 for simulation techniques.

In addition to random number generators for these distributions, there are also probability and quantile functions named by placing the letter "p" or "q" before the distribution code. (See section 7.7).

PROBLEMS

3.6. The formula for computing a sample variance on a dataset **x** of length **n** can be written as:

$$\frac{1}{n-1} \times \sum(x-xbar)^2$$

where **xbar** is the mean of **x**. The formula can also be written:

$$\frac{1}{n-1} \times (\sum x^2 - \frac{(\sum x)^2}{n})$$

(the "desk calculator" formula), Write an S expression that evaluates the variance according to the desk calculator formula. Apply it to the set of data

x ← c(1.5, 17, 2.5, 12, 19.3)

Apply it to x+5000. To x+10000. Are you surprised by the results? Numbers in S are accurate to about 6 decimal digits. How would this help to explain the observed results?

3.7.   Generate samples of size 100 from a gamma distribution with shape parameters 1, 2, 4, 6, and 10. Does the sample mean vary with the shape parameter? Regress the mean on the shape parameter.

3.5.2 *Data Manipulation*

The function **c** combines all its arguments (treated as vectors) into a single vector. It is the usual way of making small vectors inside expressions, or of entering small sets of data.

```
> mydata ← c(1:3, 5, 7, 9)
> mydata
     1   2   3   5   7   9
> 2^c(1, 12, 19)
         2     4096    524288
```

The function **rep** replicates its first argument as many times as specified by its second argument:

```
> rep(1:3,2)
     1 2 3 1 2 3
```

Use of **rep** is a good way to fill out data structures with patterned values; e.g.,

rep(c("Low","Mid","High"),c(3,6,5))

makes a character vector with 3 repetitions of "Low", 6 of "Mid", and 5 of "High". The first and second arguments are used in parallel when both are vectors.

The function **seq** is a general form of the : operator introduced earlier. The expression **m:n** is equivalent to **seq(m,n)**, but there are also one-argument and multi-argument forms. The expression **seq(x)** returns the vector **1:len(x)** or the vector **1:x** if **x** is a vector of length one. This is convenient for manipulating subsets of vectors; for example, an expression to determine which elements of **x** are negative would be:

seq(x)[x<0]

A third argument **by**, allows **seq** to step by something other than one in producing the sequence going from the first argument to the second. As an alternative, the argument **len** allows specification of the length of the resulting vector:

```
> seq(−1, 1, by=.1)
          -1.0   -0.9   -0.8   -0.7   -0.6   -0.5   -0.4
[ 8]      -0.3   -0.2   -0.1    0.0    0.1    0.2    0.3
[15]       0.4    0.5    0.6    0.7    0.8    0.9    1.0
> seq(−pi, pi, len=10)
          -3.141592     -2.443461    -1.745329
[ 4]      -1.047197     -0.3490658    0.3490660
[ 7]       1.047197      1.745329     2.443461
[10]       3.141592
```

Generally, **seq** tries to produce an appropriate sequence from whatever combination of arguments it is given.

The function **sort** returns the vector of sorted numeric or character-string data values corresponding to its argument: **sort(x)** returns the values in **x** in ascending numerical or alphabetical order. (Note: S sorts using the ASCII character set; hence, all numbers sort before all upper-case letters which sort before all lower-case letters.)

```
> x ← c(2, 6, 4, 5, 5, 8, 8, 1, 3, 0)
> sort(x)
         0    1    2    3    4    5    5    6    8    8
> sort(Weekdays)
          "Friday"     "Monday"      "Thursday"      "Tuesday"
[4]       "Wednesday"
```

The **order** function is closely related to **sort**. It returns the ordering permutation, i.e.,

x [order(x)]   # is the same as   sort(x)

Functions **sort** and **order** are very powerful in doing data analysis which depends on sorting or on finding large or small values. The **order** function is used most frequently to permute other data consistently with the result of sorting one set of data.

```
> xorder ← order( x )
> xorder
         10    8    1    9    3    4    5    2    6    7
> a ← x[xorder]          #sorted version of x
> b ← y[xorder]          #y in the same order
```

Another capability of **order** is sorting on several fields. For example,

o ← order(name, salary)

gives an ordering that is alphabetical by name. If two names are identical, the salary determines the ordering.

Notice that the sorting routines sort in ascending order. To get sorts or orderings in *descending* order, use the reversing function **rev;**

for example, **rev(sort(x))** returns **x** in descending order.

As illustrated by the example using **order**, the subset operation can re-arrange a vector. It can also be used to replicate data values in a vector or the rows and columns of a matrix:

> **x[c(1:5,4:1)]**

would contain the first through fifth values of **x** followed by **x**'s fourth through first values.

The function **rank** returns the positions of the original elements in the sorted data and takes account of ties.

> **rank(x)**
>   3.0   8.0   5.0   6.5   6.5   9.5   9.5   2.0   4.0   1.0

See section 7.6 for a statistical application of **rank**.

The function **match(x,y)** tries to find each element of **x** in **y**. The *i*th value in the result is the position of **x[i]** in **y** (or NA if **x[i]** was not found in **y**). There are two useful applications for **match**. In one the second argument is a set of possible values and the purpose is to turn the values in **x** into indices in this set. Suppose, for example, that the data in **x** was coded "L", "M" or "H" for low, medium or high. For analysis, we prefer numerical values. Then,

> **match(x,c("L","M","H"))**

is a vector that codes the characters as 1, 2 and 3. This is very similar to the operation carried out by the function **code**, described in section 5.1.3.

The second use is closely related to the **rank** and **order** functions above. In this **x** and **y** represent versions of the same data (or nearly so), but in a different order. If we do not know the relation between the two sets, then **match** can be used to find it. Specifically, if **x** and **y** have exactly the same values, **match(x,y)** is the permutation which orders **y** the same way as **x**. For example, suppose **names** is a vector of labels for the rows of matrix **mydata**. Now some new data, a matrix **update**, is obtained, but the rows of the matrix are in a different order than the rows of **mydata**, specified by **newnames**. Then the new columns of data can be put into the old order and appended to the old data by:

> **mynewdata ← cbind(mydata,**
> +   **update[match(names,newnames),])**

Notice that **match** matches **names** to **newnames**. Also, this expression has a built-in check that all the observations in **names** are included in **newnames**; otherwise, the subset expression will contain NAs and an error will be generated.

Of related interest is the function **unique(x)**, which returns all the unique values in **x**. In applications like the previous example, one would want to keep **names** as a unique list of names (e.g., so that a correct count of the number of names would be given by **len(names)**).

The functions **all** and **any** take logical vectors and return a single TRUE or FALSE according to whether any or all of the values are TRUE. For example,

> **all(x>=0)**
> **any(x<0)**

return TRUE and FALSE, respectively, if there are no negative values in **x**.

A number of special data-manipulation functions are available for time-series. If **x** is a time-series, the following data manipulation functions are available.

> **smooth(x)**   # smooth the data values
> **diff(x)**   # forward differences
> **lag(x,k)**   # same data occurring k periods earlier
> **window(x,start,end)**   # sub-series

The smoother is a robust smoother, made up of repeated simple smoothing operations (see Appendix 1). The series **diff(x)** has the same starting date, but has length **len(x)−1**. The lagged series, in contrast, has the same number of data values as **x** with the starting and ending times altered.

PROBLEMS

3.8.   Data from the New Jersey Pick-it Lottery is in the system directory under the names **lottery.number** and **lottery.payoff**.

a)   Find which winning numbers had payoffs more than $500.

b)   Find the 10 smallest payoffs and the corresponding "unlucky" numbers. *Hint:* the function **order** is helpful.

3.9.   Explain the behavior of the following set of S expressions:

> > x ← 5; seq(x)
> > x ← 2:5; seq(x)

3.10.   Try the expression

> seq(−10,10,by=.1)

Can you explain why the printed result has as many decimal places as it does?

3.11. How would you sort a set of data in reverse order? What if you couldn't use the function **rev**? Given the **name** and **salary** data, how could you produce a list in alphabetic order by name, but sorting identical names by decreasing salary?

3.12. Given a vector **name** of names, a corresponding vector **salary** of salaries, and another list of names **promotion.list**, find the salaries of the people in **promotion.list**. What happens if "Smith" is in **promotion.list** but not in **name**?

3.5.3 *Numerical Calculations*

The S language includes the standard mathematical functions:

sqrt	Square root
abs	Absolute value
sin cos	Trigonometric functions (radians)
asin acos atan	Inverse trig functions
exp log	Exponential and natural logarithm
log10	Common logarithm
gamma lgamma	Gamma function and its natural log

These functions return the values from a vector or numeric structure transformed by the corresponding mathematical function. In particular, operations on matrices and time-series yield matrices and time-series as values. By the way, note that, for positive integer values, **gamma(x+1)** is the same as x-factorial "x!". The **lgamma** function is important since the gamma function grows so rapidly.

There are also a number of functions which compute elementary numerical results:

ceiling   floor   trunc   round

ceiling and **floor** find the closest integer values not less than and not greater than the corresponding values in their arguments. The function **trunc** truncates values towards zero, while **round** rounds values to the nearest integral value. Optionally, **round** can take a second argument, giving the number of digits after the decimal to which rounding takes place. Some examples:

```
> x
     -1.90691    0.76018  -0.26556  -1.89828    0.08571
> ceiling(x)
     -1    1    0  -1    1
> floor(x)
     -2    0  -1  -2    0
> trunc(x)
     -1    0    0  -1    0
> round(x)
     -2    1    0  -2    0
> round(x,1)
     -1.9   0.8  -0.3  -1.9   0.1
```

There are also functions to add and multiply together all the elements of a vector: **sum(x)** and **prod(x)**, which return a single number giving the sum or product. The functions **max** and **min** give the largest and smallest values in their arguments (these two functions can have an arbitrary number of arguments). The function **range** returns a vector with two numbers which are the max and min of all of its arguments. That is, **range(x,y)** is equivalent to **c(min(x,y),max(x,y))**.

The function **diff** mentioned in section 3.5.2 can be used on arbitrary numerical data to produce the first differences. The function **cumsum** is almost the inverse of **diff**; that is, the i-th element of **cumsum(x)** is the sum of the first i elements of **x**. For example, if $x \leftarrow 1{:}10$, then **diff(x)** is a vector of 9 ones and **x** is equal to **cumsum(rep(1,10))**. In general, **x** is the same as **cumsum(c(x[1],diff(x)))**.

PROBLEMS

3.13.   Show that **cumsum(c(x[1],diff(x)))** gives x.

3.14.   The *geometric mean* of a set of **n** numbers is the **n**-th root of the product of the numbers. Show how to compute this in S.

3.15.   Suppose you have some data that might have very large or very small numbers. You want to find the geometric mean, but the product of the numbers might be too big or too small for the machine to represent. How would you get the answer?

3.16.   How would you compute the factorial function if there were
no **gamma** function?

3.6 Graphics

The combination of interactive computing with graphics provides a powerful environment for data analysis. In S there are a
number of high-level plotting functions, along with other facilities for
adding to plots or controlling their appearance in detail.

S is a *device-independent* graphics system—it can produce output
on any one of a large number of graphics devices, once the user
informs S of the target device. After S knows the device name, all
graphical output is tailored to that device. Thus, before using any of
the graphics functions, the user must specify the type of graphics terminal or device which will be used for plotting. The interactive terminals supported by S, and the S functions which specify the devices
are given under **devices** in Appendix 1.

The **printer** device will work on any terminal, but gives considerably less attractive plots than the true graphics terminals. A device function must be given once before doing any plotting. The device may be changed during the terminal session (for example, if
changing paper size on the pen plotters) by giving another device
function later.

The most frequently useful high-level plotting routine is

plot(x,y)

which produces a scatter plot of the two sets of data†. For example

> **plot(corn.rain, corn.yield)**

produces the plot in Figure 1. The function **plot** has an optional argument, **type**, which can be used to produce various other styles of plot;
for example, for a plot with lines connecting the points in sequence,
rather than characters plotted at the points:

---

† All S functions doing plotting follow the convention of giving the arguments **x,y** to specify data to plot, even though one might say "Plot y vs x." The
consistency (which is shared with model-fitting functions like **reg**) is convenient and easier to remember.

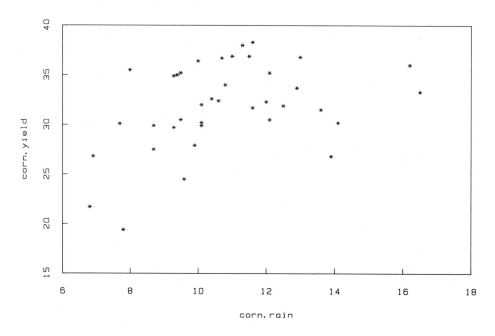

Figure 3.1. Yield of corn in bushels per acre as a function of rainfall in inches in six states, 1890 to 1927.

plot(x,y,type="l")

(The character inside the quotes is the letter *ell* standing for "lines", and not the digit one.) Other values for **type** include "b" (both) to plot both points and lines. The default value is "p" (points), for a scatter of points.

Logarithmic transformations (to base 10) of either or both axes may be done automatically:

> **plot(corn.rain,corn.yield,log="y")**

will use a log scale for y. This differs from

> **plot(corn.rain,log10(corn.yield))**

since in the first case the y-tick labels are on the original, not the log scale. Similarly, values of "x" give a log scale for x-axis, and "xy" or "yx" give a log scale for both.

The **plot** function can also be used to produce a time-series plot, if only one dataset is given. In this case, the x-axis will show time, and the y-axis the data values. (If the argument is not a time-series, the x-variable will be **1:len(x)**.) While **plot** will plot a single time-series, as above, there is a special high-level routine which plots any number of time-series:

tsplot(y1,y2, ...)

The x-axis will be a range of times which includes the range of all the arguments. For this plot, the default type is "l", connected lines. Other types may be given as above. When there are several series given, the function uses changes in line texture (dashed lines), in color, and in the plotted character to differentiate the series.

A similar function plots one or more columns of a matrix **y**, against corresponding columns of another matrix, **x**:

matplot(x,y)

The type is "p" by default, and different plotting characters, line textures and colors are again used. If the arguments differ in number of columns, the function will reuse columns cyclically. In particular, either of the arguments may be a vector.

A different high-level plotting function plots a histogram of a single set of data:

hist(x)

Arguments are available to control the number of bars, scaling, shading, etc.; see Appendix 1. There is an analogous routine for non-graphics terminals:

stem(x)

which prints a stem-and-leaf display. (There are examples in sections 3.7 and 4.1.2.) This is essentially a histogram with bars shown horizontally. In addition, the bars are made up of digits to give (to limited accuracy) the individual values in the data. Other summary information is also printed: see Appendix 1.

A powerful graphical tool for looking at probability distributions is the *probability* or *quantile-quantile (Q-Q) plot*. The S function

qqplot(x, y)

produces an *empirical Q-Q plot* useful for investigating whether **x** and **y** appear to be samples from the same underlying distribution. The function

qqnorm(x)

produces a *normal Q-Q plot* that plots the sorted data **x** against quantiles of the standard normal distribution. Plots with relatively large or systematic deviations from a straight line give evidence of non-normality.

Once you have a plot, it can be titled by

title(main,sub,xlab,ylab)

which places a main title on top, a sub-title at the bottom, and labels on the axes. (Any of the arguments may be omitted.) For example:

> **title("Iris Data Analysis","Figure 3",**
+ **"Sepal Length","Petal Length")**

Many of the high-level functions will automatically generate x and y labels from the names of the plotted datasets. In this case, simply omit **xlab** and **ylab** from **title**. You can also specify **main**, **sub**, **xlab**, and **ylab** as arguments in **name=value** form to any high-level plotting function.

Points, lines and text may be added to an existing plot through the functions **points**, **lines**, and **text**. For example, to plot a time-series and add a smoothed curve to the plot:

> **plot(co2)**
> **lines(smooth(co2),col=2,lty=2)**

We specified values for the *graphical parameters* **col** and **lty** to control the color and line-type for the plotted curve. In using the **points** function, parameter **pch** may be useful in specifying an alternative plotting character; for example, **points(x,y,pch="+")**. For plots with a large number of points, either through **points** or one of the high-level functions, one may prefer a plotting dot to a character. For a centered plotting dot, use **pch="."**. Also, the argument **marks=** to the **points** function uses a set of plotting symbols rather than characters; see Appendix 1 for the set of symbols available. For a further description of graphical parameters, see section 4.3.

When using the printer device, each frame is assembled in an internal buffer. To get the buffer to be printed, the user must signal the end of this frame. This is done automatically when another high-level plotting function is invoked (but remember that you will always be one frame behind), or can be controlled explicitly by the function

show

Use of the **show** function does not preclude further modifications to the plot. For example,

> **> plot(x,y)**
> **> show**
> **> title("This is my plot")**
> **> show**

displays the plot, augments it with titles, and then redisplays it.

PROBLEMS

3.17. Try doing a printer plot on a standard display terminal (it need not be a graphic terminal). What should you do if you have a small screen? (Hint: see **printer** in Appendix 1.)

3.18. Plot a curve showing the function **sin** over the range $-\pi$ to π. Plot both **sin** and **cos** together on this range.

3.7 A Case Study

This "basic" chapter ends with an example that makes use of many of the ideas presented above in the context of a realistic problem in data analysis. The techniques described in this section should enable you to begin working with S. After reading this far and studying the example, you should do some simple analysis on data that interests you.

We want to read data from a file "mydata" that contains five variables (pop.<15, pop.>75, per-capita income, rate of growth, and savings rate) for each of 15 countries.

The file itself looks like this ...

```
> !cat mydata
29.35   2.87   2329.68   2.87   11.43
23.32   4.41   1507.99   3.93   12.07
23.80   4.43   2108.47   3.82   13.17
41.89   1.67    189.13   0.22    5.75
```

```
42.19   0.83    728.47   4.56   12.88
31.72   2.85   2982.88   2.43    8.79
39.74   1.34    662.86   2.67    0.6
44.75   0.67    289.52   6.51   11.9
46.64   1.06    276.65   3.08    4.98
47.64   1.14    471.24   2.8    10.78
24.42   3.93   2496.53   3.99   16.85
46.31   1.19    287.77   2.19    3.59
27.84   2.37   1681.25   4.32   11.24
25.06   4.7    2213.82   4.52   12.64
23.31   3.35   2457.12   3.44   12.55
```

The rows correspond to different countries, the columns are the variables.

> **m ← matrix(read("mydata"), ncol=5,**
+   **byrow=T)   # read the data and create matrix**
Read 75 items
> **m   # print the matrix**

```
Array:
15 by 5

            [,1]    [,2]      [,3]    [,4]     [,5]
[  1,]    29.35    2.87   2329.68    2.87    11.43
[  2,]    23.32    4.41   1507.99    3.93    12.07
[  3,]    23.80    4.43   2108.47    3.82    13.17
[  4,]    41.89    1.67    189.13    0.22     5.75
[  5,]    42.19    0.83    728.47    4.56    12.88
[  6,]    31.72    2.85   2982.88    2.43     8.79
[  7,]    39.74    1.34    662.86    2.67     0.60
[  8,]    44.75    0.67    289.52    6.51    11.90
[  9,]    46.64    1.06    276.65    3.08     4.98
[10,]    47.64    1.14    471.24    2.80    10.78
[11,]    24.42    3.93   2496.53    3.99    16.85
[12,]    46.31    1.19    287.77    2.19     3.59
[13,]    27.84    2.37   1681.25    4.32    11.24
[14,]    25.06    4.70   2213.82    4.52    12.64
[15,]    23.31    3.35   2457.12    3.44    12.55
```

We construct a character vector to label the columns:

> **varname ← c("% Pop.<15", "% Pop. >75", "Income",**
+ **"Growth",**
+ **"Savings")**

> varname   # print it

```
         "% Pop.<15"     "% Pop. >75"    "Income"          "Growth"
[5]      "Savings"
```

> income ← m[,3]   # third column is per-capita income
> young ← m[,1]   # first column is % population under 15
> old ← m[,2]   # second is % over 75

> middle ← 100 − (young+old)   # percent pop not young or old
> median(young); median(old)   # overall median percents
> median(middle)   # overall median percents
```
         31.72
         2.37
         65.43
```
> stem(middle)

```
N = 15    Median = 65.43
Quartiles = 54.58, 71.65

Decimal point is 1 place to the right of the colon

    5 :  122
    5 :  5679
    6 :
    6 :  58
    7 :  002223
```

> ls   # see what datasets are in working directory
```
         "income"    "m"        "middle"    "old"
[5]      "varname"   "young"
```

With the text editor, we have already created a file containing
the names of our countries, called "names":

> !cat names
```
"Australia"
"Austria"
"Belgium"
"Bolivia"
"Brazil"
```

```
"Canada"
"Chile"
"Taiwan"
"Colombia"
"Costa Rica"
"Denmark"
"Ecuador"
"Finland"
"France"
"Germany"
```

> **countries ← read("names")**
```
Read 15 items
```

Now a pretty print of our data matrix:

> **print(m,rowlab=countries,collab=varname)**

```
Array:
15 by 5

              % Pop.<15   % Pop. >75    Income   Growth   Savings
Australia        29.35         2.87    2329.68     2.87     11.43
Austria          23.32         4.41    1507.99     3.93     12.07
Belgium          23.80         4.43    2108.47     3.82     13.17
Bolivia          41.89         1.67     189.13     0.22      5.75
Brazil           42.19         0.83     728.47     4.56     12.88
Canada           31.72         2.85    2982.88     2.43      8.79
Chile            39.74         1.34     662.86     2.67      0.60
Taiwan           44.75         0.67     289.52     6.51     11.90
Colombia         46.64         1.06     276.65     3.08      4.98
Costa Rica       47.64         1.14     471.24     2.80     10.78
Denmark          24.42         3.93    2496.53     3.99     16.85
Ecuador          46.31         1.19     287.77     2.19      3.59
Finland          27.84         2.37    1681.25     4.32     11.24
France           25.06         4.70    2213.82     4.52     12.64
Germany          23.31         3.35    2457.12     3.44     12.55
```

What countries have per-capita income above the median?

> **countries[income > median(income)]**
```
        "Australia"   "Belgium"    "Canada"       "Denmark"
[5]     "Finland"     "France"     "Germany"
```

Now plot two of the variables:

> **hp7470   # declare the graphics device**
> **plot(young,income) #scatter plot**

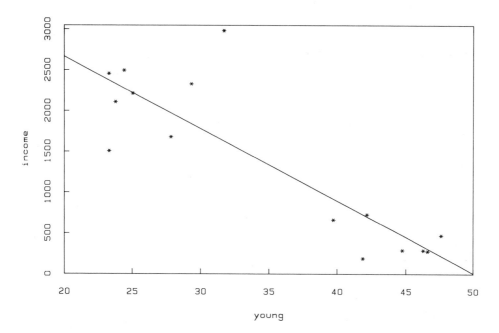

Figure 3.2. Scatter plot showing per-capita income as a function of percent of population under age 15.

Fit a straight line to the data by linear regression

> **r ← reg(young, income)**
> **abline(r)   # include line on the plot**
> **regprt(r) # print a summary of the regression**

```
                Coef       Std Err     t Value
Intercept     4439.262     505.3514     8.78451
x2             -88.6242     14.1152    -6.27864

Residual Standard Error = 516.661 Multiple R-Square = 0.752009
N = 15         F Value = 39.4213 on 1, 13 df

Covariance matrix of coefficients:
```

```
              Intercept        x2
Intercept     255379.9
x2                -6880.11   199.2387
```

```
Correlation matrix of coefficients:
              Intercept
x2                -0.964529
```

> **stem(r$resid)   # look at residuals from regression**

```
N = 15    Median = -29.1802
Quartiles = -254.477, 221.4702

Decimal point is 3 places to the right of the colon

 -0 : 95
 -0 : 3322000
  0 : 0123
  0 : 5

High: 1354.777
```

> **q   # Now finished**

PROBLEMS

The data shown in this example are available to S users under the name **saving**: **saving.x** is a matrix with 5 columns and 50 rows (countries), and **saving.rowlab** is a vector of the 50 country names.

3.19. Produce the plot showing per-capita income as a function of percent of the population under age 15. This will be just like Figure 2, except it will include 50 countries. Should either of the variables be re-expressed (e.g., transformed to a logarithmic scale)?

3.20. Fit a line to the plot from the previous problem. Is the fit better or worse than with 15 points? Show the fit to the first

15 points on the same plot.

3.21. Is savings rate related to per-capita income?

3.22. Investigate other relationships within this data set. For example, is savings rate related to any of the other variables?

4

Graphical Methods in S

The combination of interactive computing with graphics provides a powerful environment for data analysis. Plots in S can be used to look at data in a variety of ways, and also to study the results of analysis, such as the fitting of models and various statistical summaries. Plotting in S can be done in many ways, with a range of detailed control by the user:

- • S has many high-level plotting functions that produce a complete plot in one expression.
- • Users can customize the appearance of the standard plots by use of *graphical parameters* and other options within the S graphics facilities.
- • Beyond this, graphical methods can be more extensively revised, and new graphical methods constructed, by building up plots from a wide range of lower-level graphical functions.

S graphics are *device independent;* that is, users type the same expressions, regardless of the kind of plotter or terminal being used to generate the plots. Before doing any plotting, you must execute a device function. Thereafter, S automatically generates suitable plots for the particular device. A list of the graphics device functions is given under **devices** in Appendix 1. Look there for the particular devices you have available and execute the appropriate device function to enable plotting. Supported devices range from inexpensive pen plotters to high-resolution color video display terminals. In addition, some device functions have arguments that define how the device is used, e.g., the size of paper in a pen plotter.

4.1 Looking at Data

To understand the scope of graphics in interactive data analysis, it will help to organize our thinking along the lines of *what* we want to see in our plots. Summarization is an important goal of data analysis; plots use a graph, instead of numbers, to summarize the data. Plots have the advantage that much more information can be presented; the human eye and brain can perceive patterns, and exceptions to patterns, in a very sophisticated way. Further, in an interactive setting, we can respond to these perceptions by doing new analysis, leading to new plots.

In this section we will examine the most common plotting situations in data analysis and say briefly how S provides for such plots. Of course, the "data" need not be the original observations; just as often, they will be the results of earlier analysis. Readers interested in more of the data-analysis background are referred to the book *Graphical Methods for Data Analysis* mentioned in the Preface. Sections 4.1.1 through 4.1.4 correspond roughly to chapters in that book with similar names.

4.1.1 *Plotting Two-Dimensional Data*

The fundamental way of looking at the joint patterns of two variables, **x** and **y**, is the *scatter plot*. In S this is produced from

    plot(x,y)

which plots **x** on the horizontal axis and **y** on the vertical axis. For example, suppose we have observations on 21 successive experiments measuring exhaust stack loss from a chemical process, and corresponding measurements of the airflow in the process. (This is from a famous statistical example. See **stack** in Appendix 2; **airflow** is the first column of **stack.x**.)

    > **plot(airflow,stack.loss)**   # **Figure 1**

The scatter plot function has several optional arguments, to control the type of plot (points and/or connected lines) and to specify titling information. We will discuss these in sections 4.2-4.4, or you can look at **plot** in Appendix 1. For the moment, one useful argument is **log=** which controls logarithmic transformations of the axes. The expression

    > **plot(x,y,log="xy")**

will produce a plot with two logarithmic axes. The argument **log="x"**

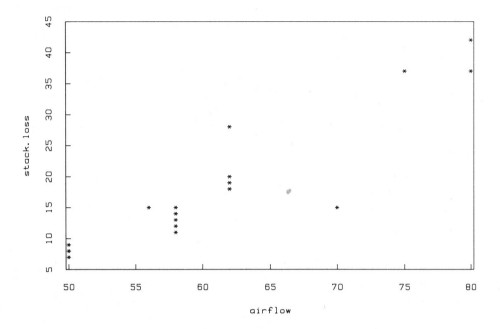

Figure 4.1. Scatter plot of stack loss and airflow.

will make just the x-axis logarithmic; **log="y"** produces a logarithmic y-axis.

A special form of a scatter plot displays a single dataset **y** in order against 1, 2, ... (or against the observation time, if **y** is a time-series). This plot, usually called a *time-series plot,* is produced by giving **plot** just the one data set. To see the stack loss in the successive experiments:

> **plot(stack.loss)   # Figure 2**

The function **tsplot** can be used to plot any number of time-series arguments. If the time domains of the arguments are not identical, the time-series plot will use a large enough time domain to include all the data. For example

> **tsplot(hstart,smooth(hstart),type="pl")   # Figure 3**

will plot the time-series **hstart** as a set of points and a smoothed version of this data as a connected set of lines. As Figure 3 shows, **tsplot** uses different plotting characters, line styles, and colors, if they are available on the graphic device.

The **matplot** function produces a scatter plot of several sets of data at once, plotting columns of one matrix against columns of

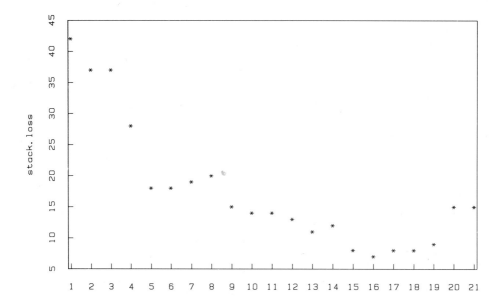

Figure 4.2. Time-series plot of stack loss.

another. By default, the sets of data are plotted as scatters using plotting characters "1", "2", ..., but the type of plot can be controlled by the **type=** argument as shown above with **tsplot**. As an example, suppose we make a plot of the sepal length and width for three varieties of iris. We can form a 50 by 3 matrix for each of the two variables, which are part of the 50 by 4 by 3 array **iris**. Sepal length and width correspond to values of 1 and 2 for the second subscript, so we can create the matrices and plot by:

> **length ← iris[,1,]**
> **width ← iris[,2,]**
> **matplot(width,length)   # Figure 4**

Many of the plotting functions try to make sense of missing values (NAs) in the data. The general philosophy is that no point will be plotted if one or more of its co-ordinates is NA, and no line segment will be drawn if either of the end points is NA. This applies to **plot**, **tsplot**, and **matplot**, as well as to the lower-level functions

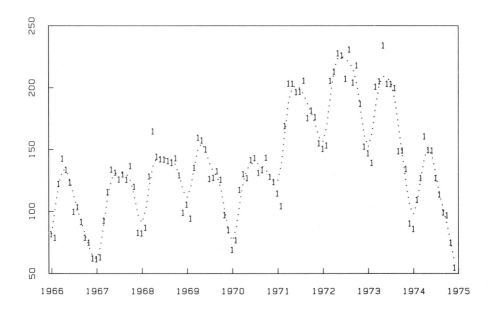

Figure 4.3. Time-series plot of US housing starts showing raw data and a fitted, smooth line.

points, **lines**, and **text** described in section 4.4.1.

PROBLEMS

4.1. Figure 1 shows that there are only a few discrete values for **airflow**. Make a *jittered* plot by adding enough random noise to the **airflow** values to prevent them from overlapping.

4.2. Data which represent mixtures of three components are basically two-dimensional, since the sum of the fractions must equal one. The macro **mixplot** produces a plot appropriate to such data. Learn to use it by studying its documentation in Appendix 1. Apply it to the **telsam** data, showing the fractions of "excellent", "good", and "fair/poor" responses obtained by each interviewer.

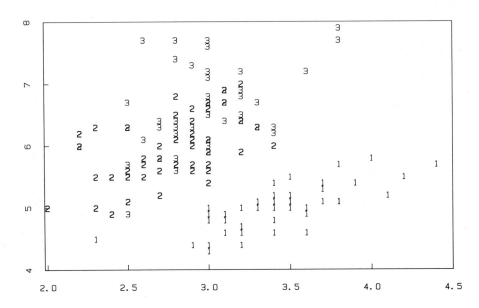

Figure 4.4. Sepal width and length for three varieties of Iris.

4.1.2 *Studying the Distribution of a Set of Data*

Plots can replace numerical summaries of location, scale, skewness, etc. for a single set of data by portraying the *distribution* of the data. A very common way of doing this is with the histogram; in S, the function **hist(x)** plots a histogram of the dataset **x**. There is also **stem(x)**, a semi-graphical form of the histogram, which prints the histogram on its side, and which allows the viewer to read the (rounded) values of the data from the digits to the left of the row, followed by a single digit in the row, one for each data value. (See Chapter 2 of *Graphical Methods for Data Analysis* for more discussion of histogram and stem-and-leaf plots.) To emphasize that plots are important when applied to the results of analysis, as well as to the original data, let's fit a regression model to the stack loss data and look at the residuals.

> stack.reg ← reg(stack.x, stack.loss)
> resids ← stack.reg$resid

> **stem(resids)**

```
N = 21    Median = -0.4550927
Quartiles = -1.711654, 2.36142

Decimal point is at the colon

   -7 : 2
   -6 :
   -5 :
   -4 :
   -3 : 10
   -2 : 4
   -1 : 97544
   -0 : 651
    0 : 9
    1 : 34
    2 : 468
    3 : 2
    4 : 6
    5 : 7
```

To show how the stem-and-leaf works, we can print the residuals, rounded and sorted:

> **sort(round(resids,1))**
```
         -7.2   -3.1   -3.0   -2.4   -1.9   -1.7   -1.5   -1.4
  [ 9]   -1.4   -0.6   -0.5   -0.1    0.9    1.3    1.4    2.4
  [17]    2.6    2.8    3.2    4.6    5.7
```

Notice that −7.2 appears as −7 in the stem (left of ":") and 2 in the leaf (on the right). Similarly, −3.1 and −3.0 appear as a stem of −3 and two leaves of 1 and 0.

Other simple ways of showing a distribution are the *one-dimensional scatter plot* and the *box plot*. The one-dimensional scatter plot just plots all the values in **x** against a constant. It hides too much information to be adequate on its own, but can be a useful extension of other techniques. We will show some uses of it in section 4.5 and in the problems. The box plot represents the data by a box showing the median and quartiles and by whiskers out from the box to show the range of the data. The function **boxplot(x)** will draw such a symbol. We will show some box plots in section 4.1.3 for *comparing* distributions, which is the box plot's strong point.

A special kind of plot is useful when we want to compare a set of data to a probability distribution; namely, the *probability* or *Q-Q plot* (*Graphical Methods for Data Analysis*, Chapter 6). For example, statistical model-fitting techniques often assume that the errors in the model are distributed according to the normal (Gaussian) distribution. Naturally, one should always check if this assumption is reasonable. A good way to do so is to use the normal Q-Q plot. In S, **qqnorm(x)** sorts the data **x** and plots it against the corresponding quantiles of the normal distribution. If the points in this plot are approximately straight, the assumption is supported; various kinds of non-straightness indicate possible anomalies in the data. The stack loss data are often used as an example of fitting a regression model, so we could examine the residuals by:

> **qqnorm(resids)**   # Figure 5

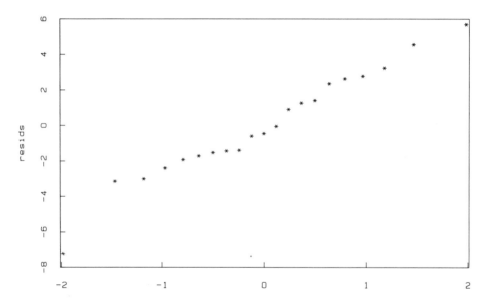

Figure 4.5. Normal quantile-quantile plot of residuals.

The data looks reasonably straight; however we ought to wonder about the smallest residual.

The normal distribution is certainly the most common reference distribution for probability plots, but the same kind of plot can be made against other distributions as well. The system macro **qqplot** does a probability plot against one of several distributions. The macro takes the dataset to be plotted, the name of the distribution (unquoted) and whatever other parameters must be supplied to define the distribution, such as shape or degrees-of-freedom. (By the nature of probability plots, location and scale parameters do not need to be given.) For example, suppose we have a set of data, **t.stats**, that are thought to be independent observations from a *t* distribution with 9 degrees of freedom. A plot to test that assumption is generated by:

> > ?qqplot(t.stats,t,9)

The name of the distribution, **t** in this case, and the definition of the other parameters must be consistent with the way quantiles are computed for that distribution. (See the table of distribution names in section 3.5.1.)

PROBLEMS

4.3.   The dataset **iris** contains 3 sets of 50 observations for 4 variables. Look at the distribution of the 150 values of sepal length (**iris[,1,]**) and sepal width (**iris[,2,]**) to see if they look normally distributed.

4.4.   Use **rnorm** to generate some data from the normal distribution, and plot it against quantiles of the *t* distribution with various degrees of freedom.

4.1.3 *Comparing Data Distributions*

Given two separate sets of data (of the same or different lengths), it is often useful to compare the distributions graphically, by the *empirical Q-Q plot*. This compares the distribution of **y** to that of another dataset, **x**, rather than to a theoretical distribution. In S, **qqplot(x,y)** plots a scatter of the sorted values of the two data sets. If **x** and **y** have different number of data values, points are interpolated in the larger set and plotted against the data in the smaller set. If the data have similar distributions (up to arbitrary location and scale parameters) the resulting plot will be roughly linear. Again, discrepancies from linearity can be analyzed to show how the two distributions differ. As an example, we can consider the lottery data used as a case study in chapter 1. Since we have several sets of data from the New Jersey lottery on the system data directory, we may want to compare the distributions of payoffs from the earliest and the latest sets of data:

```
> qqplot(lottery.payoff, lottery3.payoff)
> # Figure 6
```

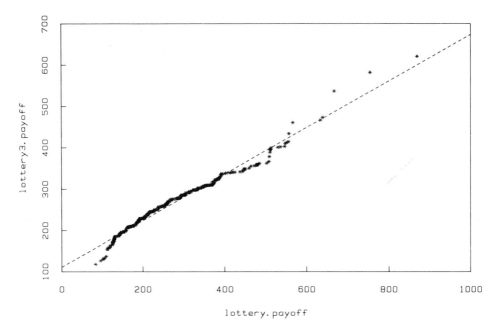

Figure 4.6. Empirical quantile-quantile plot of lottery payoffs for two different time periods.

The pattern of the plot is fairly linear over most of the range, with some deviation in the tails, suggesting a basically similar distribution for the two datasets. We also show on the figure a line, fitted roughly to the pattern of the data. This could be done by eye, but can also be done numerically. (It's a little more advanced than we want to present in this section: the technique is presented as a problem in section 4.4.) Once plotted, the line is helpful in seeing departures from linearity.

Empirical Q-Q plots are useful for comparing two distributions and analyzing their differences. Another way to compare distributions is to use a *box plot*, with a schematic box for each set of data. A horizontal line is drawn through the box at the median of the data, the upper and lower ends of the box are at the upper and lower quartiles, and vertical lines ("whiskers") go up and down from the box to the extremes of the data. (Actually, points that are very extreme are plotted by themselves.) While the box plot shows less information than the Q-Q plot, it is an easy summary to look at and has the great advantage of applying to more than two datasets. For example, we can compare the three sets of lottery payoffs by

> **boxplot(lottery.payoff, lottery2.payoff, lottery3.payoff)**
> **# Figure 7**

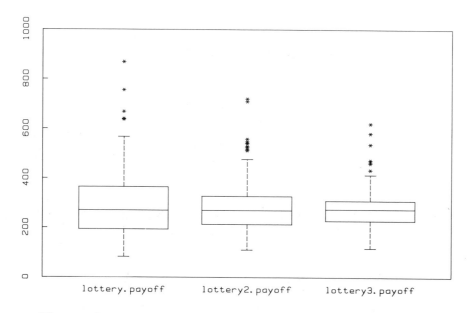

Figure 4.7. Boxplots showing distribution of lottery payoffs for three different time periods.

You can see here that the **lottery3.payoff** dataset has shorter tails than **lottery.payoff**.

Sometimes the sets of data we want to compare are subsets of one dataset grouped by values of some other variable. In the analysis of the lottery data, for example, we noticed that the payoffs seemed to depend on the first digit of the winning number. We can draw boxes for the ten subsets of payoffs in a single plot to study this phenomenon graphically. Rather than extracting each set separately, we use the S function **split** to split **payoff** into groups based on the first digit of **number**, and then give the result to **boxplot**.

> **prefix("lottery.")**
> **digit ← trunc(number/100)**
> **boxplot(split(payoff, digit))   # Figure 8**
> **title(xlab="First Digit of Winning Number",ylab="Payoff")**

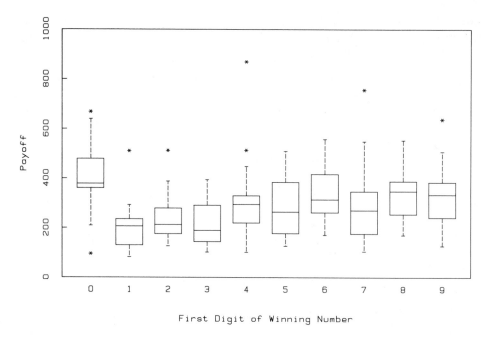

Figure 4.8. Boxplot of payoffs grouped by leading digit of the winning number. Notice the high payoffs for the first box.

PROBLEMS

4.5.   In Chapter 1, we noticed that the distribution of payoffs for winning numbers with a leading zero digit was different from the rest of the payoffs. Use a Q-Q plot to investigate this difference.

4.6.   For the **iris** data in problem 3, compare the distributions of sepal length among the three groups. Do the same for sepal width.

4.7.   In Figure 8, what produces the labels under the boxes? Hint: look at the value returned by **split(payoff,digit)**.

4.1.4 *Plotting Multivariate Data*

Plotting two variables in a scatter plot is very natural: a viewer can easily understand data represented by horizontal and vertical position. For three or more variables there is no equally obvious way to plot data. We present here some easy-to-use plots that work fairly well in many cases.

The case of three variables is special, in that we are used to visualizing the real world in three dimensions. Some modern graphics workstations are capable, for example, of simulating three dimensions by real-time motion of two-dimensional projections of three variables. S has been used experimentally to generate such displays, and we expect this to be a very useful graphical method as suitable workstations become more widely available. For the purposes of this book, however, we will stay with static, two-dimensional plots. Three-dimensional data can be represented in several ways with such plots. A simple approach is to do all three scatter plots of pairs of variables and lay these out on a page, in the style of the draftsman's drawing of an object. The function **pairs(x)** does just this, assuming that **x** is a matrix with three columns, corresponding to the three variables to be plotted. The data set **stack.x** is such a set of data. We can use the optional argument **label** to label the variables with names from the dataset **stack.collab**.

> > **pairs(stack.x, label=stack.collab)** # **Figure 9**

The **pairs** function can be used for any number of variables; if **x** is a matrix with p columns, $p(p-1)/2$ scatter plots will be drawn. However, as the number of variables goes up, it becomes harder to interpret the set of plots and get an overall sense of the data configuration.

Other graphical methods for multivariate data abandon the scatter plot and instead represent each observation with some graphical symbol. One then plots the symbols for all the observations. Such plots are useful for seeing clustering effects and overall trends (if the observations are ordered). They are less useful for seeing patterns among the variables. The S functions **stars** and **faces** are examples of such plotting techniques.

> > **stars(stack.x)** # **Figure 10**

Notice the time trends—all three variables seem to get smaller from observation 1 to observation 17. The star symbols then grow in size from observations 18 to 21.

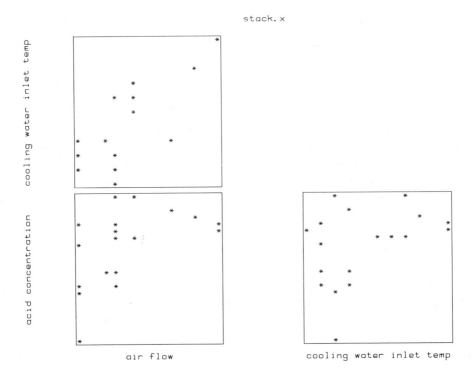

Figure 4.9. Draftsman's plot of three variables of the stack loss data.

A semi-graphical function for displaying multivariate data is

smatrix(x)

This prints a matrix of characters representing the data values in **x**. Larger values are represented by "darker" characters. Overstriking can be used for each character position to obtain a richer set of darkness levels. Optional arguments control the choice of characters and the number of overstrikes. The symbolic matrix is a relatively crude plot, but is able to represent much larger datasets effectively than most of the methods mentioned previously.

One can, of course, combine a scatter plot showing two variables with the symbolic representation of other variables. The function **symbols** takes two sets of data for a scatter plot, just like **plot**, and then one of several arguments to define a set of symbols to plot at the coordinates defined by the first two datasets. For example,

> **symbols(airflow,stack.loss,circles=abs(resids))**   # **Figure 11**

plots circles whose radius is proportional to the absolute value of the

stack.x

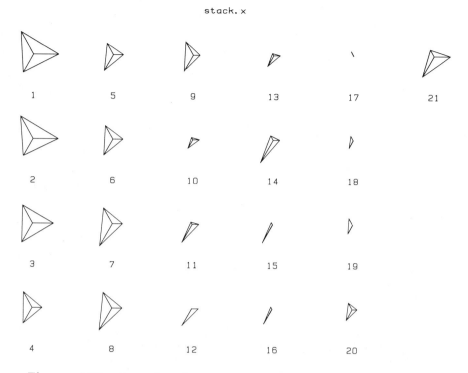

Figure 4.10. Star plot showing twenty observations on the three-variable stack loss data.

residuals from the stack loss regression, if these have been assigned the name **resids**.

PROBLEMS

4.8.   Produce a **matplot** as in Figure 3, but use the plotting characters "S", "V", and "C" to represent the species Setosa, Virginica, and VersiColor.

4.9.   Compare the iris sepal lengths and widths using two boxplots.

4.10.   Extract the 50 by 4 matrix corresponding to the first group of iris flowers. Carry out two of the multivariate plots (**pairs, stars, smatrix,** or **faces**) for this data.

4.11.   Do a stem-and-leaf display and a histogram of the iris sepal

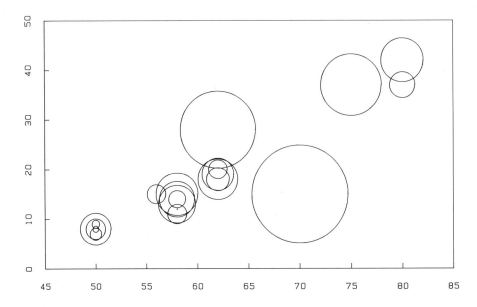

Figure 4.11. Size of residuals from stack loss regression is encoded in circle radius. Circles centers are determined by value of airflow and stack loss.

width data. Compare them. Which do you like better? Why?

4.12.   Produce a jittered **matplot** of **width** and **length** by adding random noise to the data (look up **runif** in Appendix 1). How does your plot compare to Figure 4?

4.13.   Try using the **faces** function, described in Appendix 1, to plot the **stack.x** data. How much does it help you understand the data? Try various assignments of variables to face features. How does this affect the plot?

4.14.   Construct data which lie on the surface of a sphere in three-dimensions:

$$x \leftarrow \text{runif}(100)$$
$$y \leftarrow \text{runif}(100)$$
$$z \leftarrow \text{runif}(100)$$
$$s \leftarrow \text{sqrt}(x\hat{\ }2+y\hat{\ }2+z\hat{\ }2)$$
$$xyz \leftarrow \text{cbind}(x/s,\ y/s,\ z/s)$$

Try using the **pairs** function on **xyz**. What can you see?

4.2 After Plotting: Interacting with the Plot

Much of the power of interactive graphics lies in the ability of the user to see a plot, react to it, and potentially to ask for more information or to augment the plot. This section describes some simple ways to identify information on a scatter plot or to add additional information to the plot.

Frequently, we would like to identify interesting points in a scatter plot, marking them in some way and perhaps finding out which data values produced them. This can be done interactively in S by:

> **plot(x,y)**
> **identify(x,y)**

The function **identify** causes the graphics device to prompt you to point at the interesting values. (To point, you might use a joystick, cursor buttons, or mouse. See Appendix 1 under the name of your graphics device function.) The selected points are marked on the plot by their indices in the data. When you finish pointing (how to signal this is also described in the device documentation), **identify** returns the indices of the identified points as its value. The indices can be useful, for example, if you want to pick off some unusual data points, to be treated specially in a model or other analysis. In Figure 12, we plot absolute residuals from the regression model in section 4.1.2. We notice a few unusual points in the plot, and identify some of them interactively.

> **plot(stack.loss, abs(resids))**
> **i←identify(stack.loss,abs(resids))   # Figure 12**
> **i**
        2 1   4   3   1   2

The identified points can be marked on the plot with any label, by giving **identify** a character vector of the same length as **x**. To suppress marking on the plot, the optional argument **plot=F** can be supplied. Once **i** has been computed as above, one could, for example, use **stack.loss[−i]** to look at the remaining data. See also the discussion of the **text** function in section 4.4 for other ways to identify and mark points.

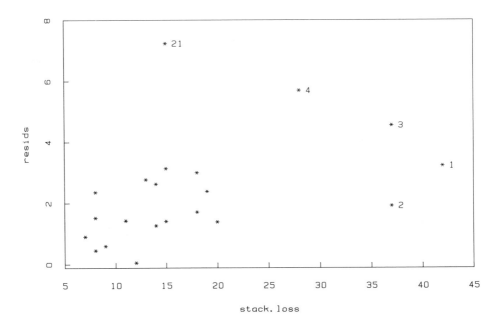

Figure 4.12. The result of interactive point identifications on a plot of absolute residuals against stack loss.

Another useful graphical technique for linear regression and other model-fitting situations is to plot a fitted line on the data from which it was derived. This is done by the function **abline(a,b)**, which plots the line whose equation is $y = a + b*x$. Also, **abline** recognizes the structure returned by the model-fitting functions (like **reg**) and draws the line defined by the coefficients of a simple regression; for example,

> > **reg1 ← reg(airflow, stack.loss)**
> > **plot(airflow, stack.loss)   # Figure 13**
> > **abline(reg1)**

To add an annotation to a region of the plot where there is empty space we can use **rdpen**, which just returns the co-ordinates on the plot at which we point. The result returned by this function can be given as plotting co-ordinates to any S function; in particular, the function **text** will plot text on the current plot at the co-ordinates (see section 4.4 for a discussion of **text**).

> > **text(rdpen(1),"Regression Line")**

With **rdpen** you can ensure that the label does not overwrite anything

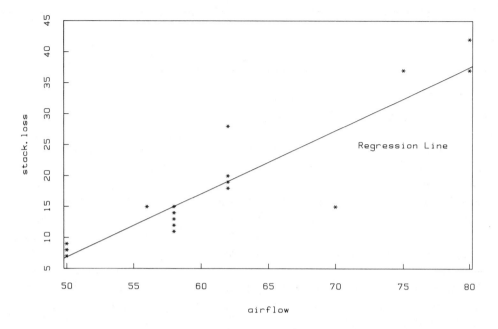

Figure 4.13. Scatter plot of airflow and stack loss, with superimposed least-squares regression line.

else on the plot.

PROBLEMS

4.15. The residuals from the smoothed housing start data are computed by

   hstart − smooth(hstart)

   Produce a scatter plot of the residuals against the raw data.

4.16. Use **identify** to point at outliers on the residuals vs. **hstart** plot from the previous problem.

4.17. Why might you want to use the **plot=F** argument to **identify**?

4.18. Construct a line plot of **sin(x)** for **x** ranging from $-\pi$ to π. Use rdpen to look graphically for the values of **x** that maximize the function. How close did you get to the actual value?

4.19. Use **rdpen** to digitize a simple shape. Hint: use the **type=** argument.

4.3 Customizing Plots

High-level graphics functions can generate a complete plot from a description that says only *what* to plot, without detailed instructions about *how*. For data analysis, particularly analysis at an exploratory stage, this is what we want: to get a picture with a minimum of effort, and then to get on with the process of learning from that picture. However, as we look in more detail at the plot, or when we generate plots for communication with others as well as for learning, we will typically want to control the plot in more detail. The previous section started us off, by showing how to add some information to a plot interactively. In this section, we will explore another form of control: using *graphical parameters* to change the effect of the S graphical functions. Section 4.4 discusses a third technique: *building up plots* from low-level functions in addition to, or instead of, using high-level plotting functions.

4.3.1 *Graphical Parameters*

Graphical parameters control many aspects of the appearance of plots. They can be used in two ways:
1. as arguments to a graphics function, making *temporary* parameter settings for color, line type, plotting character, etc. The parameter settings stay in effect through execution of the graphic function and are then reset to their previous value.
2. as arguments to the special function **par**. Parameters set in a call to **par** stay in effect throughout all plotting until reset in another call to **par** or until re-initialized to their default values when a new graphic device function is invoked.

As arguments either to graphics functions or to **par**, the parameters should be specified in the **name=value** form. For the complete list of graphical parameters, see **par** in Appendix 1.

We have already seen a number of these graphical parameters. For example, a special set of graphical parameters are recognized by most high-level functions, including **plot** and other functions that produce plots with a surrounding box and titles.

xlab=, ylab=
> labels for the x or y axis. These labels default to the names of the data sets plotted on the x or y axis.

main= a main title plotted in enlarged characters above the plot.

sub= a sub-title for the bottom of the plot, below the x-axis label.

type= the type of plot: "p" (points), "l" (lines), "b" (both), "o" (over-struck), or "h" (high-density). See Figure 14. (Section 4.4 introduces **type="n"**, used to set up a plot that will later be filled in.)

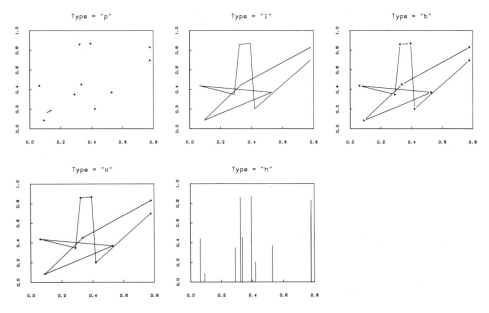

Figure 4.14. Five versions of the **type=** graphical parameter.

> Another set of parameters changes the appearance of points and lines drawn on the plot:

pch= plotting character. The character given is used for scatter plots, in calls to **plot** and similar functions. As we have seen, the default plotting character is an asterisk: for plots with many points, one might prefer **pch="."**, which changes the plotting character to a (properly centered) dot.

cex= character expansion, relative to the standard character size for the graphical device. For example, **cex=0.5** gives half-size characters, **cex=1.5** gives characters 1.5 times normal size. Characters are drawn using the hardware character generator in the device, so beware of the funny results that happen if you specify a character expansion that the device doesn't have— positioning and centering of text will not be correct.

lty= line type. A value of 1 gives solid lines. For devices that support multiple line-types, values 2, 3, ... will select a non-solid line type.

col= color. Devices which can automatically change color will change to one of a set of device-dependent colors.

Some examples of these parameters are shown in figure 15:

```
> plot(2:10,pch="X")
> abline(0,1,lty=2,col=6)   # line with slope 1, intercept 0
> text(6,3,"This is text",cex=2.5)
```

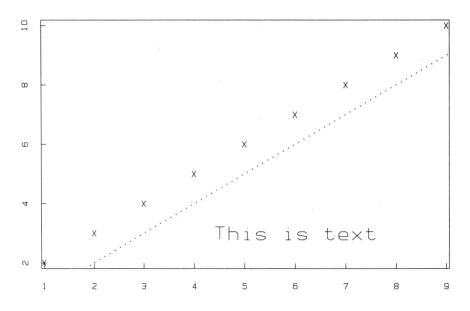

Figure 4.15. A plot showing a number of modified graphical parameters.

These parameters are mostly useful with functions that build up a plot, since they allow different vectors of data on the plot to be distinguished. High-level functions, such as **tsplot** and **matplot**, that produce several lines or sets of points generally try to vary these parameters automatically. Such functions also allow users to specify sets of values to override the defaults. The parameter **type** mentioned above is dealt with similarly. For both **type** and **pch**, the user can supply character strings with each character in turn taken as the parameter value for successive sets of data. Thus,

> **tsplot(hstart, smooth(hstart), type="pl", col=c(1,3))**

plots the series **hstart** as a set of points in color 1, and adds the series **smooth(hstart)** as a connected line in color 3 (see Figure 3).

To generate a printed list of current values for the graphical parameters, use the function **pardump**. The function prints those parameters whose (quoted) names are supplied as arguments; e.g.,

> **pardump("pch","cex")**

will print out the current plotting character and character size. With no arguments, **pardump** prints out the names and current values of all graphical parameters. Since the total printout takes about 60 lines, supplying arguments is a good idea in interactive use.

4.3.2 *Layout; Multiple Plots per Page*

Further description of S graphics will be helped by the introduction of some terminology. A simple graphical display including titles and axes is called a *figure*. A figure is composed of a *plot* surrounded by *margins* as shown in Figure 16. The plot contains the graphical information (the plotted points of a scatter plot, the bars of a histogram, etc.) and is addressed by a *user coordinate* system; that is, by the coordinates determined by the data. The margins surrounding the plot are used for tick labels, axis labels, and titles. Locations within the margins are specified by *side* (1 for bottom, 2 for left, 3 for top, and 4 for right) and the number of lines of text away from the edge of the plot.

The function **par** is the only way to control graphical parameters that deal with the *size* and *layout* of plots, margins, and figures. The parameter **mar=c(m1,m2,m3,m4)** controls the size of the margins which surround the plot. The margins by default are set large enough to accommodate 5, 4, 4, and 2 lines of text on the bottom, left, top, and right of the plot. Looking at a plot shows that the numerical tick labels are spaced out 1 line from the edge of the plot. Since the labels themselves take 1 line, we need at least 2 lines of margin on the bottom and left for the tick labels generated by scatter plots. Therefore, if we want to make the margins as small as reasonable, assuming we won't generate any titles, we could use:

> **par(mar=c(3,3,1,1))**

(Of course, if we don't suppress axis labels in scatter plots, for

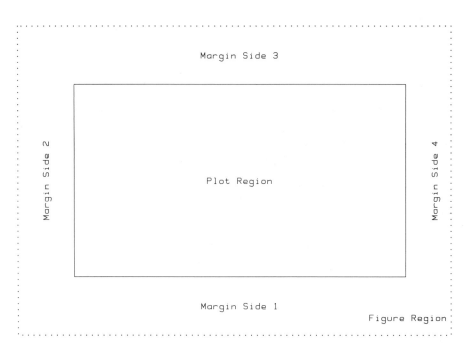

Figure 4.16. Graphical displays consist of a figure region, made up from a plot region (for points, lines, etc.) surrounded by margins which normally contain text.

example, by **axes=FALSE**, there will be warning messages from the plotting functions.)

The parameter **pty** is a character string which describes the plot shape. The default, **pty="m"**, sets up the maximum size plot possible on the device, after allowing for the margins. The other possibility is **pty="s"**, which will force the plot to be square, regardless of the shape of the device.

A more complicated situation involves the positioning of an array of figures on a single page of output as shown in Figure 17. In this case, each figure contains its own plot and margins, and the entire page has a set of *outer margins* surrounding the figures.

A parameter **oma=** specifies four values for the size of the outer margins, just as **mar=** does for margins. By default, there are no outer margins (**oma=rep(0,4)**).

Graphical parameters used with **par** provide the ability to set up an array of figures on an output device. To have 6 plots appear on a page (2 rows of 3 plots each), use

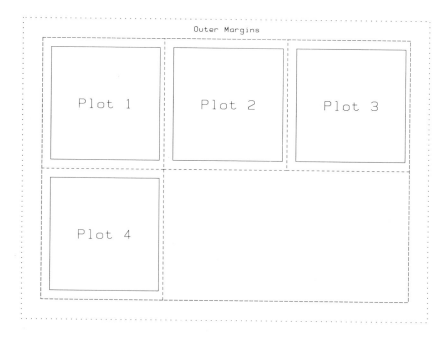

Figure 4.17. An array of figures positioned on the page, surrounded by outer margins.

par(mfrow=c(2,3))

Each successive high-level plotting function will automatically advance to the next available figure region on the page, in a row-first order (first the top row first column, then the top row second column, and so forth). When a page is exhausted, a new one is started, still in the 2 by 3 configuration. Similarly, the graphical parameter **mfcol** sets up an array of figures and accesses them in a column-first order. In addition, **mfrow** and **mfcol** automatically reduce character and margin size to half the standard size if there are more than two rows or columns.

Plotting can be returned to the normal one-per-page form by

par(mfrow=c(1,1))

This also restores standard character size.

4.4 Building Plots

So far, we have shown how to create plots using some of the high-level S functions, and then how to customize these plots by adding some information and by setting graphical parameters. The next step in using S to make customized graphical displays is to build up a plot, composing it out of text, lines, and other symbols that can be added to the basic plot. The paradigm for doing this consists of two steps:

1. set up the plot, typically by using the **plot** function to generate axes and labels (either plotting some data or with **type="n"**);

2. repeatedly invoke S functions to add whatever kind of information you would like to the plot.

4.4.1 *The Data Part*

For a first example, suppose we want to make a scatter plot of **airflow** and **stack.loss** as we did in Figure 1, but instead of plotting the same character at each point, we want the points labelled by the observation number (**1:len(airflow)**) to show the pattern of recording the data. The **text** function can produce this (or any other sequence of numeric or character values). The first step of our paradigm is done by giving the argument **type="n"** to **plot**, to tell it to set up the plot, but not to plot any points. The complete expression to generate our numbered scatter plot is then:

> **plot(airflow,stack.loss,type="n")   # Figure 18**
> **text(airflow,stack.loss,label=1:len(airflow))**

Next, suppose we want to put a "+" where the residuals from the stack loss regression were positive and a "−" where the residuals were negative. (Assume we saved the residuals in **resids**.)

> **plot(airflow, stack.loss, type="n")**
> **text(airflow[resids>0], stack.loss[resids>0],"+")**
> **text(airflow[resids<=0], stack.loss[resids<=0],"−")**

(As you might expect, there are are more compact ways of doing the above: see Problem 24 below).

There are several graphical parameters that affect text strings, whether plotted with **text** or with other text-plotting functions:

adj= justification. The value **adj=0** will cause the string to be left-justified at the specified position, **adj=1** will right-justify, and **adj=.5** (the default) centers the string.

srt= string rotation, measured in degrees counterclockwise from horizontal.

crt= character rotation, measured in degrees counterclockwise from horizontal. By default, whenever **srt** is specified, **crt** is changed to the same value. However, it is possible to change **crt** afterwards. Thus,

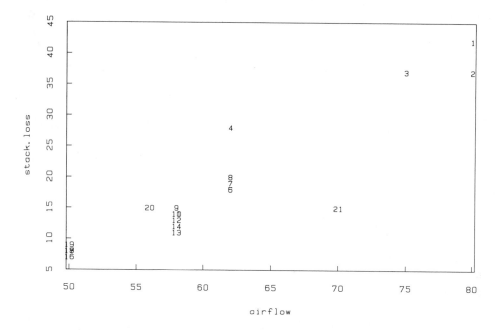

Figure 4.18. Plot of airflow and stack loss where each point is labelled by its observation number.

text(x,y,"My Label",srt=−90,crt=0)

writes a vertical string from top to bottom, with characters oriented horizontally.

Using S interactively, we can label interesting points in a plot using **text**, as an alternative to the use of **identify**. If the interesting points can be described by a logical expression, **text** will be convenient. Since **text** plots labels centered on the co-ordinates given, we will want to offset the labels when adding them to a scatter plot. One method is to include **adj=0** to left-justify the text and stick a couple of blanks on the front of the labels. As an example, in Figure 12, the unusual points could be described by the condition:

> **odd** ← **abs(resids)>4 | stack.loss>25**

To label them:

> **text(stack.loss[odd], abs(resids)[odd], adj=0, " odd")**

The **text** function can take numerical or character data for its **label** argument.

Other useful functions for building up plots include **points,**

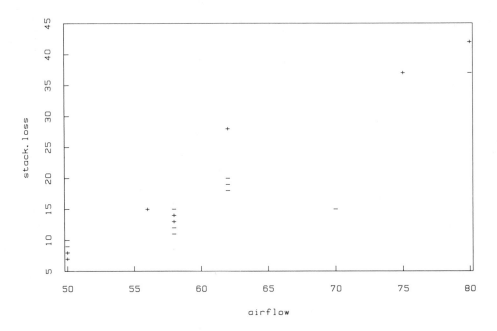

Figure 4.19. Plot of airflow and stack loss, with points labelled by the sign of the residual from a least-squares fit.

lines (for connected lines), **segments** (for unconnected line segments), and **arrows**. The functions that add points, lines, or text to a plot (with the exception of **abline**) all recognize whether the previously set up plot had standard or logarithmic axes. If a logarithmic axis was set up, the data are transformed to the log scale prior to plotting, and should be given in the normal, untransformed scale. The functions **rdpen** and **identify** also work correctly in the presence of logarithmic axes.

The function **symbols** (used in section 4.1.4 to generate a plot) can also add to an existing plot if given the optional argument **add=TRUE**. There are also some functions that generate the data for a plot, but do no plotting. Instead, their result becomes the argument to one of the functions **points**, etc. For example, **lowess(x,y)** returns a smoothed versions of two variables. Thus

```
> plot(airflow,stack.loss)   # Figure 20
> lines(lowess(airflow,stack.loss))
```

generates a smooth curve on the scatter plot of Figure 1. (Here we wanted both the scatter plot and the curve, and so did not use the **type="n"** argument to **plot**.)

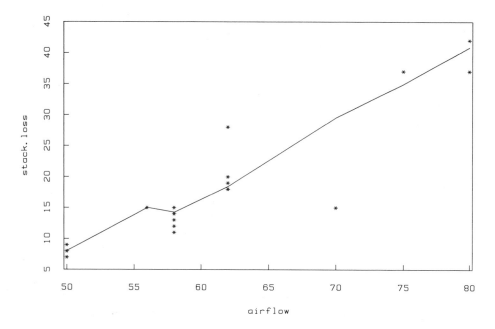

Figure 4.20. A smooth curve superimposed on the airflow and stack loss scatter plot.

This use of **lowess** illustrates a *graphical data structure*. The function **lowess** returns a structure with components named **x** and **y**. The **lines** function, and most other functions that expect two arguments **x** and **y** can, instead, receive a single graphical data structure. The usefulness of the graphical data structure is that it may be generated as the result of many different S functions. Functions that generate a plot structure have no need to do actual plotting.

Other functions returning graphical data structures are **density** (smoothed probability density estimates) and **rdpen** (graphical input). We can construct an empirical density plot by:

```
> plot( density(resids), type="l")   # Figure 21
> points( resids, resids*0 )   # a 1-dim scatter plot
> title("Empirical Density Plot", xlab="resids")
```

Some high-level plotting functions can be used with an argument **plot=FALSE**. In this form, they do no plotting, but instead return a data structure which may be used subsequently as an argument to other functions. For example, **qqplot** and **qqnorm** will return a graphical data structure representing the ordered quantiles and data values which they would have plotted. If **boxplot** is called with

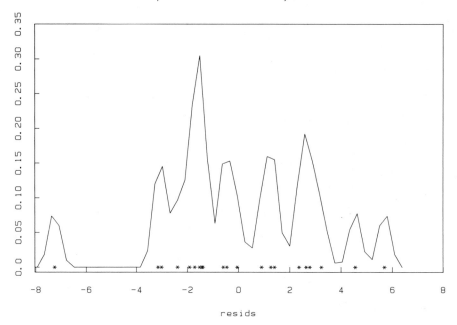

Figure 4.21. A smoothed probability density plot of the residuals from the linear fit to the stack loss data. A one-dimensional scatter plot appears at the bottom.

plot=FALSE, the structure returned is a matrix giving 5 statistics for each of the data sets given as arguments. (Function **bxp** produces a box plot when given one of these structures.) The function **hist** will return a data structure suitable as an argument to the **barplot** function. The function **plclust** returns a structure suitable for an argument to the function **labclust**.

We can use the idea of building-up a plot to combine powerful computational facilities with graphics. For example, suppose we want to join selected cities into a network. We have a dataset in the shared directory (see **city** in Appendix 2) giving locations for various cities, so we will use those cities for the basis of the network. We can draw a map of the United States by using the **usa** function introduced in section 2.6. We can use the **mstree** function (see Appendix 1) to compute the *minimum spanning tree*, that is, the tree with the minimum total length that joins the cities. We can then superimpose the tree on the map. The following S expressions produce Figure 22:

```
> usa(states=F,lty=2)   # dotted map – no state boundaries
> points(city.x,city.y,pch="o",col=2)  # plot the cities
> tree ← mstree( cbind(city.x,city.y*.8), plane=F )
>    # a degree of latitude is approximately
>    # 80% as large as a degree of longitude
> s ← seq(tree)
> segments(city.x[s],city.y[s],city.x[tree],city.y[tree],col=3)
```

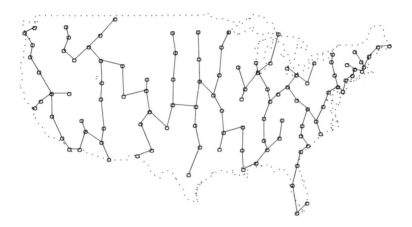

Figure 4.22. A minimal spanning tree connecting selected US cities.

PROBLEMS

4.20. Using the datasets listed in Appendix 2 under **state**, produce a plot of illiteracy vs. income, plotting the state abbreviation at the appropriate position on the scatter plot.

4.21.* Using the function **arrows**, produce a plot showing the sequence of observations on the **airflow, stack.loss** scatter plot. That is, draw an arrow pointing from observation 1 to observation 2, another from observation 2 to observation 3, etc.

4.22. For a set of two-dimensional data of your own choosing (you can always make up data using random number generators), produce several scatter plots using different plotting characters. Which plotting character do you like best?

4.23.   The function **text** allows the arguments **col**= and **cex**= to give vectors of colors and character sizes. Use this to produce the airflow, stack.loss plot using one color for the "+" characters and another color for the "−" characters. Also try to make the character size proportional to the size of the residuals.

4.24.   Look up the functions **matpoints**, **matlines**, **tspoints**, and **tslines** in Appendix 1. When might they be useful?

4.25.   The macro **idplot** uses a combination of the **plot** and **text** functions in order to produce a scatter plot with sequence numbers. Use **mprint** to print the definition of macro **idplot** and comment on its features.

4.4.2 *Annotation*

In addition to data-based information, we often want to add special explanatory information to graphs. S functions can be used to add text (**title, mtext, text**), to generate extra axis ticks and labels (**axis**), and to create legends with sample points and/or lines (**legend**).

The **title** function can plot a main title (at the top), a subtitle (at the bottom) and text to label the **x** and **y** data. However, for more precise control of the location of information in the margins of a figure, the function

mtext(side, line, label)

can be used. The argument **side** specifies whether the text is to be on the bottom (1), left (2), top (3), or right (4) of the plot. The character string in **label** is plotted parallel to the specified side **line** lines of text out from the edge of the plot.

The optional argument **at**= to **mtext** allows text in the margin to be associated with specific user coordinates. For example, if **side** is given as 1 or 3, the argument **at** is interpreted as a set of x-coordinates at which the vector of labels is to be plotted. (The y-coordinates of the text are controlled by the **line** argument). This feature can be used to construct special user-defined axes, etc.; however, the **axis** function (section 4.4.6) is more flexible for this purpose.

Another option to **mtext** allows the overall labelling of a set of multiple figures. Remember the *outer margins* mentioned in section 4.3.2? The optional argument **outer=TRUE** allows the positioning of labelling information in the outer margins instead of in the normal figure margins. For example,

> **par(mfrow=c(3,2),oma=c(0,0,4,0))   # multiple figures**
> **   # with 4 lines of outer margin at top**
> **mtext(side=3,line=0,cex=2,outer=TRUE,**
> **   "This is an Overall Title for the Page")**
> **plot(x,y,main="This is a Title for the Figure")**
> **...**

plots an array of figures with an overall title for the page as well as titles for each figure.

Graphics functions that deal with character strings have the added capability of utilizing embedded *newline* characters that may occur in strings. Whenever this character is encountered, the current plotting position is moved down one line (perpendicular to the string orientation), and the rest of the string is plotted there. In addition, each part of the string between newlines is adjusted independently according to the parameter **adj**. The newline character is entered into a string by using the escape sequence "\n":

> **title("This is the first line \nand this is the second")**

A useful application of **text** is to label a curve or line on the plot. In interactive computing, it is often easiest to select the position for the text using **rdpen(1)** to point at the plot. An example of this process was given at the end of section 4.2. If we are not working interactively, or want to label a curve as part of the plot we are building, some systematic labelling is needed. Generally, we need some point (x,y) on the graph that is a good position to put a label (say, the mid-point of the curve). Given this, **text** can put on the label as shown in section 4.2.

PROBLEMS

4.26. System datasets **corn.rain** and **corn.yield** give amounts of rainfall and corn yields for a number of years. Produce a scatter plot of this data, and draw the **lowess** fit on the plot.

4.27. Compute residuals from the fit in the previous problem, and plot absolute residuals as a function of rainfall. Does the variance of the residuals appear to be constant? Hint: the function **approx** is useful for finding the fitted values based on **lowess** output.

4.28. Use **tsplot** to plot manufacturing shipments (dataset **ship**) and a smoothed value of shipments. Use the function **legend** (look it up in Appendix 1) to annotate the plot.

4.4.3 *Control of Plotting Range*

So far, we have built up plots by using the high-level functions to generate the plotting range, axes and other paraphernalia, and then filling in with whatever we really wanted to plot. This is the preferred approach, since it leaves S to do the necessary calculations about ranges, pretty tick labelling, and so forth. Occasionally, we may be obliged to control the range for one of the coordinate axis to include some specific values. In this case, the range for plotting can be given by the optional arguments **xlim** and **ylim** to the **plot** function. These should be 2-element vectors containing the lower bound and upper bound for data values to be plotted on that axis. For an example, let's turn to the **auto** data and suppose the 3 datasets **price**, **mpg**, and **repair** have been created from the corresponding columns of **auto.stats** (from the shared data directory, see Appendix 2). We want to plot a scatter of **price** and **mpg**, superimposing arrows that point to lower or higher prices for good or bad repair records (assuming $500 for each level of the 5-point repair scale above or below 3). We compute the adjusted price **p2** and draw the desired plot (shown in Figure 23):

```
> p2 ← price−500*(repair−3)
> plot(price,mpg,xlim=range(price,p2),type="n")
> arrows(price,mpg,p2,mpg)
> points(price[repair==3],mpg[repair==3],mark=0)
```

Because arrows of length zero disappear, we marked the corresponding points. See **lines** in Appendix 1 for pictures of the available marks.

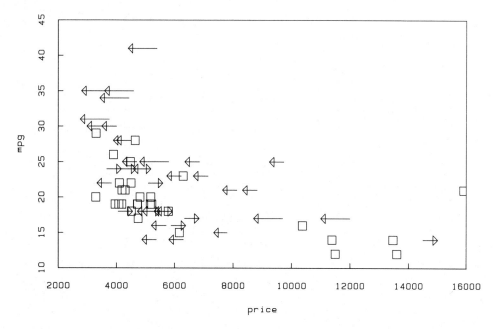

Figure 4.23. On a plot of price and miles-per-gallon, the arrows show price adjusted for repair record.

4.4.4 *User Coordinate Systems and Axis Parameters*

Before any plotting can be done, S must set up a *user coordinate system* which specifies the way that the user's **x** and **y** values will be mapped onto the plotting region. In addition, before axes can be drawn to show the coordinate system, it is necessary to determine where tick marks should be placed. The high-level plotting routines that were introduced in section 4.1 perform both tasks: they determine coordinate system limits from the range of the data values and position the tick marks at "pretty" values.

Occasionally, more control is needed. Perhaps several plots are to be drawn using a single set of axes or the data being plotted are integers so that ticks should be placed only at integral values.

Producing a plot with axes is really a two-stage process: choosing the labelling values and plotting the labels. Table 1 shows the important graphical parameters in the process. The "set-up parameters" are used to compute the "critical parameters"—the user coordinate system and tick labels. The "drawing parameters" affect how the axis is drawn.

To create a plot, the user or the high-level function must

Table 1: Graphical Parameters for Axes	
Set-Up:	
xaxs,yaxs	Axis style.
lab	Number of tick intervals desired.
xlim,ylim	Initial data limits.
Critical Parameters:	
xaxp,yaxp	Axis parameters: min tick, max tick, number of intervals.
usr	User coordinates: xmin, xmax, ymin, ymax.
Drawing Parameters:	
tck	Tick length as fraction of plot. Positive points inward, negative outward.
las	Label style: horizontal or parallel to axis.

determine the range for the coordinate axes. (For now, we will just consider the x-axis, since the y-axis can be treated identically.) These axis limits are initially the range of the data or the values given by a user-supplied **xlim** argument. Using the data range and the graphical parameters **xaxs** and **lab** (see Figure 24), a high-level graphics function will set the critical graphical parameters: **usr** and **xaxp**. The suggested number of tick intervals on the x and y axes is controlled by the parameter **lab=**. For example, to specify approximately 2 intervals (1 tick mark) on the x axis and 5 on the y axis, use **lab=c(2,5)**. Default values are c(5,5). Various axis styles also influence the way in which high-level functions pick pretty values at which to put ticks and labels: see Table 2.

You should hardly ever attempt to set the critical parameters explicitly, since the values are altered by high-level plotting functions. The **usr** parameters are 4 values containing the minimum and maximum values of the user coordinate system for the x axis and minimum and maximum values for the y axis. The **xaxp** parameters are three numbers giving the minimum and maximum coordinates for tick marks, as well as the number of intervals between tick marks (one less than the number of ticks).

Once the critical parameters have been set up, plotting can begin. A number of parameters affect the way axes are drawn after they have been set up. They influence the generation and plotting of numerical labels in the margins of scatter plots and similar displays.

las=   label style. If **las=1**, all axis labels will be horizontal. By default (**las=0**), they are oriented parallel to the axis on devices that have the ability to rotate characters.

tck=   tick mark length. Its default value is 0.02, i.e., 2% of the shorter side of the plotting region. This parameter can be negative to

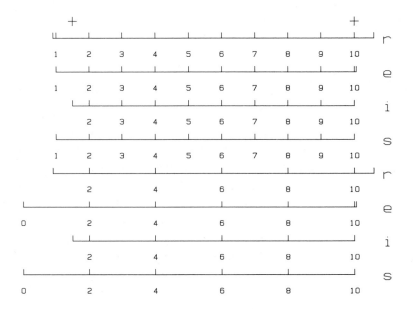

Figure 4.24. The influence of the axis style and number of ticks on an axis set up to accommodate a range from 1.5 to 10 (marked with "+" symbols). The top four axes were based on a nominal 10 tick intervals (parameter **lab**), the bottom four on 5 tick intervals. The letter to the right of each axis identifies the axis style (**xaxs**).

cause ticks to point out of the plot. If tck is set to 1.0, grid lines will be drawn instead of ticks.

Another parameter controlling tick labels is sometimes useful:
xaxt=, yaxt=

> axis type. Ordinarily, the axis type is determined by a high-level graphics function, and takes the values "s" (standard, linear), "l" (logarithmic), or "t" (time). However, a user specification **xaxt="n"** will suppress the drawing of the x-axis. Thus, the expression

> > **> plot(x,y,xaxt="n")**

> produces a scatter plot, but does not generate x-axis ticks or labels.

> You might want to suppress the axis in order to draw a specialized version with the **axis** function (section 4.4.6).

Table 2: Axis styles (**xaxs=** or **yaxs=**)	
Style/Name:	"s" Standard
Coordinates:	Initially: xmin − xmax. May be enlarged to hold the pretty values of tick labels.
Ticks:	Pretty values surrounding xmin/xmax.
Advantages:	Ends of axis are labelled.
Disadvantages:	Fraction of the plot utilized by xmin−xmax varies depending on the particular pretty tick values chosen. Points at xmin/xmax overplot the box if xmin/xmax happens to be a pretty value.
Style/Name:	"e" Extended
Coordinates:	Initially xmin − xmax. May be enlarged to hold the pretty values of tick labels and further enlarged to make xmin/xmax miss the coordinate limits by at least ½ a character width.
Ticks:	Pretty values surrounding xmin/xmax.
Advantages:	Ends of axis are labelled, except where actual data values at xmin or xmax might overplot the box, in which case the axis is extended slightly. Designed to improve upon standard axis.
Disadvantages:	Fraction of the plot utilized by xmin−xmax varies depending on the particular pretty tick values chosen. The extended axis may confuse some people.
Style/Name:	"i" Internal
Coordinates:	xmin − xmax
Ticks:	Pretty values within the range of xmin/xmax
Advantages:	The entire plot area is occupied, regardless of actual xmin/xmax values.
Disadvantages:	Ends of axes are not at pretty values, hence unlabelled. Points at xmin/xmax overplot the box.
Style/Name:	"r" Rational
Coordinates:	xmin − xmax, extended by 7% on each side.
Ticks:	Pretty values within the range of the extended min/max.
Advantages:	Constant fraction of the plot is occupied, regardless of actual xmin/xmax values. No points overplot the box. Attempts to combine the advantages of internal and extended.
Disadvantages:	Ends of axes are not at pretty values, hence unlabelled.

4.4.5 *Low-Level Axis Control*

Another use of the axis style parameter is to specify an axis as "direct". The direct style allows an axis to be "locked in" so that it cannot be changed by later high-level plotting functions. For example

```
> plot(x,y,xlim=c(0,1),ylim=c(0,100))  # set up special axes
> par(xaxs="d",yaxs="d")    # lock the axes in
> plot(x2,y2)   # plot will have same axes as previous plot
```

will cause a special set of axes to be propagated to future plots. Another use of the **par** function is needed to "unlock" a direct axis, i.e.,

```
> par(xaxs="s",yaxs="s")   # set back to standard style
```

The values used by the graphics functions in labelling axes can be generated explicitly by the function **pretty(x,nint)** which returns approximately **nint** values (5 by default) that cover the range of values in **x** and are "pretty" in the sense that their difference is a simple decimal number. These values are useful, for example, as contour levels to the **contour** function.

4.4.6 *Building up Axes*

Many high-level graphic functions allow an argument **axes=F** to tell them to compute the critical parameters, but not draw the axes or box surrounding the plot. This capability, in conjunction with the functions **axis** and **box**, allows the user to build up plots with appropriate coordinate systems and non-standard axes.

The function **axis** provides precise control over the characteristics of an axis. When used with the simple argument **side=**, it will draw an axis on that side of the plot according to the current critical parameters. However, **axis**, through other arguments, allows tick and tick label positioning, character strings for tick labels, control over tick drawing, transformations for probability axes, and drawing in the outer margins.

We can now produce a result just like

```
plot(x,y)
```

from component steps:

```
plot(x,y,type="n",axes=F)   # coordinate system and axes
points(x,y)   # plot the points
axis(1); axis(2)   # draw the axes
```

```
box( )  # draw the box
```

The advantage of the stepwise approach is that, at any of these steps, we can modify the graphical parameters to get a custom plot.

For example, it is common in time-series work to plot two sets of data on the same axes; the two series are normally related but may be measured on quite different scales. A plot could be constructed with number of housing starts (in thousands) and manufacturing shipments (in millions of dollars). Given two time-series, **hstart** and **ship**, measured at identical times, the following will plot them on the same picture with a housing axis on the left and a shipment axis on the right (Figure 25):

```
> # allow 6 lines of margin information on each side of plot
> par(mar=rep(6,4))
> tsplot(hstart)  # plot with x and left y axis
> # declare that the current plot is new so that S will overplot
> par(new=TRUE)
> # dotted line, suppressing axes and y axis label
> tsplot(ship,lty=2,axes=F,ylab="")
> axis(side=4)  # draw y axis on right
```

PROBLEMS

4.29. As mentioned earlier, the iris data comes in three groups. Produce three separate plots, one for each group, of the sepal length and sepal width. Make the three plots appear on the same page. For ease of comparison, all three should have the same x-axis scale, and all three should have a common y-axis scale.

4.30. Produce a scatter plot of your favorite x-y data. At the top and right edges, draw one-dimensional scatter plots of the x- and y-data.

Next, at the top and right edges, draw *jittered* one-dimensional scatter plots of the x- and y-data. Jittering is accomplished by adding small amounts of random noise to the data to avoid exact overplotting. For one-dimensional scatter plots, jittering can be done perpendicular to the data axis.

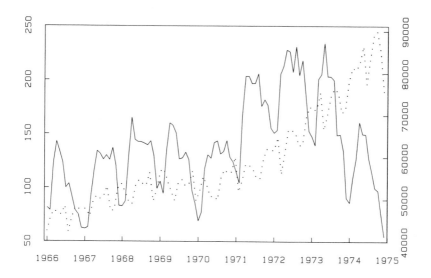

Figure 4.25. A time-series plot showing two sets of data with very different scales. The left axis is for the housing starts series, the right axis is for manufacturing shipments (dotted line).

4.31.   Produce a normal Q-Q plot, and use the **axis** function to add an axis labelled with a probability scale.

4.32.* Produce a figure like Figure 26 to display various device-dependent graphical parameters.

4.33.   Produce your favorite scatter plot using the various axis styles given in Table 2. Which style do you like best?

4.34.* Fit a line using robust regression to the empirical Q-Q plot of Figure 6. Show the line on the plot, using a non-solid line style (if your device has one). Hint: use the **plot=F** argument to **qqplot**.

4.35.   Produce your favorite scatter plot, with a y-axis, but no x-axis. Do this in two different ways, once using the **xaxt** parameter and once using the graphical parameter **axes** in the call to **plot**.

4.36.   Produce an empty plot in which the x-axis has tick marks at

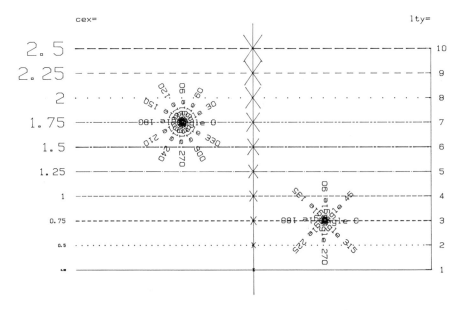

Figure 4.26. A figure that shows various device-dependent graphical parameters.

c(−15,−5,5,15) and the user coordinate system runs from −17 to 17.

4.37.   Produce your favorite scatter plot overlaid with dotted grid lines.

4.38.   If your plotting device has a surface 8 inches wide and 10 inches long, the expression

par(usr=c(0,8,0,10))

will set up a coordinate system in inches. How could such a coordinate system be useful when trying to draw diagrams?

4.5 Specialized Plots

This section deals with graphical methods tied to special analytic procedures (such as two-way table models), or to specialized forms of data (e.g., surfaces in three dimensions).

4.5.1 *Two-Way Tables*

S has some plotting functions designed for matrices that are treated as two-way tables. The functions are **barplot** (discussed in section 8.1), **plotfit** (section 7.2), and **mulbar**. If the matrix is regarded as (signed) counts of some phenomenon, indexed by the rows and columns, a simple way to show the counts graphically is with **barplot(height)**, using its ability to plot divided bars. If **height** is a matrix, the plot will have one bar per column of **height**, with the total bar height the sum of the values in the corresponding column of the matrix. For example consider the system dataset **telsam.response**. If we take the first 5 rows of the data, we can display the results that the first 5 interviewers obtained:

```
> counts ← t( telsam.response[1:5,] )   # transpose so that
>         # columns correspond to interviewer
> barplot(counts,names=encode("Interviewer",1:5))
```

The plot is shown in Figure 27.

The function **plotfit** is another high-level plotting function, which is designed to display the results of a two-way analysis of a data matrix. The rows and columns can be labelled, and residuals can be displayed with or without the fitted grid. For details and a sample plot, see section 7.2 or Appendix 1.

The **mulbar** function lays out a rectangular array of bars of varying width and height, and as such is a very general way of representing a two-way layout of data. For example, we could fit a model to the interviewer data which postulates that responses are independent of interviewer. A χ^2 test might be used to determine adequacy of the model; however, we can use **mulbar** to provide a graphical version of that test, as shown by Figure 28.

```
> counts ← telsam.response[1:5,]
> fit ← loglin(counts,c(1,0,2))  # fit independence model
> resid ← counts − fit  # compute residuals
```

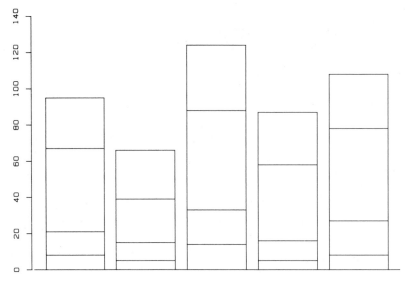

Figure 4.27. Bar plot of responses received by 5 telephone interviewers. Bar divisions, from bottom to top, correspond to "Poor",
"Fair", "Good", and "Excellent".

```
> mulbar(
+    sqrt(fit),
+    resid/sqrt(fit),
+    collab=telsam.collab,
+    rowlab=encode(telsam.rowlab[1:5]),
+    ylab="Interviewer",
+  )
```

4.5.2 *Maps*

A rather specialized high-level plotting function is **usa**, which
produces maps of the United States. Just typing **usa( )** gives a map of
the coast line and state boundaries. The argument **states=FALSE** will
omit the state boundaries; **coast=FALSE** omits the coastline. In addition, it is possible to specify a vector giving lower and upper limits for
latitude (**ylim**) and longitude (**xlim**) to get a rectangular section of the
map, e.g.,

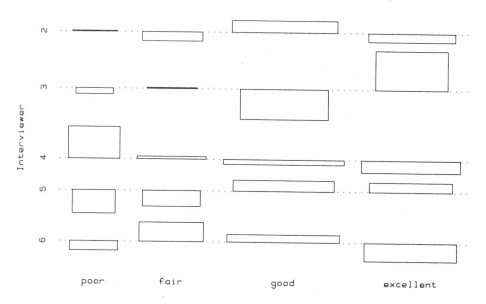

Figure 4.28. A plot showing a fit to the interviewer data, assuming independence of responses. Height of each bar is proportional to the signed contribution of that observation to the Chi statistic; width is proportional to square-root of the fitted value; area is proportional to the residual.

usa(xlim=c(65,100),ylim=c(35,50))

for the Northeast. Unfortunately, the resolution of the map does not increase as smaller regions are displayed, so maps of small areas are likely to look coarse. The coordinate system for the map places latitude on the y-axis and *negative* longitude on the x-axis (so that coordinates increase from left to right). The system data directory contains (under the prefix "state.") several datasets with coordinates appropriate for the map. For example, the dataset **state.center** is a plot structure, giving coordinates near the center of each state, so

usa( )
text(state.center,state.abb)

produces a map with states identified by their 2-letter abbreviations.

4.5.3 *Plotting Surfaces in Three Dimensions*

Several S functions are concerned with the display of surfaces defined as a function of two variables. The first,

> **contour(x, y, z, v)**

produces a *contour plot* of the surface which shows lines of constant height. The vectors **x** and **y** give positions along the x- and y-axes and the matrix **z** gives the height of the surface at the corresponding points. That is, **z[i,j]** is the surface height at **(x[i],y[j])**. Contours are drawn at the heights given by **v**, and if **v** is omitted, approximately 5 equally spaced contours are drawn.

The function

> **persp(z, eye)**

provides a 3-dimensional perspective view of a surface with hidden line removal. The matrix **z** is assumed to have come from an underlying regular grid of equally spaced **x** and **y** values. It is also assumed to be centered at the origin and to extend ±1 in the **x** and **y** directions. The vector **eye** gives the (x,y,z) coordinates of the viewpoint in this coordinate system. The default viewpoint is $(-6, -8, 5)$. Because of the implied coordinate system and because the surface is considered as a 3-dimensional object with comparable x-, y-, and z-coordinates, values in **z** should ordinarily be less than one.

The function

> **interp(x, y, z)**

provides a facility for creating the grid of surface heights needed by **persp** and **contour** from irregularly spaced (x,y,z) data points. The surface is approximated by triangular patches joining groups of 3 data points, and the regular grid of surface values is interpolated. The output structure contains components **x**, **y**, and **z**. The first two are vectors defining the x-y grid, and the last is the matrix of interpolated values. Since **interp** produces NAs rather than extrapolating, **contour** will not draw lines outside the convex hull of the data; no segment of a contour line will be drawn to an edge if either of the vertices of the edge is NA in the **z** matrix.

Let us now tie together the functions **interp**, **persp**, **contour**, and **usa** by means of an example. We can display the results of the 1976 United States presidential election, first on a map with contour lines, and next as a perspective view. First, let's extract the data from the **votes.repub** dataset, omitting Alaska and Hawaii:

```
> forty.eight ← seq(50)[ NA(match(state.abb,c("AK","HI"))) ]
> vote.1976 ← votes.repub[forty.eight,31]
```

The data is irregularly spaced, therefore we will need to use **interp** to generate a grid of data. Let's assume that by plotting the percent vote at the state center, we will get an adequate map.

```
> x ← state.center$x[forty.eight]
> y ← state.center$y[forty.eight]
> i ← interp(x,y,vote.1976)
```

Lets now plot a map and superimpose the contours (Figure 29).

```
> usa(state=F)
> text(x,y,round(vote.1976))   # the actual data
> contour(i$x,i$y,i$z,add=T,labex=0,col=3)
> # contour plot with unlabelled contour lines
```

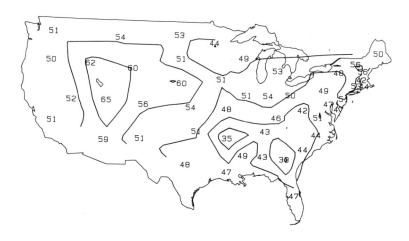

Figure 4.29. Percent vote for Republican candidate in the 1976 Presidential election, with contours.

We can show the election results in a perspective plot, too. However, we must first eliminate the missing values generated by **interp**.

```
> p ← i$z/50   # scale
> p[NA(p)] ← min(p)   # replace NAs
> persp(p)
```

The result is shown in Figure 30.

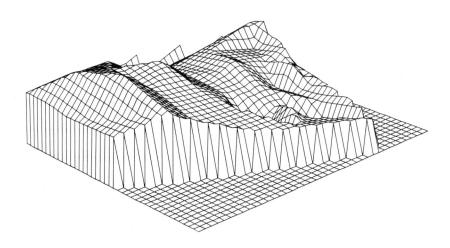

Figure 4.30. Percent vote for Republican candidate in the 1976 Presidential election, shown by height in a perspective grid plot.

Notice how it was necessary to decide on a replacement value for the missing values, and how the **i$z** values were scaled to make a surface with an appropriate range of values. (Remember that the height matrix is treated as a physical object with sides ranging from −1 to 1).

4.6 Deferred Graphics

Users with many plots to produce or without access to graphical terminals may find it inconvenient to use the interactive graphics capability in S. S provides a *deferred graphics* facility that allows replotting, either interactively or in a batch mode. This is carried out by a two-part process. First, a deferred graphics file is produced during a normal session with S. This can be done with or without interactive graphics. Once the S session has ended, the deferred graphics file can be replotted on any batch or interactive graphics device.

Deferred graphics is controlled by the function

defer(on, ask)

The **defer** function should be used after executing a device function. Argument **on** tells whether to turn deferred graphics on (TRUE, the default) or off (FALSE). In addition, if **ask** is TRUE, the user will be asked at the end of each frame whether it should be saved on the deferred graphics file. By default, all output is saved on the file; no questions are asked.

During a deferred graphics session, a symbolic form of the plotting is saved on the deferred graphics file, (which is named "sgraph".) This file contains lines describing the graphical operations being carried out; details of the format of the file are in Appendix 1 under **defer**. Any further S sessions that utilize deferred graphics will append their output to the file.

Suppose that the only graphics device available is a printer, and that the user will have access to an interactive device at some later time. Then,

```
> printer
> defer(ask=T)
Deferred Graphics on - ask about each frame
> plot(height,weight)
> title("Sample plot of Weight vs Height")
> show
    ... printer plot produced here ...
> qqnorm(weight)
Save last frame? yes    refers to height,weight plot above
> ...
```

This session produces the deferred graphics file. Later, on an interactive device, e.g., a Hewlett-Packard 2623 display terminal,

```
> hp2623
> replot    # re-plots all frames from save file
```

(The plot may not look as good as if it were done originally using **hp2623**, since character position, size, and rotation are based on the capabilities of the **printer** device.)

The **replot** function turns off deferred graphics to prevent appending to the same file it is reading. Deferred graphics mode is also turned off whenever a new device is specified or by the expression **defer(FALSE)**. In both cases, the message "Deferred Graphics turned off" will be printed as a warning.

5
Advanced Use of S

This chapter describes some of the more advanced features of the S language and support environment. It is divided into sections describing data structures, iterative computations, the S language environment, data management, and advanced language features.

5.1 Data Structures

In section 3.4, we introduced a number of the basic data structures used in S: in particular, vectors, matrices, and time-series. The following sections describe arrays, vector structures, categories, and general structures as well as a number of functions that operate on them.

5.1.1 *Arrays*

The generalization of a matrix is a multi-way array. Where matrices have two subscripts, arrays may have one, two, three, or more. The array is a structure which has a component named **Dim** (the dimension) and a component named **Data**. For example, a three-way array (like **iris**) with range 1:50 for its first subscript, 1:4 for its second subscript and 1:3 for its third subscript, would be created by the **array** function as follows:

> `> x ← array( read("afile"), c(50,4,3) )`

The data values on the file should be ordered so that the first subscript varies fastest, then the second, etc. The first value read from the file becomes **x[1,1,1]**, the second **x[2,1,1]**, and so forth. (See the function **aperm** to rearrange the subscripts of a multi-way array.)

Once created, arrays are used primarily as input to functions like **loglin**, which fits a loglinear model to an array, or to **apply** (see section 5.2.1) which can apply S functions to subarrays.

5.1.2 *Vector Structures*

Arrays, along with the matrices and time-series introduced earlier, are examples of *vector structures*. A vector structure in S is any structure which has a vector component named **Data**. A vector structure can be treated as a vector when suitable, even though it has more structural information than a simple vector. In particular, arithmetic, logical comparisons, and numerical calculations can operate on vector structures. Consider arithmetic or logical operations; e.g.,

> **x + y;   x >= y**

First, suppose one of the operands, say **x**, is a vector structure and the other is a simple vector. Then the result is a vector structure with the same structural information as **x** but with data determined by the expression. Thus if **x** is a matrix, **x+1** is a matrix with the same number of rows and columns, but with data values increased by 1. In nearly all cases, the result is what one would expect intuitively. (See also section 5.5.2.)

When both the operands are vector structures, the situation is more complicated. If both arguments are arrays, then the two dimension vectors must match in all elements, in which case the result is an array of identical dimensions. If the two operands are both time-series, the result is a time-series with time domain the intersection of the two time domains (if possible). The number of observations per period must be the same in both series. For all other cases, the arguments to the operation are treated as vectors, and the result of the operation is a simple vector.

Most numerical functions operating element-by-element (e.g., **abs, sin**) do the obvious: they produce a vector structure with the data values changed and the structure unchanged. However, some functions produce a vector without structure, regardless of the structure associated with the argument. All the sorting functions have this property. For example, sorting the data values in a matrix should not (and does not) produce a matrix.

Occasionally, it is helpful to extract the components of a vector structure. Whatever the type of structure, the component called **Data** is the vector of data values. Thus, in the (50 by 4 by 3) three-way array **x**, **x$Data** is a vector of 600 data values. For a matrix or array **x$Dim** is the dimension vector. For a time-series **x$Tsp** is the vector of time-series parameters (the starting date, the ending date and the number of observations per unit time).

5.1.3 *Categorical Variables*

A *category* is a vector structure for data which is either qualitative, discrete-valued, or which is to be analyzed by breaking quantitative variables into finite ranges. The key difference between a category and a simple vector is that the set of possible levels is explicit in a category; namely, it is given by a component called **Label**, a vector of character labels for the levels. As a result, not all possible levels need be present in a specific vector, and compatible results can be obtained by analyzing different datasets corresponding to the same set of levels.

Functions **code** and **cut** form categorical variables from discrete and continuous data values. For example, suppose **occupation** is a character vector with values like "statistician". Then

> `> code(occupation)`

returns a category with **Label** component containing an alphabetical list of the different occupations and **Data** component coding the values as 1, 2, ...

Additional arguments to **code** specify levels and corresponding character labels. Suppose **sex** is a vector that contains 1's and 2's representing men and women. You can construct a category **csex** by executing:

> `> csex ← code(sex,1:2,c("Male","Female"))`

The advantage of the category is that the relationship between the numbers 1 and "Male" and 2 and "Female" is now explicit.

The **cut** function creates categories from continuous variables. For example, if **income** is a numeric vector,

> `cut(income,4)`

creates a category with 4 levels by cutting the range of the income values into four equal-length intervals. It is also possible to give **cut** an explicit set of interval boundaries. Any values in the numeric vector outside the range of the boundaries given will be coded as NA. One simple procedure is

> **cut(income,pretty(income))**

using a set of "pretty" numbers over the range of **income** to define the intervals.

Once categories have been constructed, there are many ways in which they can be used for data analysis; a number of these uses are presented in section 7.1. The simplest use is to form a table of the counts in a multi-category set of data:

> **table(csex, cincome)**

for example, produces a two-way table of counts corresponding to the two arguments, which should be categories corresponding to the same set of observations.

PROBLEMS

5.1.   The function **dput** prints its argument, showing all components. Look at the output of **dput** for an array and for a time-series.

5.2.   The system data set **votes.repub** is an array whose columns are the percent vote for the Republican candidate for president in a number of elections. Form categories with two levels corresponding to whether the percent was greater than 50, for each of the 3 latest elections in the data. Make a table of counts for these categories.

5.3.   Look at the documentation for the **liver** data in Appendix 2. Form a coded category for the number of cells injected and compute tables describing the experimental design (how many observations were taken for each combination of cell number, experiment and section).

5.4.   Suppose you are given a character vector **occupation** which contains the occupations of a number of individuals. The expression

> > **occ ← code(occupation)**

will create a category named **occ**. What would be the result of the expression:

> > occ$Label[occ]

5.5. Given a category named **age** which was originally broken down into 10-year intervals, how could you re-code it for 20-year intervals? Would it be easier or harder if the original (uncoded) data were available?

5.6.* Suppose you are given two equal-length integer vectors, **current** and **new** which describe how the current integer values of a category **old.cat** are to be mapped into new integer values of a re-coded category **new.cat**. Explain how the following expression carries out the re-coding.

> > new.cat ← code(new[match(old.cat,current)],
> + lab=new.lab)

Make up an example yourself and try it.

5.1.4 *General Data Structures*

Aside from vector structures, the commonly occurring data structures in S are each shared by a set of functions that analyze some data to produce a structure, display the results of the analysis or do some post-processing to the analysis. The user of these functions does not typically have to know the internal organization of the structure, so long as the functions do. Table 1 gives a useful summary of the common data structures, along with their role in S.

Occasionally, one may want to generate or modify structures explicitly, by naming their components. (If you wanted to write a set of macros that did some analysis and produced a new kind of data structure, for example.) The functions **cstr** and **mstr** can be used to create hierarchical structures and modify existing structures:

cstr(arg1,arg2, ...)

creates a structure whose first component is the value of **arg1**, whose second component is the value of **arg2**, etc. If the arguments are given in the **name=value** form, **name** becomes the component name. Otherwise, if **arg***i* is a dataset, the name of the dataset becomes the component name. If **arg***i* is not a dataset, the component is unnamed.

To modify an existing structure:

Table 1: Common S Data Structures		
Structure Name	**Components**	**Comment**
Vector Structure	Data	Recognized throughout S
Matrix	Data, Dim	Special case of Array
Array	Data, Dim	First subscript varies most rapidly
Time-Series	Data, Tsp	Operators align time
Category	Data, Label	Discrete (categorical) data with labels
Table	Data, Dim, Label	Multi-way table; see **table**, **login** functions and section 7.1
Plot Structure	x, y	Recognized by most plotting functions
Model Structure	coef, resid	Produced by **reg**, etc., interpreted by **abline**
Index Structure	Data, Label	returned by **index**; used by **tapply**
Cluster Structure	height, merge, order	Used by **hclust**, **plclust**, etc.
Distance Structure	Data, Size	Generated by **dist**; recognized by **hclust** and **cmdscale**

mstr(z,arg1,arg2, ...)

The component of structure **z** with the same name as **arg1** is replaced by the value of **arg1** and so forth. If no component of this name exists, the argument is added, as in **cstr**. For example, **mstr** can be used to modify a vector structure **mydata** to contain a new component named **Comment**.

> mydata ← mstr(mydata, Comment="The original numbers")

Because the resulting dataset will still be a vector structure, we will still be able to operate upon it with S functions. When printed, the comment will also show up.

At some point, we may tire of having the comment tag along with the dataset. We can delete a component by giving an empty named argument:

> **mydata** ← **mstr(mydata, Comment=)**

Certain functions give structural information for general data structures. The function **ncomp(z)** returns the number of components of the structure **z**, while **compname(z)** returns a character vector containing the names of the components.

5.2 Iteration

One of the most important concepts in S is that of implicit iteration—S expressions operate on entire data structures, implicitly iterating over all of the values in the structure. Conceptually, a given function acts simultaneously on all the data values in its arguments. To get the full power of the language, it is important to think of computations in terms of the data structures as a whole, in contrast to lower-level programming languages in which computations are on single data values.

Occasionally, such implicit iteration is difficult to apply in a particular analysis. In this case, various explicit forms of iteration must be used. This section discusses iteration in S, and provides a number of examples of the techniques.

The techniques of this section allow S functions to be applied over subsets of data and provide for iterative execution of expressions. These techniques are useful, particularly in conjunction with the macro facility, in developing new analyses. Users who come from a background in lower-level programming languages, however, need to remember that iteration in an interactive, high-level language is intrinsically less efficient than non-iterative use of the language facilities, especially when the iteration is applied to individual data values. With experience, it is often possible to recast the computation into a natural non-iterative form.

5.2.1 *Apply and Implicit Iteration*

The function **apply** allows an arbitrary S function to be applied to sub-arrays of an array, usually to the rows or columns of a matrix. The matrix case is conveniently handled by calls to two system macros,

>   **?col(x, fun)**
>   **?row(x, fun)**

In both cases, **x** is a matrix and **fun** is the (unquoted) name of an S function (not an expression or the name of a macro). The macro **?col(x,fun)** applies **fun** to each column of **x**. If **fun** returns a result of length **n** for each column, the result is a matrix with **n** rows and **ncol(x)** columns. If **n** is 1, the result is a vector of length **ncol(x)**.

Similarly, **?row(x,fun)** returns the result of applying **fun** to each row of **x**; in the form of an **nrow(x)** by **n** matrix or a vector. In either case, if **fun** returns a null result, so does the macro. Additional arguments, if any, are passed directly through to **fun.** For example,

>   **?col(x, mean)**

returns the vector of column means of **x**. To generate 10% trimmed column means, give an additional argument to **mean**:

>   **?col(x, mean, trim=.1)**

for a trimming argument of .1.

For non-matrix arrays, **apply** must be used directly. In this case, the arguments are a multi-way array **x** (see section 5.1.1), a vector **margin** specifying which subscripts of **x** should appear in the result, and a character string, whose *value* is the name of the S function to be applied. The following example, using a 3-way array, illustrates this more general form of use. The dataset **iris** is an array of dimension **c(50,4,3)** (see Appendix 2). Suppose we want a 4 by 3 matrix of medians for all values of the second and third subscripts.

```
> medians ← apply(iris,c(2,3),"median")
> medians
Array:
4 by 3

           [,1]   [,2]   [,3]
   [1,]    5.0    5.90   6.50
   [2,]    3.4    2.80   3.00
   [3,]    1.5    4.35   5.55
   [4,]    0.2    1.30   2.00
```

Here **median** is invoked on 12 sets of 50 numbers each.

The function **sweep** goes along with **apply** in many applications. It takes values produced (usually) by a use of **apply**, and "sweeps" them out of an array. By default, the values are subtracted from the array. System macros **?rsweep** and **?csweep**, analogous to **?row** and **?col**, perform row and column sweeps. To remove column medians from a matrix:

> > **resids ← ?csweep(x, ?col(x,median))**

The result of **?col** is a vector of **ncol(x)** medians; **?csweep** then subtracts each of these from all the values in its corresponding column. Optionally, **sweep** and the related macros can take the name of an operator or function for the sweep process; for example, "/" to cause the marginal values to be divided into the array. Suppose we want to divide the residuals produced above by the standard deviation of each set. In one expression:

> > **standardized ← ?csweep(resids, sqrt(?col(resids,var)), /)**

As with **apply**, the general form of **sweep**, for use with multi-way arrays, takes a vector of margins and an optional character string to identify the sweeping operation. To sweep out the medians in the 3-way array example of **apply**:

> > **resids ← sweep(iris,c(2,3),medians)**

The **apply** and **sweep** operations work on regular arrays. Analogous problems arise with data which is indexed irregularly, by one or more categorical variables. The function **tapply** carries out the analogous calculations to **apply** with categorical variables. See section 7.1 for **tapply** and for analysis of categorical data in general.

A specialized member of the **apply** family is the **sapply** function. It is used much less frequently than **apply** or **tapply**. To apply a function to the components of a structure:

> **sapply(z,fname)**

with **z** a general structure (see 3.4.7) and **fname** a character string giving the name of the function (remember to quote it). The result of **fname** can be any structure; the result of **sapply** will be a structure with the successive results as components. The structure returned will, however, be turned into an **n** by **ncomp(z)** matrix if the function always returns **n** values, and into a vector if **n** is 1. For example, to find the 10% trimmed means of each of the components of **z**,

> location ← sapply(z,"mean",trim=.1)

As with **apply**, additional arguments may be passed through to the function being applied.

5.2.2 *Explicit Iteration; For Loops*

The various ways of applying functions provide an iterative facility suitable to many applications. However, they can only apply a single S function. The S language itself has an iterative facility for general situations, although where the previous techniques are applicable, they will be much more efficient.

Explicit iteration in S takes the form:

for(name in expr$_1$) expr$_2$

The expression **expr$_2$** is executed repeatedly; normally **expr$_2$** contains **name**. Before the first execution, the dataset **name** is set to the first value in **expr$_1$**, before the second evaluation it is set to the second value, etc. Execution terminates when all the values in **expr$_1$** have been used. As an example, suppose we want to regress each of the columns of **y** against the matrix **xmat**, then plot the residuals on a normal probability plot:

> resid ← reg(xmat,y)$resid
> for(cy in 1:ncol(y))
+   qqnorm(resid[,cy])

This will use the implicit iteration in the **reg** function to carry out the entire regression operation, assign the component containing residuals the name **resid**, and will then loop, giving successive columns of matrix **resid** to **qqnorm**. More typically, the expression being iterated is a compound expression; that is, several S expressions enclosed in braces and separated by semicolons or newlines. (See section 5.5.1 for a further description of compound expressions). For example, suppose we also want to plot the absolute values of the residuals against the fitted values (y minus the residuals) each time:

```
> ?T(resid) ← reg(xmat,y)$resid
> for( cy in 1:ncol(y)) {
+       ?T(label) ← encode("Regression of column",cy,"of y")
+       qqnorm(?T(resid)[,cy],main=?T(label),ylab="Residuals")
+       plot( y[,cy]−?T(resid)[,cy], abs(?T(resid)[,cy]),
+             xlab="Fitted Values",
+             ylab="Abs. Residuals",
+             main=?T(label)
+             )
+       }
```

Notice that the example is careful to label each of the plots distinctly. The use of **?T(resid)** and **?T(label)** as the names for intermediate datasets follows the practice we recommend for macro definitions (see section 5.3.1). In fact, the user of iterative loops will usually want to put expressions such as the above into macro definitions. The chance of making typing mistakes is too large.

The **name** in a **for** loop is a built-in S dataset. These built-in datasets are quite special. Although they can be referred to in expressions like ordinary datasets, they are not actually stored in a directory and their values cannot be modified. When the iteration is complete, they disappear.

5.2.3 *Using NULL Data in Loops*

Often, when using explicit iteration, it is necessary to collect together values that are computed at different iterations. Suppose, for example, that each iteration of a loop computes some values, and that you want a vector of all values computed during the loop. In a case like this, it is often convenient to be able to start with an empty collection prior to the first iteration. The built-in dataset **NULL** represents the absence of any data values, i.e., a *null* value.

For example, to produce a list of all datasets accessible on any of the first 3 data directories,

```
> all ← NULL
> for(i in 1:3)
+       all ← c(all, list(pos=i))
> all ← sort(unique(all))     # sort and omit duplicates
```

In this case, NULL serves as a place holder until some values are produced. (The c function throws away NULLs).

Null values are also possible results from certain functions. For example, if there are no data sets on the working directory, then the value of the **ls** function is NULL:

```
> ls()
NULL
```

As always, remember that explicit looping should be used only if an equivalent computation cannot be carried out by the looping implicit in functions such as **apply** and **tapply**.

PROBLEMS

5.7. Construct a matrix **small** from the first 10 rows of the system dataset **telsam.response**.

a) Find the overall mean, **M**, and subtract it from **small**.

b) Find the row means, **R**, of the mean-adjusted **small**. Subtract each row mean from the corresponding row, giving a row-adjusted matrix.

c) Compute column means, **C**, from the row-adjusted matrix. Subtract them from the columns, giving a residual matrix.

d) We have now decomposed **small** into pieces **M**, **R**, **C**, and residuals. Glue the whole thing together to produce an 11 x 11 matrix with the last row containing **C**, the last column containing **R**, with **M** in the [11,11] position. Print it with row/column labels 1, 2, ..., 10, and "Total".

5.3 The Language Environment

This section describes the S language environment, including techniques for reusing expressions by means of source files and macros, keeping track of the steps when performing data analysis, running S as a background process in the UNIX system, and executing special expressions whenever S is invoked.

5.3.1 *Reusing Expressions: Source Files and Macros*

There are times when an analysis performed with S must be carried out repeatedly; for example, each time new observations are added to a dataset. One may then wish to save the effort of typing the expressions each time. The simplest approach is to use a text editor to put the expressions onto a file. The function **source** then causes S to read expressions from the file:

> **source("cmdfile")**

causes expressions to be read from the file "cmdfile". When the whole of the file has been read, S resumes reading expressions from the user's terminal. An error or user interrupt will also terminate reading from the source file. Source files may themselves contain **source** expressions; on terminating reading from the second file, S resumes reading the first.

Until you are sure the expressions in a source file are correct, it is good to have S echo the expressions to the terminal before it evaluates them. The argument **echo** to the function **options** controls this.

> **options(echo=1)** # echo expressions from source file

We have already shown the **source** feature in a disguised form. The automatic dump and use of **edit** described in section 3.3.2 actually prepares a file which is then executed by a hidden source function.

The simple use of **source** is limited since the file contents are completely fixed. Any changes in the expressions to be executed must be edited into the file before execution. Usually, we would like to generate similar, but not identical, expressions; for example, if we want to analyze different datasets in the same way. The need is for something which acts like a new S function, taking arguments which name the datasets to be used in expressions. We can achieve this in S by defining a *macro* which expands into a source file, with argument text substituted into the macro definition. † The rest of this subsection introduces briefly the use and definition of macros. Further

---

† It is also possible for users to write their own S functions which are written in a programming language, and are compiled and loaded into a form exactly like built-in S functions. This is the most efficient way of extending the language and is more general than macros, since it is not restricted to combinations of existing S functions. However, writing new S functions requires much more expertise from the user than does macro writing, and macros are usually simpler and quicker to set up or modify. Information about writing new S functions is available in a Bell Laboratories technical report *Extending the S System*, (R. A. Becker and J. M. Chambers, 1984).

information on writing macros is in chapter 6.

Macros are invoked in S by giving the macro name preceded by the character "?" and followed by the arguments to the macro, if any, in parentheses. As with S functions, the arguments are separated by commas and may be given either positionally or by name. The macro processor looks up the definition of the macro as a dataset stored in an S data directory, and substitutes the arguments into the macro body. The result is a file of text which is passed to S to be evaluated, as a hidden source file.

Macros are defined with the S function **define**. This takes an optional file name as an argument and reads one or more macro definitions from the file, or from the user's terminal if no file is given. The form of a macro definition is shown by the following example:

> **MACRO uscale(x)**
> **(x−min(x))/(max(x)−min(x))**
> **END**

This macro returns its argument rescaled to the range 0 to 1. The MACRO statement gives the name of the macro, **uscale**, and specifies the arguments, here just **x**. The END statement terminates the definition. In between comes the body, consisting of any S expressions, possibly including other macro calls. When the macro is called, occurrences of arguments in the macro body are replaced by the corresponding actual arguments. For example:

```
> myx ← 1:5
> options(echo=1)
> ?uscale(myx)
> (myx-min(myx))/(max(myx)-min(myx))
        0.0    0.25   0.50   0.75   1.00
```

(Note how the expression **options(echo=1)** also caused S to print the expanded version of the macro. This is very useful for debugging macros.) Arguments to macros may be omitted. If default text is to be substituted for a missing argument, this text is specified between slashes, following the argument name, in the MACRO statement. For example, changing the first line of the definition to

> **MACRO uscale(x/myx/)**

would cause the macro to substitute **myx** for the argument **x** in the macro text if the user typed **?uscale** without an argument.

As a side-effect of **define**, a dataset is stored in the save directory for each macro defined. The macros are all stored under the prefix "mac.": the dataset corresponding to macro **uscale** is **mac.uscale**.

To print a macro definition, call the function **mprint** with the macro dataset name as an argument:

> **mprint(mac.uscale)**

```
MACRO uscale(x)
($1-min($1))/(max($1)-min($1))
END
```

Occurrences of the argument names in the text are replaced, during definition, by $1, $2, etc. for the first, second, etc. argument. To edit the macro, call the function **medit**, also with the dataset name as an argument. After completing the edit, write the edited version and quit from the editor; the macro will be redefined. The UNIX shell variable **EDITOR** determines which editor will be used by **medit**; the default editor is **ed**.

Suppose the user invokes the macro **uscale** by typing

> **?uscale(sqrt(abs(myx)+1))**

The expanded macro text would contain four instances of the expression **sqrt(abs(myx)+1))**, and would evaluate the expression four times.

```
> ?uscale(sqrt(abs(myx)+1))
> (sqrt(abs(myx)+1)-min(sqrt(abs(myx)+1)))/
+ (max(sqrt(abs(myx)+1))-min(sqrt(abs(myx)+1)))
        0.0        0.307007  0.565826  0.793850  1.000000
```

Assigning argument values to intermediate datasets avoids repeated evaluation. Using

$$?T(x) \leftarrow x$$

as the first expression of the macro definition evaluates the argument and creates the dataset **$T1x**. Using **?T(x)** in the rest of the macro ensures that the argument is evaluated only once. The "$" character overrides any prefix (see section 5.4.1) and the "T" is a convention to denote a dataset for temporary use. The "1" differentiates nested uses of "?T(x)"—if a macro calls another macro which also uses **?T(x)**, the second macro will generate the name **$T2x**.

Macros will generally involve several expressions; by making the text into a compound expression (section 5.5.1), the value of the macro will be the last subexpression. The macro **uscale** could thus be written:

> **MACRO uscale(x)**
> **({**
>
>   ?T(x) ← x
>   ?T(r) ← range(?T(x))
>   (?T(x)−?T(r)[1])/diff(?T(r))
>
> **})**
> **END**

The use of (**{** at the beginning and **}**) at the end of the text ensures that **uscale** can be used wherever an S function could be used (see section 6.2).

It is important to understand the distinction between *macro time*, when the expressions involving macro calls are being expanded, and *evaluation time*, when the expanded S expressions are being parsed and evaluated by the S executive. Macro time involves the substitution of text into macro bodies, and the generation of text as output. Expressions at macro time are subject to the rules of the macro system, not to the rules of S. In particular, expressions are not evaluated numerically or in terms of the contents of datasets referred to. Again, liberal use of

> **> options(echo=1)**

is a good way to see, just prior to their evaluation, the expressions that result from macro processing.

Chapter 6 discusses a number of additional topics concerning macros, including: predefined macros, macros with arbitrarily many arguments, conditional expansion, and interaction with the user during expansion.

PROBLEMS

5.8.   With the first definition given for macro **uscale**, what would be the result of executing **?uscale(runif(20))**?

5.3.2 Keeping Track of an Analysis: Sink and Diary

In a project that requires an extensive amount of data analysis, it is often difficult to keep track of what procedures have been carried out; it may also be difficult to remember details of how an analysis was performed. In addition, if S is used at a display terminal, there is no hard-copy record of the S expressions that carried out the analysis.

The function **diary** provides assistance in this area. Executing **diary( )** activates a diary-keeping facility in S which appends the date and time to the file "diary" in the current UNIX directory. This is followed by all expressions subsequently typed to S. In addition, any expressions created by macros or source files (section 5.3.1) are also recorded on this file. This provides a record of the actual definition of macros and the content of source files at the time they were used.

Because the diary file contains all input typed to S, it also contains comment lines that the user may type. Thus, thoughts can be recorded by typing comments:

> **diary()   # turn diary-keeping on**
> **plot(x,y)**
> **#Three outliers appear on the plot -- investigate**

Also, the lines that came from macros or source files appear on the diary file as indented comments. This makes clear the distinction between typed input and macro or source file input, and means that the diary file itself can be used as a source file to recreate an analysis at some later time.

Diary keeping can be turned off by typing

> **diary(FALSE)**

A second argument to **diary** is taken to be the name of the file for output, if you do not want to use "diary". The expression

> **diary(TRUE,"today")**

writes the diary to the file "today" (overwriting, not appending).

The **diary** function records expressions that you type to S; the **sink** function can be used to record the results that S ordinarily prints on the terminal. Typing

> **sink(TRUE)**

will cause normal terminal output (aside from prompts) to be appended to the diary file. This allows the diary to contain selected portions of output along with the expressions typed by the user. When a **sink** function is in effect, no printed output appears on the terminal.

The **sink** function can also be used to direct standard S printing to a UNIX file rather than to the terminal:

> **> sink("outfile")**

This is useful for editing printed output, preparing examples or archiving results.

To have results printed again on the terminal, type **sink()**.

The function **stamp** is also a useful adjunct to diary keeping. When **stamp()** is executed, the current time and date is put on the diary file, printed on the terminal and, if a graphics device is active, drawn in the lower right-hand corner of the plot. (Optional arguments allow you to modify these actions.) This can help you associate printed output, graphical output, and the expressions that generated them.

5.3.3 *Batch Execution of S*

There are occasions when interaction with S is unnecessary, e.g., when a large computation is to be carried out utilizing a source file or macro. At these times, *batch* execution of S is desirable since it frees the interactive terminal for other tasks such as editing, etc.

The utility function, typed to the UNIX shell

> **$ S BATCH infile outfile**

initiates batch execution of S. Input commands to S come from **infile**, and output is placed on **outfile**. The input file is equivalent to a source file, as described in section 5.1.1. (The **source** function is a good way of interactively testing a file of expressions before using BATCH.) The output file contains a listing of the expressions from **infile** along with the printed output that results from evaluating those expressions. Any datasets created or modified during batch execution are available for use in subsequent S sessions. Also, a deferred graphics file (section 4.5.3) can be produced in batch mode, and later replotted.

5.3.4 *Executing Expressions on Invoking S*

Users may wish to set options in S or do other operations automatically, whenever they invoke S. This can be done by defining a **profile** macro. When S is invoked, it examines the user's save directory for a macro with the name **profile**. If such a macro exists, it is automatically invoked before prompting the user for expressions.

For example, suppose you were working on a data analysis project with Joe, and want to keep track of what you have done, and also have access to the raw data which is kept in Joe's save directory. Then you could define a profile macro:

MACRO profile
attach("/usr/joe/sdata")
diary
END

Other uses of **profile** might be to invoke a graphics device function or to set system options by the **options** function (see Appendix 1).

5.4 Data Management

Data management in S concerns the organization and searching of data directories, getting data into and out-of S, as well as documentation for datasets.

5.4.1 *Prefixes*

It is sometimes useful to make sure that related data sets have similar names. For example, all data generated in a particular analysis may be made to begin with the same initial characters. This can be done automatically in S by specifying a **prefix**; that is, a character string which will then be automatically prefixed to all names used in assignments or references to datasets. For example,

prefix("iris.")

causes an assignment, $x \leftarrow 1:2$, to create a dataset in the working directory with the name **iris.x**. The use of prefixes reduces confusion between data generated in different analyses, and guards against accidentally replacing a useful dataset that has a simple name (like **x**). The empty prefix can be restored by giving **prefix()** without an argument.

To refer to a dataset by its exact name, regardless of any prefix in effect, precede the name by "$":

> `> x ← 1:5`
> `> prefix("iris.")`
> `> x ← 1:2`
> `> $x`
>   `1 2 3 4 5`
> `> x`
>   `1 2`

The current prefix applies to all data directories; for example, names created in the **save** function will carry the current prefix. Be careful with long prefixes. Some versions of the UNIX system restrict the length of file names (and therefore of S datasets). If the prefix is long, even names that are distinct after a few letters may not lead to distinct file names under the prefix.

The function **prefix** returns as its value the prefix which was previously in effect. This is useful (particularly in macros) to allow restoration of the existing prefix:

oldpref ← prefix("temp")

allows one to restore the prefix by **prefix(oldpref)**.

Another effect of specifying a prefix is that the function **ls** will, by default, only list datasets whose names begin with the current prefix. To list all dataset names, it is necessary to give a *pattern* as an argument. The pattern that generates a list of all dataset names is "$*", e.g.,

ls("$*")

This is a special case of patterns that end in the character "*". Such patterns cause the function to recognize the subset of data names which match the pattern up to the "*". For example, when executing without a prefix,

ls("iris.*")

returns the names of all the datasets in the working directory which begin with "iris.". However, patterns in **ls** do not have any other special characters except "*" at the end, and are *not* general regular expressions.

The function **rm** can be used in an extended form which is also useful in this context. If a special argument **list=v** is given to **rm**, **v** will be taken as a character vector whose elements are the names of datasets to be removed. This form has two advantages. The vector of names can be used to remove the datasets without first having to evaluate them. This is more efficient in the case of many or large datasets. Second, the character string vector can be generated as the value of **list** with a pattern, with the effect of clearing all the datasets with a given prefix.

> **rm(list=ls("iris.*"))**

removes all data whose names start with "iris."

5.4.2 *Attaching Data Directories; Search Lists*

When users invoke S, they have available automatically three data directories. Other directories may be included explicitly; for example, another user's directory or a special directory shared by a group of users. To attach a directory, use

> **attach(name)**

where **name** is the name of the directory to be attached. Note that **name** must completely identify the directory. For example, to attach the save directory of a user who runs S in his login directory, "/usr/joe", say

> **> attach("/usr/joe/sdata")**

The **attach** function puts the new directory onto a *search list* used in looking for a dataset. The three original directories plus the directories named in calls to **attach** are searched, in order, for a specific dataset name when the dataset is used in an S expression. By default, **attach** puts the new directory onto the end of the search list. It is possible to specify an argument, **pos**, to **attach**. In this case, the new directory becomes position **pos** in the search list. Directories that were in position **pos** or beyond are pushed down in the search list. In particular, one may replace the working or save directories by attaching a directory with position 1 or 2:

> **> attach("newwork",pos=1)**

changes the working directory to "newwork".

Data directories may be detached from the search list once they are no longer needed, by supplying either the name or the position. For example, after attaching the new working directory, one can detach the default working directory by either of the following:

> **> detach("swork")**
> **> detach(pos=2)**

It is also possible to examine the current search list. The function

search()

returns the search list, as a character vector.

5.4.3 *Moving Data between Computers*

The **read** function in section 3.4.3 can read a vector of data values from a file. Additional information about the mode and length of the data is inferred from the data items or is supplied by the user. As the example involving **matrix** in section 3.4.5 illustrated, data structures more complex than a simple vector cannot be read directly in this way. Similarly, there is a **write** function which writes only data values; other information is not part of the output file.

There are, however, S functions which provide input and output for complete S structures, including the dataset name, mode and length, as well as data values. Although these functions read and write files of character items in human-readable form, they are fully general interfaces for transmitting any S data structures. The applications of these functions are mainly transmitting datasets between computers and providing archival storage, in readable form, of S datasets.

The function

dput(x,file)

writes the S dataset **x** to the external file, in a self-describing form. If **file** is omitted, the output will be to the user's terminal. Conversely,

dget(file)

reads an S dataset in self-describing form from the file. For output or input of several datasets, the functions **dump(list,file)** and **restore(file)** may be used. The argument **list** is a character vector whose values are the names of the datasets to be written to the file. Typically, this argument is itself the result of the **ls** function, possibly with a pattern argument or with a prefix in effect. The corresponding call to **restore** reads all the datasets from the file and assigns them to a data directory, using the dataset names read from the file. Either function can take an optional **pos** argument to cause the datasets to be dumped from or restored to the save directory or any other attached data directory. For example,

> **dump(ls(pos=2),"dbfile",pos=2)**

dumps all the user's save directory to the file "dbfile". The contents of "dbfile" may then be transmitted to another computer on which S runs, either over a data network or, if necessary, through copying to a tape. If the transmitted data were written to a file "newdata" on the new machine,

> **restore("newdata",pos=2)**

would restore the data to the save directory on the new machine. Note that **dump** and **restore** ignore the current prefix; that is, the **list** argument to **dump** contains complete dataset names, and **restore** creates datasets with the corresponding complete names.

5.4.4 *Nonstandard Data Input*

Frequently, a file containing data to be read into S will have a more complicated format than the simple "data items separated by white space" that **read** can accommodate. Many of these files can be handled by the S macro **extract**, which is able to deal with character files in which each line (record) represents an observation, and where all lines are in an identical form. Each line can consist of fields that begin and end in fixed character positions (as would be the case with FORTRAN formatted output), or it can contain fields that are delimited by a field-separating character, such as a tab or by white space. Both character and numeric fields can be present in each record.

The first step in reading such a file into S is to describe the layout of each record. This description is placed in another file, typically by use of a text editor outside of S. The format of the descriptor file is best shown by example. Suppose a data file named "state" has a state name in character positions 1 to 30 of each line, followed by population in positions 31 to 40, and crime rate in positions 41 to 50. We create a descriptor file named "state.des" that contains:

```
name         CHAR   1    30
population    INT    31   10
crimes        REAL   41   10
```

Each line of this file describes one field of the data file, and contains the field name, the mode (CHAR, REAL, INT), the starting position in the line, and the length of the field, all separated by white space.

Once a descriptor file is constructed, the **extract** macro can be invoked.

> **?extract(state)**

After execution, three new datasets, vectors named **name, population**, and **crimes**, will be present in the working directory, containing the data described by the "state.des" file.

A similar process is used if the lines of the data file consist of variable-width fields separated by a distinctive character. In this case, the descriptor file should begin with a line

```
-f%
```

where "%" is the desired field separator character, e.g. a *tab*. If "%" is omitted, fields are assumed separated by white space (tabs or blanks). Also, each line of the ".des" file should contain name, mode, and field number. Otherwise, the processing through **extract** is identical. The following "state.des" file would be appropriate if state name, population, and crime rate were the first, third, and fourth fields in each record:

```
-f
name          CHAR    1
population    INT     3
crimes        REAL    4
```

The operation of the **extract** macro is as follows. The macro first invokes a utility function to process the data and descriptor files. The utility function is designed to be usable as a stand-alone utility:

$ S extract state

A file named "state.ext" is created by the utility, with the format of a **dump** file. The data in the file can be read into S by

> restore("state.ext")

Only the positions or fields specified in the descriptor file are processed. The contents of the rest of each record are irrelevant. Fields beyond the end of a record are set to NA or the null string ("") if they are numeric or character, respectively.

It is possible to use the **extract** utility function on one machine, and then to ship the resulting ".ext" file to another machine for processing by the **restore** function. This can be useful when extracting small amounts from large data files, since it avoids shipping the entire data file.

Since one may often be unsure about details of format on a file of data, there is a utility, usable at shell level, to summarize the contents of such a file. The UNIX command

$ S scandata <myfile >summary

writes on the output file, **summary**, a summary of the characters appearing in the file. For each position (column), **scandata** reports which characters appeared, on how many lines and what percent of the lines. The **summary** file is intended to be human-readable and is often useful in determining how many fields are on a set of records, where the fields start and stop and what kind of data they contain.

There will still be kinds of data not easily handled by any of the methods so far described. For these we have a general fall-back facility: if, by any means whatever, you can get the set of data into an ordinary program (say, written in the C language), a set of C-callable subroutines are available to generate S datasets directly. This requires somewhat more programming background than anything else mentioned so far, and should probably only be used by those fairly proficient in writing C or FORTRAN programs in the UNIX environment. The basic idea, however, is simple. For example, suppose we managed to put together a vector of real (floating point) data in the C vector **mydata** of length 100. Then the C-language statement

```
cvreal("x",mydata,100);
```

will create and write an S dataset named **x** in the *current* UNIX directory. This is an *internal* S dataset, ready to use. Therefore, once the C language program is ready to run, either it should be run in the user's **swork** directory, or the file "x" created should be moved to that directory. Use

$ S MAN cvwrite

to get further details.

5.4.5 *Documenting Datasets and Macros*

Good documentation can be the key to the usability of macros, and is also important for shared datasets. The function **help** can print user-written documentation for user datasets and macros in the same form as documentation for the system macros and datasets listed in Appendices 1 and 2. Associated with each directory on the search list is an (optional) collection of documentation. The **help** function searches through these collections whenever macro or dataset documentation is requested. Thus, your macro and dataset documentation is available to anyone who attaches your save directory.

A macro and several S utilities are available to assist in the creation of this on-line documentation. The first step is to use the system macro

?prompt(name)

to create a *documentation file*, named "**name**.d". The argument **name** can be the name of a macro, or the name of a dataset or dataset prefix (section 5.2.2). Obviously, **?prompt** cannot compose the documentation, but it will make note of macro arguments or dataset names within a prefix. Once the documentation file is created, it should be edited with a text editor so that it adequately describes the macro or dataset. The file will contain instructions (preceded by the character "~") describing what should appear at various places in the file.

Once the documentation file has been edited, it can be installed in the documentation collection by the utility

$ S NEWDOC name.d

At this point documentation is accessible via **help**, and the documentation file may be removed.

If the documentation needs to be changed, the documentation file can be retrieved by the utility

$ S EDITDOC name

which writes a documentation file "**name**.d" This file can be modified and then re-installed by the **NEWDOC** utility.

To provide printed copies of documentation, use the utility

$ S PRINTDOC [name ...]

By default, all documentation is printed at the terminal in the same form as the S detailed documentation. If selected **name**s are given, only that documentation is printed.

5.5 Advanced Language Features

This section describes some of the more advanced features of the S language. They include new syntactic constructs, the concept of *coercing* data from one mode to another, the precedence of S operators, and functions that perform like S operators.

5.5.1 *Syntax: Conditional and Compound Expressions*

Conditional expressions allow the evaluation of expressions to be *conditional* on the logical result of some other expression. The conditional expression is

if (e_1) e_2 else e_3

Each of the e_i is an arbitrary expression. The first expression is evaluated as a logical vector. If the first element of e_1 is true, expression e_2 is evaluated and returned as the value of the conditional expression. Otherwise, e_3 is evaluated and becomes the value of the conditional expression. The **else** part of the expression may be omitted; in this case, the whole expression has a null value if the condition is false. If the **else** part is present, it should start on the same line as expression **e2**. Otherwise, S will evaluate the whole expression as an **if** without an **else**.

Notice that this is a sequential process: e_2 and e_3 are not evaluated until the condition is examined, and then only one of them will be evaluated. In contrast, there is a conditional *function*, which operates on its arguments in parallel:

$$\text{ifelse} (e_1 , e_2 , e_3)$$

This evaluates all of its arguments and returns elements from either the second or third, depending on whether the corresponding element of the first argument is true. Both forms are useful, but they have different effects. The expression

$$x \leftarrow \text{if(min(y)} > 0) \ \log(y) \ \text{else} \ z$$

checks whether the smallest element of **y** is positive. If so, the logarithm of **y** is returned; otherwise, the data structure **z** is returned. Notice that **y** and **z** do not need to have similar structure at all. Also, **log(y)** is not evaluated at all if the condition fails.

On the other hand, one might write

$$x \leftarrow \text{ifelse(} y > 0 , \log(y) , 0)$$

This returns a data structure identical to **y** except that it will contain the logarithm of all positive values of **y**, but with zero wherever **y** had non-positive elements. However, the logarithm of **y** is always evaluated, and a warning message is printed if any elements of **y** are not positive.

Generally, one may want to control more than one expression in a conditional expression or iteration: *compound expressions* allow several expressions to be grouped into one by separating them by semi-colons or new-lines and enclosing the list in braces. For example:

$$\text{if(min(y)} > 0) \ \{ \ y \leftarrow \log(y); \text{label} \leftarrow \text{"Log of response"} \}$$

Wherever an expression can be used in S, one can use instead a compound expression (which acts as a single expression). There are three common applications for compound expressions: in conditional expressions, in iterative expressions, and in macros. The following details about compound expressions are sometimes important:

- The value of a compound is the value of the last subexpression computed.
- The value of an individual sub-expression within a compound expression is never printed automatically; use the **print** function with the sub-expression as an argument.

PROBLEMS

5.9.   What happens if you type the expression:

        if(pi<0) "This is unexpected"
        else "This is correct"

exactly as shown here? Why?

5.10.   What value will **x** have after the following expression?

        x ← { y ← sqrt(9); y+2 }

Why? How is this related to macros that return values (section 5.3.1)?

5.11.   Write an expression that determines whether **x** contains any missing values, and if so, will print a message to that effect and also recreate the dataset **x** without the missing values.

5.12.   What do you think of the expression:

        if(x==TRUE) "X is true"

Can you think of a simpler way to express the same idea?

5.5.2 *Arithmetic and Operators: Coercing; Precedence*

In evaluating arithmetic or logical expressions, S attempts to make the result meaningful, using knowledge of the structures of the operands. If two operands are of unequal length, values from the shorter will be reused cyclically, to make the result the same length as the longer. We already saw the most common case of this, when one argument has only a single value, as in **x+1**. Somewhat less obviously:

```
> 1:7 * 1:2
  1  4  3  8  5  12  7
```

Whenever a logical vector is used where the context requires numerical values, the logical values are **coerced** to numbers by the definition that TRUE is equivalent to 1 and FALSE to 0. This occurs not only in arithmetic, but in any function which expects numbers and gets logical values. As an example, suppose we have is a logical expression for a set of observations and we want a corresponding vector of labels (perhaps as input to the **text** function). One way to get this is to subscript two labels, using the subscript 1 for people under age 65 and 2 for those over 65:

> **c("Young","Old")[(age>65)+1]**

(However, the function **code** gives a clearer way of doing the same thing.) Conversely, numerical values will be coerced to logical by mapping all zeroes to FALSE and any other values to TRUE.

S functions will not generally coerce from numerical or logical data to character data, or conversely. This can be accomplished by an explicit function or operator; equivalent forms are

> **x %m mode   # operator**
> **coerce(x,mode)   # function**

The mode provided will usually be one of the built-in modes, REAL, INT, LGL, or CHAR. Conversions not involving character data, however, are done automatically and do not require the **coerce** function. The **encode** function is a more flexible way of coercing numeric data to character mode. This function creates a character vector out of any combination of character vectors and numeric vectors. These are frequently useful as labels for printing or plotting.

> **print(votes.repub,collab=encode("Election of",votes.year))**

The corresponding elements of all the arguments to **encode** are concatenated to form a single character string. The vector returned has as many character strings as the longest argument; shorter arguments are reused cyclically (as with the first argument in the example). Arguments which are not character strings are converted in **encode**. By default, a single blank is inserted between values from successive arguments. The special argument **sep=string** allows the user to specify any separating string desired; most usefully, **sep=""** jams successive arguments together without blanks.

When several operators appear in an expression, as in **1:7 * 1:2**, some uncertainty may exist in your mind as to which operation is done first. In this case, both occurrences of ":" are evaluated first, and the results then given to "*". This is usually described by saying that ":" has *higher precedence* than "*". Similar questions about how any expressions involving S operators are evaluated may be answered by Table 2. Operators higher in the table are evaluated before operators lower in the table. Operators with the same precedence are listed on the same line, and are evaluated left to right, as they appear in the expression.

Table 2: Precedence of S Operators		
Operator	Name	Precedence
$	component selection	**HIGH**
%x	special operator	
−	unary minus	
:	sequence operator	
^ **	exponentiation	
* /	multiply/divide	
+ −	add/subtract	
< > <= >= == !=	comparison	
!	not	
& |	and/or	
← _ →	assignment	**LOW**

Remember, when in any doubt, that parentheses can and should be used to make explicit the order of evaluation desired:

> **(1:7)*(1:2)**
> **n:(m−1)**
> **(y−mean(y))*.01**

In general, the precedence of operators in S follows conventional rules for arithmetic and logic.

5.5.3 *Numbered Components of a Structure*

It is possible to address the components of a structure by position rather than by name; that is, **z$[i]**, where **i** is taken as a single integer value. The main use of this expression is for looping over all the components of a structure (the function **ncomp(z)** gives the number of components).

If you want to apply a function to each of the components of a structure, it is usually preferable to invoke the function **sapply** (see section 5.2.2), rather than referring to the numbered components directly. However, given a hierarchical structure named **z**:

```
> for(i in 1:ncomp(z)) {
+       stem(z$[i])
+       qqnorm(z$[i])
+     }
```

The loop does a stem-and-leaf display and normal probability plot of each component of **z**.

5.5.4 *Functional Form of Operators*

Some operations in S can be used in an optional functional form which is more general, but less convenient. These include the functions **get** and **assign** to get and assign datasets, the function **select** to select a component of a dataset, and the function **sys** to invoke an operating system command.

The function **get(name)** gets the dataset whose name is the *value* of the argument **name**. This is the chief advantage of the **get** function; it can be used to access (usually in a **for** loop) datasets from a list of the relevant names. For example, suppose we have a number of S datasets which contain character vectors appropriate for the **vu** function. If the datasets are named **vu.1, vu.2, ...,** then the expression

> ``` > for(i in ls("vu*")) vu(get(i)) ```

will invoke **vu** on each of the datasets.

Another reason for the use of **get** is the possibility of constructing dataset names systematically. This tends to be used in situations where there is a sizable amount of data, so much so that we do not want to operate on all of it at once. Constructed dataset names might be convenient, for example, when setting up a large number of datasets using the **extract** utility (section 5.4.3).

The function **get** may take an arbitrary number of arguments; the effect is to select the first-level component of the dataset given by the second argument, the second-level component given by the third argument, and so on. Again, this might possibly be useful if the component names were systematically the same over a collection of datasets. It is also possible to use an argument **pos=** to **get** to restrict the search for the datasets to a single directory (see Appendix 1).

The **assign** function is the symmetric partner to **get**:

> **assign(name,expression)**

assigns the value of **expression** to a dataset whose name is the value of **name**. The applications and relative advantages are similar to **get**; for example, suppose that we have a number of UNIX files containing printable data, perhaps constructed by some UNIX utilities, and suppose further that the files are all in a UNIX directory named "data.dir". We can use the UNIX *ls* command to get a list of the files in "data.dir":

```
> !ls data.dir >files
> files ← read("files")   # create dataset of file names
> for(i in files)
+     assign(i,read(encode("data.dir/",i,sep="")))
```

The **encode** function creates the names of the files that we want to read (e.g., "data.dir/abc"); **assign** constructs datasets of the same base-name ("abc").

It is possible to give an argument **pos=** to **assign** if the assignment should be to a directory other than the working directory. In the previous example, we might have wanted to save our datasets in the save directory.

The function **select(str, comp$_1$, comp$_2$, ...)** selects the component of **str** defined by **comp$_1$** at the first level, **comp$_2$** at the second level, etc. The reasons for using **select** are not as strong as for **get** or **assign**. One may want to use a systematic set of component names, as with **get**.

The function **sys(cmd)** executes the UNIX system command which is the value of **cmd**. Again, the advantage of the functional form, over using the "!" escape, is the possibility of constructing the command with **encode**. For example, suppose **files** is a character vector with a list of file names. We can construct and execute a sequence of invocations of the UNIX system command **wc** as follows:

sys(encode("wc",files,">>counts"))

(in this case, we wanted the blank to separate the command from the file name). The **sys** function executes each element of **cmd** as a separate UNIX system command. The counts are appended to a file named "counts", which can then be read and analyzed.

PROBLEMS

5.13. Use the UNIX *ls* command to create a file giving the names of several UNIX text files. Read this file to construct a character dataset. Try the word-count example given above, and read the resulting output file.

6
The S Macro Processor

The S macro facility provides a means for extending the S language. It allows new function-like operations to be built up from S expressions, and allows commonly used expressions to be saved and executed with a minimum of typing.

Section 6.1 describes the macro concept. Basic techniques for defining macros are covered in section 6.2. After this section, you should be able to write simple S macros. The remaining sections discuss more advanced topics in macro design.

6.1 Macros in S

Macro calls in S expressions look much like ordinary S functions. Macros are introduced by the character "?", followed by the macro name and optional arguments in parentheses. For example,

> `> colmean ← ?col(x,mean)`

The expression in which the macro call appears is passed to the S macro processor. Finding **?col** in the expression causes the macro processor to search for a macro named **col**. Macro definitions are stored in an S data directory; the definition of macro **col** is expected to be in a dataset named **mac.col**. (This particular macro is available to all users, since **mac.col** is in the shared directory.)

Once the definition of **col** has been found, the macro processor scans the arguments in the call to the macro. The text for the arguments is substituted into the body of the macro, and the output resulting from the substitution is rescanned, to look for further macro calls.

If the scan proceeds without finding any more macro calls, the resulting text is written to a file. This file is executed by S through a hidden
use of the **source** function.

Keep in mind that the whole process of expanding macro calls
and of substituting arguments into the macro text takes place without
evaluating any S expressions. In particular, macros operate only on
the text of their arguments, without knowing the contents of datasets
or anything else about the effect of the S expressions.

Some distinctive features of the approach to macros in S are
relevant. Macro definitions in S are kept in individual datasets which
are read only when the specific macro is invoked. There is essentially
no penalty (other than disk space) for having many macro definitions
of which only a few are likely to be used in any expression. Although
macros in directories can be treated like any other S data structures,
there are special S functions **define, mprint,** and **medit** to generate,
print and edit macro definitions. Macros are stored under the special
prefix "mac.". While legal, it is probably unwise to use names beginning "mac." for anything other than macro definitions. When specifying a macro to **medit** or **mprint** use the dataset name.

```
> mprint(mac.col)
MACRO col(x,fun)
apply($1,2,"$2",$*)
END
```

6.2 Defining Macros

Here is the definition of a simple macro, **myreg**, which does a
regression and then produces two diagnostic plots (a normal Q-Q plot
of residuals and a plot of the absolute values of residuals against fitted
values):

```
MACRO myreg(x,y)
?T ← reg(x,y)$resid
qqnorm(?T)
plot(y−?T,abs(?T)) #abs resid vs fitted
END
```

The definition consists of a MACRO statement giving the name of the
macro and its arguments, then the S expressions forming the macro
body, and finally the END statement. The body generally involves
the arguments to the macro.

A macro can be created initially by typing (with a text editor)
the definition as shown into a file, say "regmac", and then using the S

function **define**:

> **define("regmac")**

The macro definitions on the file will be processed and saved; in this case as a dataset **mac.myreg** in the user's save directory (an optional argument, **pos,** to **define** can be used to save the definition in another directory; e.g., **pos=1** for the working directory). Notice that the name of the *macro* determines the name of the dataset: the name of the *file* from which the definition is read has no effect. Any number of macro definitions can be read from a single call to **define**. As they are read from the file, the *dataset name* for each macro is printed. If no file argument is given to **define**, the macro definitions will be read from the user's terminal. In this case, the user will be prompted by "N>" for the line containing the MACRO name and with "D>" for lines containing macro definition. When the user hits carriage return in response to "N>", **define** finishes. When the macro(s) to be defined are extensive or complicated, it is better practice to prepare them with a text editor, on a file, and use **define** when you are satisfied with the appearance of the macros.

The macro **myreg** is an example of a useful, simple macro. The advantages of the macro to the user are that all the S expressions can be invoked by one macro call, saving most of the typing and possible errors. For now only the **resid** component produced by the S function **reg** is assigned, since the other portions of the returned value are not used.

The use of **?T** as the name for the intermediate dataset is a suggested convention. In this chapter, we shall follow the convention of naming all intermediate datasets by using the **?T** macro. This macro provides consistent, unambiguous names for intermediate datasets and avoids accidental name conflicts. For example **?T(x)** produces the name **$T1x**. Even if the user has a prefix in effect (section 5.2), all of the intermediate datasets will be placed under the prefix "T". Therefore, users should not give important datasets names like "T1x", "T1y", etc. We will show later on how these intermediate datasets can be removed at the end of the macro. The "**1**" in the dataset name varies with the nesting of macros within macros, to prevent conflicts.

Once defined, the body of a macro can be printed by the function **mprint**:

> **mprint(mac.myreg)**

```
MACRO myreg(x,y)
?T ← reg($1,$2)$resid
qqnorm(?T)
```

```
plot($2-?T,abs(?T)) #abs resid vs fitted
END
```

The only substantial difference from the original is that argument names in the macro body have been changed to the positional form: **$1** for the first argument and **$2** for the second argument. This is accomplished in the **define** function, which scans the macro definition for all occurrences of the argument names, and replaces these with the positional symbols. (It is also perfectly legal for you to use the **$1, $2** form in the macro body.)

It is possible to give **mprint** a *list* of macro names. For example

> **mprint(list=ls("mac.*",pos=2))**

prints all the macros in the save directory. An optional argument **file=** to **mprint** allows macro definitions to be written to a file.

To edit a macro, use the S function **medit**:

> **medit(mac.myreg)**

medit invokes a text editor with the body of the macro as the file to be edited. Writing an edited version of the file and quitting saves the new version. Since the new macro body is reprocessed through **define**, the edited changes can refer to the arguments by name. To create a *new* macro with the edited body (and leave the original available) just change the name of the macro on the MACRO statement.

By default, **medit** uses the **ed** editor. You can specify which editor should be used by setting the UNIX shell variable **EDITOR**; for example

$ **EDITOR=vi; export EDITOR**

selects the visual editor **vi**. You may want to include this setting as part of your default UNIX login procedure. (The **EDITOR** shell variable also selects which editor is used by the **edit** function.)

Let us edit the macro **myreg** to illustrate some other possibilities with macros. We can make the second plot optional by introducing an argument **plotfit** as the third macro argument, and changing the last expression to

if(plotfit)plot(y−?T,abs(?T))

If we want the plot to be done by default, **plotfit** can have a default value of TRUE. Default values for macro arguments are strings supplied after the argument name in the MACRO statement, enclosed in slashes:

> MACRO myreg(x, y, plotfit/TRUE/)

Default values can contain other macro calls. They cannot, however, contain references to the other arguments; a use of **x** or **y** would be left untranslated (see section 6.3).

So far, our **myreg** macro has thrown away the regression data structure (except for residuals). A better approach is to return the regression structure as the value of the macro, so that the macro then acts as an extended form of the S function **reg**. The technique to use here is the *compound expression* (see section 5.5.1). If the body of the macro is made into a compound expression, the value of the last subexpression becomes the value of the compound expression and, therefore, the value of the macro. To be even more sure that the macro can be used as a function, the compound expression in turn should be enclosed in parentheses. Among other effects, this ensures that the macro body will not be broken up by its context. For example, a macro **abc** with body **$1+$2** would not produce the intended result in the expression

> ?abc(x,y)*2

since the expanded text would be

> x+y*2

ALWAYS REMEMBER: macros are expanded as text and only then are evaluated by S. A good technique for debugging macros is to use

> **> options(echo=1)**

to instruct S to print each expression from the expanded macro prior to executing it.

The general rule in defining a macro which should act like an S function is to begin the body with "({", end with "})" and return the intended value as the last sub-expression. However, there are some exceptions. First, do not use parentheses and braces if the macro does not return a value. Second, since braces make the entire macro into a single expression, do not use them until the macro is debugged. Otherwise, **options(echo=1)** will simply print the entire macro body. During debugging, assign the intended value to an intermediate name, say ?T, and look at that dataset after the macro finishes execution.

If an argument appears more than once in the body of a macro as it does in **myreg**, it is generally desirable to evaluate it once and save the value in an intermediate dataset. Not only may this save considerable execution time, but some uses of arguments might otherwise lead to illegal S expressions.

In our regression example, we can assign the regression to **?T** and return this as the value. Here is the macro, with the changes we discussed edited in.

```
MACRO myreg(x, y, plotfit/TRUE/ )
({
        ?T(y) ← y
        ?T ← reg(x,?T(y))
        qqnorm(?T$resid)
        if(plotfit)plot(?T(y)−?T$resid, abs(?T$resid))
        ?T
})
END
```

The macro now returns **?T** as its value. Unfortunately, **?T** and **?T(y)** remain as clutter in the working directory. We have a conflict between the desire to have **?T** as the final expression and the desire to use **rm** as the final expression in order to get rid of the intermediate datasets.

The following trick will remove the intermediate datasets and simultaneously return a value from a macro. By using the special argument **value=**, **rm** can return any desired expression. For example, in **myreg**, the last line of the macro definition could be

> **rm(?T,?T(y),value=?T)**

One more change will illustrate some of the flexibility of the use of macro arguments. Suppose we want our plots to have an informative title; for example, the call to **qqnorm** could be changed to:

> **qqnorm(?T$resid,sub="Normal Q-Q plot",**
> **main="Regression of y on x")**

The point to note is that **x** and **y** are substituted from the actual macro arguments into the character string used as the main title. The call to the macro of the form

> **?myreg(pred[,1:4],log(response))**

would produce as its main title on the plot:

> **Regression of log(response) on pred[,1:4]**

In this way, macros used to generate plots and printing can be made considerably more informative than simple function calls. The macro is now:

```
MACRO myreg(x, y, plotfit/TRUE/ )
({
    ?T(y) ← y
    ?T ← reg(x,?T(y))
    qqnorm(?T$resid,sub="Normal Q-Q plot",
        main="Regression of y on x")
    if(plotfit)plot(?T(y)−?T$resid, abs(?T$resid))
    rm(?T,?T(y),value=?T)
})
END
```

You should now be able to use the macros in the shared directory, and to define and use some macros of your own. Documentation for shared macros is given on-line by the S function **help**; e.g.,

> > help(macro="col")

for the documentation of macro **col**. Section 5.4.4 describes utilities that can be used to construct documentation of your own macros. You should now do some experimenting, or try some of the problems below.

PROBLEMS

6.1.   Create a macro to compute the logarithm base 2 of its argument. (Hint: log base 2 is log base 10 divided by **log10(2)**.)

6.2.   Create documentation for your macro. Use the **help** function to print it.

6.3.   Tukey defines a function **flog** to compute a *folded logarithm* for fractional data as:

   .5*log(fraction that do)
      − .5*log(fraction that do not)

(Fractional data ranges from zero to one).

Write a macro to compute flogs. Should you use intermediate datasets? Produce a graph of **flog(x)** for values of x ranging from 0 to 1. What are the properties of flogs?

6.4.*   Suppose you write the following macro to "Winsorize" a set of data: to set all values of **x** which are less than **limit** to **limit**.

> **MACRO winsor(x,limit)**
> **x[x<limit] ← limit**
> **END**

Create a dataset **mydata** containing 100 random Gaussian variates. Try **?winsor(mydata,0)**. Does it do what you expect? Re-initialize **mydata** and try **?winsor(mydata,−1)**. Is there any problem now? If so, how could you fix it?

6.5.* You have a friend that uses S on a different computer. How could you send all of your macros to your friend?

6.6.* In **myreg,** why can the expression **rm(?T,?T(y),value=?T)** remove the dataset **?T** and still return it as a value?

6.3 Literals; Conditional Expansion

Sometimes it is necessary to hide some text from the macro processor. The most common case is that the name of one of the macro arguments must be used literally, perhaps because it is the same as the name of an argument to an S function. In the **myreg** macro, if we had named the third argument **plot** rather than **plotfit**, there would be trouble calling the S function **plot**, since **define** would turn all occurrences of "plot" into "$3". A built-in macro construction is used to avoid this:

> **?**(*anything*)

passes *anything* on without scanning it. These literals are recognized both when the macro definition is processed by the **define** function and when the macro body is processed to substitute the arguments. Literals can protect characters in the default text for macro arguments (principally, this allows the character "/" to appear in the default string), and they can protect argument names or other special characters, such as "?" or "," within the macro body. In our example,

> **if(plot)?(plot)(...)**

would solve the problem. (Macro expressions need to be legal S expressions only *after* expansion. The strange use of parentheses above does no harm.)

The literal in this example protected the name "plot" during the definition process. Using a literal to protect a special character during expansion is a little more subtle. The contents of the literal are passed on to the macro output without processing, but with "?(" and ")" stripped off. In particular, attempting to protect characters in arguments to other macros can be tricky. (See the problems at the end of the section.)

The **if** statement in the **myreg** macro is evaluated at *function execution time*. This has the advantage that the argument **plotfit** may be an arbitrary expression, but the disadvantage that more interpretation of the macro output is needed when the resulting expression is evaluated. An alternative is to use the conditional macro **IFELSE**. In its simplest form, this has four arguments. If the first two are equal (as strings at macro time), it expands into the third argument; otherwise, it expands into the fourth argument. In our example, suppose the plotting is to be done only if the **plotfit** argument is the string "TRUE":

> ?IFELSE(plotfit,TRUE, plot(?T(y)−?T$resid, abs(?T$resid)).)

Here, the fourth argument to **IFELSE** is implicitly empty. When **?IFELSE** is used, equality is determined at macro time, so that the test cannot depend on the value of an S expression. It is a character-by-character test for equality: "T" would not be equal to "TRUE", even though at execution time they both stand for the same logical value. If we replace the **if** statement in the example with the **?IFELSE** form, the user of the macro could not type

> ?myreg(x, y, rss<.8)

and expect the macro to expand reasonably. Generally, execution-time tests are preferable, unless one is explicitly concerned with the *text* of an argument. Again, see the problems below for an example.

A more powerful example of conditional expansion solves a problem mentioned in the previous section: how to make default values depend on other arguments. Suppose we want the default test for **plotfit** to be **len(y)>50**. This would not work as the default text in the MACRO statement because default text is not scanned on substitution. However, we can generate the same result by not supplying a default string, but instead testing for a null argument:

> MACRO myreg(x, y, plotfit)
> ?T(pl)← ?IFELSE(plotfit,,len(y)>50,plotfit)
> if(?T(pl)) plot(...)

Now ?T(pl) represents either the macro argument or the default test.

We want the default test to depend on data values; therefore, the test must be done at execution time by **if**, not at macro time by **?IFELSE**.

There is a more general form of **?IFELSE** which takes an arbitrary number of arguments, in sets of 3. If the first two of each set match as strings at macro time, the expansion of **IFELSE** is the third argument in the set. Otherwise the first two of the next set are compared, and so on. If one final argument is given, this is the default expansion if all matches fail. For example, to allow "YES" and "NO" as values of **plotfit**:

```
?T(pl) ← ?IFELSE(plotfit,YES,TRUE,
    plotfit,NO,FALSE,
    plotfit,,len(y)>50,
    plotfit)
```

Notice that leading white space in arguments to macros is ignored, so that the macro calls can have a natural layout. Trailing white space is retained; see the example at the end of section 6.4.

PROBLEMS

6.7. Write a macro that takes an *unquoted* string, and uses that as the title in invoking the **plot** function on two fixed datasets, say **always.x** and **always.y**. (Perhaps these get updated and we want to replot them, providing a new title.)

How would you call the function to produce the title "Species counts, adjusted"? How would you rewrite the function if it invoked another macro, not the **plot** function, and the second macro also wanted the title as an argument?

6.8. Write a version of **myreg** that does the plot of fit if and only if the first character of the **plotfit** argument matches one of "Y", "y", "T" or "t". (Look at the definition of the **SUBSTR** built-in macro in section 6.8 below.)

6.9.* Write a macro which produces some S expressions which execute and which then invoke another macro.

6.4 Macros with Many Arguments

As with S functions, some macros will want to have arbitrarily many arguments. This is managed by including in the MACRO statement only as many arguments as should be recognized specially, and then matching all the remaining arguments by the special construction **$***. For example, here is a form of the **myreg** macro which takes one y variable and an arbitrary number of x variables:

```
MACRO myreg( y )
({
        ?T(x) ← cbind($*)
        reg(?T(x),y)
})
END
```

(For the moment, we have omitted the plotting.) The occurrences of **$*** in the text of the macro are replaced by all the actual arguments to the macro (if any) which have not been matched to any of the arguments in the MACRO line. In the body, **$*** is replaced by the remaining arguments, separated by commas. In this case, for example, calling the macro in the form **?myreg(resp,var1,var2,x)** expands the first line of the body into

```
?T(x)←cbind(var1,var2,x)
```

This form of construction is useful when one wants to call S functions with arbitrarily many arguments. Also, a macro calling a graphics function may wish to include graphics parameters (section 4.3) in the macro call. The macro **symplot** below generates a plot of the sorted values of a vector against the same values in the reverse order (a plot which can be used to test for symmetry):

```
MACRO symplot(x)
?T(x) ← sort(x)
plot(?T(x),rev(?T(x)),xlab="x",ylab="rev(x)",$*)
        #plot with graphics parameters
END
```

The macro call

```
?symplot(resid,pch="+",col=3)
```

passes the graphics parameters **pch** and **col** to the **plot** function. Since this macro returns no useful result, the enclosing (**{** and **})** are not used.

One problem with having arbitrarily many arguments is how

to allow special arguments with default values. How does one reintroduce the argument **plotfit** into the **myreg** macro, without having it match the first of the x variables? In both S functions and macros, the usual solution is to require the argument to be supplied in the form **name=value**. To restore optional plotting to the regression macro we reintroduce the argument with a trailing =:

> **MACRO myreg(y, plotfit=/TRUE/)**

with the same sort of tests as discussed in the previous section to determine whether to do the plotting. With the macro defined in this way, the user must use

> **plotfit=FALSE**

in the macro call to suppress the plotting. Lets have a complete look at **myreg** now:

```
MACRO myreg( y, plotfit=/TRUE/)
({
     ?T(y) ← y
     ?T(x) ← cbind($*)
     ?T ← reg(?T(x),y)
     qqnorm(?T$resid, sub="Normal Q-Q plot",
         main="Regression of y on ($*)" )
     if(plotfit) plot(?T(y)−?T$resid,abs(?T$resid))
     rm(?T,?T(y),value=?T)
})
END
```

Note: the remainder of this section describes some advanced features of the macro processor, and can be skipped on first reading.

The expansion of **$*** can be generalized to use the list of arguments in any way at all, through the built-in macro **LOOP**. This macro concatenates instances of a string (its first argument) in which the remaining arguments are substituted successively. It is effectively a macro-within-a-macro, in that the string is a (limited) form of macro definition and the remaining arguments provide instances of this definition. Here is a simple application: suppose we want to create a vector which has all the macro arguments combined as quoted strings. The purpose is to create a vector of character names in S from the list of arguments to the macro. The **LOOP** macro does this if its first argument contains "$%1". Each of the other arguments to **LOOP** is then substituted for $%1 in the first argument, and the results concatenated together, with commas in between:

```
?LOOP("$%1",x+y,−3,log(resp))
```

for example, expands into

```
"x+y","−3","log(resp)"
```

The desired trick of making the vector of names is then accomplished by the macro **names**, defined as follows

```
MACRO names
c(?LOOP("$%1",$*) )
END
```

We could use **names** in **myreg** for example, to add a labelled printout of regression results:

```
regprt(?T,names=?names($*))
```

More complex applications of **LOOP** are also possible. Suppose we want to generate separate expressions for each of the arguments. For example, a macro to produce separate scatter plots of all other arguments against the first argument would be:

```
MACRO mplot(x)
?T ← x
?LOOP( plot(?T,$%1,xlab="x",ylab="$%1")
   ,sep=,$*)
END
```

In this case, the first argument to **mplot** is **x**. Each of the remaining arguments to **mplot** is substituted for both instances of **$%1** in the **plot** call. Notice that the first argument to **LOOP** is "plot(.....)" which contains a final newline character. The special argument **sep=** to **LOOP** gives a separator different from the default comma. Here the separator is null because we put a newline at the end of the first argument to **LOOP** (the preferred way to separate statements generated by **LOOP**). The output is more readable on separate lines, and the danger of overflowing text buffers during parsing is reduced. Thus

```
?mplot(pred[,1],resp,log(resp),sqrt(resp))
```

expands into

```
?T ← pred[,1]
plot(?T,resp,xlab="pred[,1]",ylab="resp")
plot(?T,log(resp),xlab="pred[,1]",ylab="log(resp)")
plot(?T,sqrt(resp),xlab="pred[,1]",ylab="sqrt(resp)")
```

In this example, we have again taken advantage of the macro

processor to generate informative labels, and have copied the argument **x** into **?T** for efficiency, since it is used repeatedly.

6.5 Interaction with the User; Prompting

In some circumstances it may be important for a macro to communicate with the user. Messages can help make the macro easy to use. The built-in macros **MESSAGE**, **FATAL** and **PROMPT** all take a message to be printed on the user's terminal when the macro is expanded. In addition, **FATAL** induces an error which terminates the evaluation of the expression containing the macro call. **PROMPT** produces, as its expansion, a line of input read from the user's terminal in reply to the prompt. **MESSAGE** produces no expansion text.

The **MESSAGE** macro can be used just to print a message to the user:

> **MACRO learn**
> **?MESSAGE(Hello, this is the learn macro)**
> **...**

More typically, all three macros will be used conditionally to report special situations or establish values for arguments. The **FATAL** macro is one way to make arguments obligatory, or to check other problems:

> **MACRO myreg(x/?FATAL(Must supply argument x)/, ...)**
> **?T(x) ← x**

The result is that, at macro expansion time, the user who omits the argument **x** will see the error message. As a general approach, this is better than just allowing expansion without an essential argument. The best that can happen in the latter case is an S syntax error. Not only does this take longer, but the error is usually obscure to the user, particularly if it comes far into the expansion of a complicated macro. Even worse, the S expression may turn out to be syntactically correct but produce ridiculous results.

Some conditions cannot be checked at macro expansion. For these, the S function **fatal** can be used to print a message and force an error (which in turn causes an exit from any macros or source files):

> **if(min(y)<0) fatal("Negative values make no sense here")**

The argument to **fatal** can also be constructed from data values:

> **?T ← sum(y<0)**
> **if(?T>0) fatal(encode("y contains",?T,"negative values"))**

While **FATAL** is better than nothing, one may give the macro user a more friendly response to missing arguments by the use of **PROMPT**. The message given to **PROMPT** is printed on the terminal. The user replies with a line of input, which, not including the new-line, becomes the expanded form of **PROMPT**.

> **MACRO myreg(x/?PROMPT(X matrix for regression:)/,...)**
> **?T(x) ← x**
>
> **...**

Thus, a user who had totally forgotten how to use this version of the macro **myreg** might go through:

> `> ?myreg`
> `X matrix for regression:` **pred[,1:5]**
> `Y vector for regression:` **log(resp)**
> `Plot fitted vs abs. resid (T or F)?` **T**

It is a question of style or psychology as to how far such prompting is desirable. So long as users can get out of the prompting session (by hitting interrupt), the use of prompts often makes casual use of the macro easier. See also section 6.6 for a general menu interface to S expressions.

In order for more complicated macros to be usable, there is a need for documentation. See for example, the documentation for system macros in Appendix 1. Section 5.4.4 describes how you can write documentation for both datasets and macros. In addition, macros may be made self-documenting through the **SYS** built-in macro which executes an operating system command. For example

> **?IFELSE(?PROMPT(Do you want Instructions?),**
> **yes,?(?SYS(cat /usr/joe/instruct)))**

will print the file "/usr/joe/instruct" if the user requests instructions.

The style advocated earlier, of assigning all arguments to intermediate datasets if they appear more than once in the macro text, becomes obligatory if **PROMPT** is used to supply default values. Otherwise, the user will be prompted with the message *every* time the argument appears (and the macro processor will cheerfully substitute different responses each time!). Occasionally, it may be impossible to assign arguments, as when the argument is part of a function or dataset name. In this case, the use of **PROMPT** requires that the macro come in two layers. The outer layer contains the arguments and the **PROMPT** defaults, but its text consists only of the invocation of the inner layer. The following is a general probability plot macro, **qqplot**, which takes the coded name of the distribution as an argument and constructs from it the name of the appropriate S function.

```
MACRO qqplot( x,dist/?PROMPT(What distribution?)/)
# create quantile-quantile plot against arbitrary distribution
?qqplot2(x,dist,$*)
END

MACRO qqplot2(x,dist)
?T(x) ← x;
plot(?(q)dist(ppoints(?T(x)),$*), sort(?T(x)),
     xlab="Ordered quantiles of dist $* distribution",
     ylab="Sorted values of x")
rm(?T(x))
END
```

The prompt will occur only in preparing the macro call to **qqplot2** so that each message will be issued no more than once. See section 6.7 for another means of accomplishing this end.

Note: the remainder of this section contains advanced material which can be skipped on first reading.

Readers who have come this far in the macro discussions may find it useful to understand more details of how macro substitution works. Suppose the macro processor receives the line

```
?qqplot(abc)
```

It begins by reading the dataset containing the definition of **qqplot**, and extracting the body, argument names and argument.default values as processed by **define**. If the macro name is followed by a parenthesized argument list, the arguments in this list are now scanned. If another macro call is encountered during the scan of the arguments, this definition is read in and the process of argument scanning now proceeds on this lower level. Eventually, for some macro, the actual arguments will be pure text. These are matched to the argument names, and the default text is used for any missing arguments. But (this is critical here) the default text is treated as part of the macro definition, and is not scanned at this stage.

The argument text is substituted into the macro definition text, and the result is pushed onto an internal buffer. It is not output yet, because the text of **qqplot** is free to contain other macro calls. If the current macro was part of an argument to another macro, the process of expansion continues. When the top level expansion is complete, the macro processor rescans the pushed-back text, and at this point will pick up macro calls in the definition. In particular, the macro calls to **PROMPT** are now part of the actual arguments to **qqplot2**. Therefore, they will be evaluated (once) before the text of **qqplot2** is expanded.

This detailed knowledge of how the macro processor works is not needed for most applications, but occasionally it becomes necessary to understand such techniques as delaying macro expansions or protecting strings with "?(...)". For further reading, see the macro processor discussed in *Software Tools* by Kernighan and Plauger (Addison-Wesley, 1976), on which the macro expansion part of the S macro processor is based.

PROBLEMS

6.10. Why did the example

> ?IFELSE(?PROMPT(Do you want Instructions?),
>   yes,?(?SYS(cat /usr/joe/instruct)))

use **?(...)** around **SYS**?

6.11. Given the macro

> MACRO double(x/?PROMPT(What do you want doubled)/)
> x+x
> END

What happens if you use **?double(runif(20))**? What about **?double** without an argument? How would you fix the problems?

6.12. What does the **$*** do in the **qqplot** macro?

6.6 Menu Interface

When S is used to provide a very high-level interface to specialized analysis or graphics, the eventual users may prefer to interact via menu selections, as a partial or complete replacement for typing S expressions. The **PROMPT** macro is one approach to generating multiple-choice input from a user at macro expansion time. The S function **menu** provides a more general and pleasant interface at execution time. The technique is as follows. We define a character vector of **actions**, with each element of the vector being one possible action, in the form of an S expression. We define another vector of messages corresponding to the actions, these being what the user should see to define each of the possible actions. Then executing the function **menu(items, actions)** prints out for the user a list of the items. When the user selects one of the legitimate items, the corresponding action is written out and executed by a hidden **source**.

Consider, for example, the macro **myreg** that we have been working on throughout the chapter. This had two optional plots. We can set up a menu that offers the user of **myreg** either of these, both or neither. The item vector is set up as:

```
?T(item) ← c(
  "Normal probability plot",
  "Abs. residuals vs fitted values",
  "Both",
  "None")
```

The corresponding action vector could contain the actual S expressions that do the plots. The user of the macro sees the following:

```
> ?myreg(x,y)
1- Normal probability plot
2- Abs. residuals vs fitted values
3- Both
4- None
Which one?
```

The user responds by typing a number between 1 and 4.

A good strategy, particularly for applications that are at all complicated, is to organize the actions as a set of macros; in the case of the plots above, for example, we might define the two plots as **myplot1** and **myplot2**:

```
?T(act) ← c( "?myplot1",
  "?myplot2",
  "?myplot1; ?myplot2",
  " ")
```

This general approach allows one to build up a network of menus; that is, to handle the situation in which the first menu does not answer all the questions, but instead selects one of several possible second-level menus. As long as we keep the various action and item vectors around (in datasets whose names we can remember) and if all complicated actions are turned into equally well-named macros, we can study the current menus by looking at these datasets. In this context, remember that the expression

> **> edit(item)**

lets us edit any character vector, **item**, and

> **> medit(mac.myplot1)**

lets us edit the current definition of the **myplot** macro.

The **menu** function, used in this style, will allow quite complicated menu interfaces to be managed and used effectively.

PROBLEMS

6.13.   Define the **myplot1** and **myplot2** macros from our previous definition of **myreg** and test out the menu interface described in this section.

6.7 Temporary Definitions; Recursive Calls

Macro definitions generated through the S function **define** are permanent definitions, created outside the macro processor and saved in S datasets. It is possible, for specialized requirements, to define macros during the macro expansion itself. This is accomplished by the **DEFINE** macro:

> **?DEFINE(name,text)**

creates a macro with **text** as the body. (It is very difficult to create a macro with arguments in this fashion.) Throughout the current invocation of the macro processor, the macro may be invoked as **?name**. The macro is not saved permanently, however. Once the macro processor has expanded the expression given to it, all temporary macro definitions disappear.

Temporary definitions are useful mainly as a mechanism to save some information during the expansion of a macro, which will then be re-used later on in the expansion. For example, the use of **PROMPT** shown with a two-layered macro in the previous section could also be handled by defining the value of the prompt as a temporary macro.

Suppose, in **myreg**, we want to prompt the user for a choice of one or both of a Q-Q plot and a plot of residuals. The first character of the response is extracted, by a built-in macro **SUBSTR** (section 6.8), and defined temporarily as the macro **plotopt**. In the remainder of the definition of **myreg**, the temporary macro can be invoked every time the user's response is needed. The extended form of the **IFELSE** macro can then be used to generate one or the other or both of the plotting expressions.

```
MACRO myreg(x,y,
    plots/?PROMPT(Plotting (qq, residual,both or none):)/
    )
({
  ?T(y) ← y
  ?T ← reg(x,?T(y))
  ?DEFINE(plotopt,?SUBSTR(plots,1,1))
  ?IFELSE(
      ?plotopt,q,
          qqnorm( ?T$resid ),
      ?plotopt,r,
          plot(?T(y)−?T$resid, abs(?T$resid), xlab="Fitted"),
      ?plotopt,b,
          qqnorm( ... )
          plot( ... )
      )
  rm(?T,?T(y),value=?T)
})
END
```

PROBLEMS

6.14. Use **DEFINE** to create a version of the **qqplot** macro in section
       6.5 that does not have the inner layer (the separate **qqplot2**
       macro).

6.8 Built-In Macros and Special Constructions

Built-in macros are part of the S macro processor itself: they are not the result of the **define** function. The following built-ins are recognized:

?IFDEF(name,s1,s2)

If macro **name** is currently defined, this macro expands as **s1**, otherwise it expands as **s2** (or null if **s2** is omitted).

?IFELSE(a1,b1,c1, a2,b2,c2, ..., d)

The macro takes 3*k or 3*k+1 arguments. If the arguments **a1** and **b1** match as strings, the macro expands as **c1**. Otherwise, arguments **a2** and **b2**, if both present, are compared, and so on. If all matches fail, the macro expands as **d**, which is implicitly null if 3*k arguments were given.

?LOOP(string, a1, a2, ...)

The output of the macro consists of the contents of **string** repeated as many times as the number of remaining arguments **a1**, **a2**, ..., imply. Successive repetitions are separated by a separating character (comma by default) which may be given by the **sep=** argument. In order to separate by new lines, put a new line at the end of **string** and specify the separator as null.

The contents of **string** are scanned for the special tokens $%1, $%2, etc. Let **m** be the largest integer which appears in such a token. Then the first **m** arguments from **a1**, etc. are substituted into the first occurrence of **string** to replace $%1, etc. The next **m** arguments are substituted into the next occurrence, and so on until the list of arguments has been used. (Obviously, values of **m** are usually just 1 or 2.) However, if the argument **skip=** is supplied to **LOOP**, each substitution moves ahead by **skip**, regardless of the value of **m**.

?DEFINE(name,text)

When an invocation of this macro is encountered in the course of scanning the text of another macro, **name** is entered into the table of currently known macros, with a definition consisting of **text**. The contents of **text** are not scanned for argument names or defaults. All arguments in **text** must be in the form $1, etc., or $*, and all missing arguments have null default text. The macro so defined will normally be invoked later in the text of the macro containing the invocation of **DEFINE**.

?SUBSTR(string,start,length)

Returns the substring of its first argument, starting at character position **start** and continuing for **length** characters (character positions are indexed from 1). If the last argument is omitted, the rest of **string** is returned.

?SYS(string)

Executes **string** as a command to the UNIX operating system. Useful for printing a file of instructions, etc.

?T(name)

Returns **$T**, followed by the nesting level of the macro call, followed by **name**. Used for constructing temporary S dataset names.

$*

Notation for "all arguments to this macro that were not matched by formal parameters". These arguments are substituted for **$***, and are separated by commas.

?(string)

Quoting convention for the macro processor. Each time a string enclosed in **?(...)** is encountered, the **?(** and **)** are stripped off, and **string** is not evaluated.

6.9 More About Macros

One of the best ways to learn how to write macros is to read macros. Try using **mprint** to print macros in the shared data directory. To find what macros exist, type

> ls("mac.*", pos=3)

Remember, too, that you can use

> help(macro="abc")

to see the on-line documentation for macro **abc**.

PROBLEMS

6.15. Construct a macro that takes an arbitrary number of arguments **a,b,c,...** and produces scatter plots of adjacent pairs, **a** and **b**, **b** and **c**, etc.

7

Data Analysis Using S

Chapters 3 to 6 have told you about most of the structure of the S system. This chapter presents some new functions and new ways of using familiar functions, to demonstrate the use of statistical techniques in the analysis of data.

7.1 Analyzing Categorical Data

Arrays represent a scheme of *regular* classification in data analysis; e.g., the columns of an **n** by **p** matrix divide the **n*p** data values into **p** equal-sized groups. *Irregular* classifications are also common, and S provides several ways of dealing with these. Usually the best is by the use of categorical variables, (which in S are called *categories*, see section 5.1.3) that run parallel to the data vector(s) of interest. Each categorical variable divides the data into as many groups as there are levels to the category. The S functions **cut** and **code** create categories out of continuous and discrete values respectively. Several categorical variables relating to the same data are analogous to the several dimensions of an array. The S function **index** creates a multiway *indexing structure* from an arbitrary number of categories.

A fundamental tool in analyzing data with respect to several categories is the **tapply** function, analogous to **apply** for arrays. The arguments to **tapply** are a vector of data, an index structure and the name of a function (plus any other arguments for the function). In the typical case that the function returns a single number summary for a vector, the value of **tapply** is a *table*, which in S is a multiway array that also contains the information about the labels for the levels of the categories. Tables are very like arrays in S, and the labelling information is kept intact by operations on tables, such as subscripting and **apply**. The function **tprint** prints one or more tables with all the labels in the right place.

With this background, we can now describe the sequence of expressions required to produce a special table summarizing a set of data:

1. form all the categories needed to describe the data, normally using **cut** and/or **code**;
2. use **index** to combine the categories into an index structure;
3. use **tapply** with the summarizing function(s) of interest to produce the table(s).
4. look at the table, using **tprint** on the result.

Because one often wants to use different categories or to modify the definition of categories many times in analysis, it is usually better to think of the process dynamically as described here, rather than having permanent index structures. As we shall illustrate, the summaries can be used in combination with the index structures to generate residuals.

Here is an example worked in some detail, to illustrate. The system dataset **saving.x** is a matrix of 5 variables measured on each of 50 countries. One of the variables is rate of growth in per capita income. Suppose we want a table of the median growth rate, for various income levels and regions of the world. First, we create the income and growth data:

> **income** ← **saving.x[,3]**
> **growth** ← **saving.x[,4]**

We want to cut the income data into nice intervals.

> **stem(income)**
```
N = 50    Median = 695.665
Quartiles = 287.77, 1813.93

Decimal point is 3 places to the right of the colon

   0 : 1112222222333333444
   0 : 56667778889
   1 : 134
   1 : 557789
   2 : 12234
   2 : 556
   3 : 03
   3 :
   4 : 0
```

Three levels of "less than 1000", "1000 to 2000", and "more than 2000" seems a reasonable choice:

```
> income.level ← cut(income, c(0,1000,2000,5000),
+     c("Poor","Medium","Rich"))
```

(Since **cut** allows non-exhaustive categories, we need to put in lower and upper bounds as well as the intermediate break points. See Appendix 1 for the full definition of **cut**.) Now, for the regions we do not have any dataset to hand, so we create, with a text editor, a file giving the region corresponding to each country. For a start we choose 4 regions: Europe, North America, South America, and the rest (Asia, Africa and the Pacific). The file, named "continent", contains 50 one letter abbreviations, and looks like:

```
> !cat continent
A  E  E  S  S  N  S  A  S  S
E  S  E  E  E  E  S  S  E  A
E  E  A  A  E  E  E  E  A  S
S  S  S  A  E  A  A  E  E  E
A  A  E  N  S  A  S  S  A  A
```

We can make this into a category with **code**, which takes a character vector and forms a category with as many levels as distinct strings in the vector. By default, the labels are the strings, but here we would like something a little more readable.

```
> region ←code( read("continent"),
+     level=c("A","E","N","S"),
+     label=c("Afr/As","Europe","N. Am.","S. Am."))
```

First step complete; the other steps are easy:

```
> i ← index(income.level,region)
> grow ← tapply(growth,i,"median")
> tprint(grow)
```

```
region:          Afr/As   Europe   N. Am.   S. Am.
income.level:
   Poor           2.960    6.880       NA    2.480
   Medium         4.985    3.540       NA       NA
   Rich           2.870    3.530    2.440       NA
```

The results are quite interesting: notice that the largest growth rates are in the poor European countries and the medium Afro-Asian countries. Looking at which countries fall in these two groups will show why.

> **saving.rowlab[growth > 5]**

```
          "Taiwan"     "Greece"        "Japan"      "Korea"
[ 5]      "Malta"      "Netherlands"                "Portugal"
[ 8]      "Zambia"     "Jamaica"       "Libya"      "Malaysia"
```

Notice that we did not attempt to find directly which countries fell into the two large-growth cells, but rather found the countries with large growth.

We can compute residuals from the fitted values by using:

> **residuals ← growth − grow[i]**

Notice how the table **grow** is subscripted by an indexing structure to produce a vector parallel to the data that generated the table. This completes our example.

Several other functions are also useful for categorical data. The function **table** generates a contingency table (a table of counts) from an arbitrary number of categorical variables (all referring to the same set of observations). A table based on **k** categories gives the counts of the number of observations in the **k**-way array defined by the categories. Specifically, the table is an array with **k** dimensions, the number of levels in the **i**-th dimension being equal to the number of levels in the **i**-th category. In addition the table has a component named **Label** which is a structure whose **k** vector components are the labels from the categories.

> **counts ← table(income.level,region)**

The output of **table** is appropriate for analysis by the **loglin** function below.

Although the normal S **print** function can be used to print tables, they can be printed in a more pleasing form by the function

tprint(table1, table2, ...)

Unlike **print**, **tprint** makes use of the labels, and can also print several tables in parallel, cell by cell. For example, the table generated above can be printed by:

> **tprint(counts)**

region: income.level:	Afr/As	Europe	N. Am.	S. Am.
Poor	11	4	0	15
Medium	2	7	0	0
Rich	1	8	2	0

The **loglin** function fits a loglinear model to a table or to a multi-way array. The simple form is

> **loglin(table,margins)**

The argument **margins** specifies which marginal totals the loglinear model should fit. It is a vector of integers: subsets of the integers are separated by zeroes. Each subset specifies the factors (i.e., dimensions of **table**) to be included in one of the marginal totals. We can fit the row and column margins in our two-way table above, for example, with **c(1,0,2)** and then compute and print the fitted values and residuals with the **tprint**function.

> ```
> > fit ← loglin(counts, c(1,0,2) )
> > tprint(Fitted=fit, Resid=counts−fit)
> ```

region: income.level:		Afr/As	Europe	N. Am.	S. Am.
Poor	Fitted	8.40	11.40	1.20	9.00
	Resid	2.60	-7.40	-1.20	6.00
Medium	Fitted	2.52	3.42	0.36	2.70
	Resid	-0.52	3.58	-0.36	-2.70
Rich	Fitted	3.08	4.18	0.44	3.30
	Resid	-2.08	3.82	1.56	-3.30

Notice that the argument names given to **tprint** appear as the names of the tables in the output. Also, notice that arithmetic on tables of identical shape produces the same shape table as result.

PROBLEMS

Table 1 presents a set of data on breaking strength of plastic specimens, related to whether an additive was used, their color and whether they were exposed to cold.

7.1.   With a text editor, enter the data for the four variables in the table into one or more UNIX files. Then read the data into S and do problems 2 and 3 below. (Data entry is an important part of analysis, too often overlooked. Before you start typing, think about how to make the data easy to read, and what arrangement into S dataset(s) will simplify the analysis. After reading it in, print the data in a form that facilitates

Table 1

Break	Add've	Color	Cold	Break	Add've	Color	Cold
90.0	yes	beige	no	80.1	no	green	yes
81.4	no	black	yes	78.5	yes	black	no
79.1	no	black	no	76.8	yes	green	no
76.7	no	beige	yes	87.3	yes	black	yes
78.9	no	black	yes	78.0	yes	black	yes
71.9	no	black	no	78.3	yes	green	no
88.2	yes	beige	no	84.7	no	green	no
72.8	yes	green	no	79.9	no	green	yes
90.1	yes	beige	no	91.4	no	black	yes
79.5	yes	green	no	75.9	no	beige	yes
81.7	yes	green	no	82.6	yes	beige	no
77.7	no	beige	yes	72.8	no	green	yes
84.3	yes	black	no	78.1	yes	black	yes
79.5	no	beige	no	81.8	no	beige	no
78.8	no	green	yes	81.4	no	green	no
73.3	yes	beige	yes	70.7	yes	black	yes
79.3	yes	green	no	73.1	no	beige	yes
81.1	yes	beige	no	88.0	no	beige	yes
77.1	yes	green	yes	85.3	yes	beige	no
83.2	no	black	no	85.9	no	beige	no
85.1	yes	beige	yes	72.0	no	green	yes
69.1	no	green	yes	79.9	yes	green	yes
81.2	yes	black	yes	73.9	no	green	no
75.2	no	green	yes	86.0	yes	beige	yes
78.9	yes	black	yes	81.5	yes	green	yes

proofreading against the Table 1.)

7.2.   Using **tapply**, calculate the 25% trimmed mean of the breaking strength for each cell of a table indexed by additive, color, and cold.

7.3.   Print a table that shows these trimmed means and the number of samples that were present in each cell.

7.2 Regression and Models

Section 3.5.1 introduced three functions to fit linear models: **reg**, **l1fit** and **rreg**. This section discusses some additional techniques for analyzing models. We begin by discussing **rreg** in more detail. The simple call is similar to least-squares regression:

> > z ← rreg(x,y)

As with **reg**, x is a matrix whose columns are the predictor variables and y is a vector to be fitted to the linear model. The model is estimated by an iterated, weighted least-squares algorithm, in which the weighting attempts to reduce the influence of observations in the data which, on the previous iteration, had unusually large residuals. The method is useful when the usual assumptions of linear modeling have been violated (e.g., because there are a few unusual observations or because the normal distribution is not a good model for the residuals). It is a good practice to compare the robust fit to a least-squares fit from **reg**. If the two are in reasonable agreement as to coefficients and residuals, the use of the least-squares analysis is supported. If there is significant disagreement, the user should study the data and the residuals of the two regressions to understand what is happening. For example, probability plots of the residuals may point out a few unusual observations. The structure returned by **rreg** also includes the final weights, w; studying the observations with small weights may shed some light.

There are a number of optional arguments to **rreg** to control the fitting and/or to study the intermediate results. While these may be helpful, their use sometimes requires an understanding of how robust regression is done. See **rreg** in Appendix 1.

A different problem with some linear models is that the number of x variables may be too large, perhaps because you want a simpler model for explanatory purposes. The S function

leaps(x,y,nbest)

implements a technique (known as "leaps and bounds", see the reference in Appendix 1 to learn details of the algorithm) to find **nbest** or more of the "best" regressions against a *subset* of the x variables, according to a criterion of good fit. The **leaps** function is actually a generalization of the *stepwise regression* methods that are more widely known. Stepwise regression begins with a model consisting of either all of the x variables or none of them. Then, at each step, a *single* variable is deleted or added.

The **leaps** function, unlike stepwise regression, is not limited to operating with a single variable at each iteration. Instead, it examines all possible subsets of **x** variables, using clever computational techniques to avoid examining obviously bad subsets. For each subset, a statistic is computed to evaluate the goodness of fit for the model. By default, the criterion is the C_p statistic, measuring the residual mean square against the mean square for the total regression. Other arguments may be supplied to **leaps**, including selection of alternative methods for judging the goodness of fit. The subsets are defined by a component, **which**, of the value of the **leaps** function. This is a logical matrix with as many rows as there are subsets returned and the same number of columns as **x**. Each row defines a subset of the **x** variables. Other components returned are a vector, **label**, of labels for each of the subsets and a vector giving the values of the criterion for each of the subsets returned. This component has the same name as the method used, "Cp" by default.

A common use of **leaps** is to produce a C_p plot of the results. This shows the criterion against the size of the subset (which is another component of the value returned by **leaps**). The system macro

?Cp(x,y,wt,int,names,identify)

does this plot and optionally (if **identify=TRUE** is specified) allows the user to identify points on the plot. The full **regprt** output for the identified points is then printed. For example, we can look at the C_p plot for the Longley data:

> > ?Cp(longley.x,longley.y)

The results here are not extremely clear-cut, but suggest that all the regressions including at least variables 3, 4 and 6 do fairly well.

The **leaps** function has a limit of 30 **x** variables. For larger problems, you should probably first attempt to reduce the dimensionality of the problem using techniques such as principal components.

The **hat** function supplies diagnostic information concerning linear regression. It returns a vector that tells how much *leverage* each observation had in the regression. A point with high leverage has a large effect on the fit because of its position among the **x**-values. While the leverage values cannot by themselves say what effect a single point has on the various regression coefficients or on the predictive use of a regression model, data points with large values are candidates for further study. Consider as an example, the regression of the system data sets **stack.x** and **stack.loss**. We compute the **hat** values,

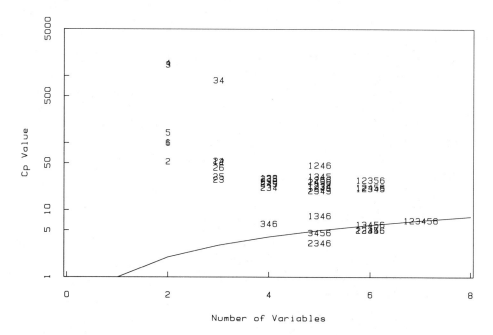

Figure 7.1. Plot produced by the **Cp** macro. The letters plotted at each point represent which columns of the **x** matrix were involved in a particular subset. In general, points with the smaller y-values represent "better" fits.

look at them and find out which data points gave large values:

```
> stem( h ← hat(stack.x) )
N = 21    Median = 0.1254351
Quartiles = 0.097474, 0.163433

Decimal point is 1 place to the left of the colon

   0 : 55688
   1 : 002223344
   1 : 567
   2 : 2
   2 : 7

High: 0.2980961 0.3176047

> rev(order(h))[1: 3]
     2   1   21
```

Now one could investigate points 1, 2 and 21 on other plots, or try the regression again omitting some or all of these points, to see what effect they had on the analysis.

In a multiple regression situation, we are often interested in assessing the effect of a single **x** variable on the response variable **y**. One way of doing this is by producing an *adjusted variable plot*. This plot shows the selected x-variable and the response variable, each adjusted linearly for all of the other x-variables. (This process is also carried out by the **adjust** macro.)

Suppose we wish to look at **stack.loss** and air flow, adjusted for the other two **stack.x** variables.

```
plot(
     reg(stack.x[,−1],stack.x[,1])$resid,
     reg(stack.x[,−1],stack.loss)$resid,
     xlab="Adjusted Air Flow",
     ylab="Adjusted Stack Loss"
) # Figure 2
```

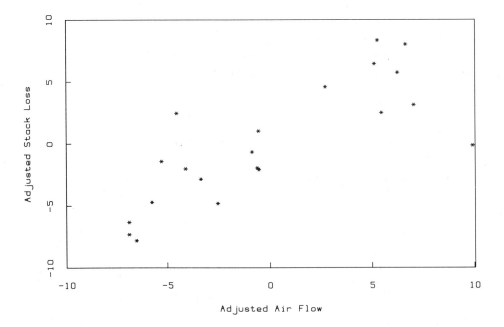

Figure 7.2. Adjusted variable plot for stack loss data. Both the stack loss and the air flow variables were adjusted for linear effects of water temperature and acid concentration.

The slope of a line fit to this plot is identical to the slope of the air flow variable in a multiple regression of stack loss on the other three variables.

Another broad class of models are those applied to multi-way arrays, regarded as tables. Section 7.1 introduced log-linear models for such data. The function

twoway(x)

introduced in section 3.5.1, provides a means of looking at two-way tables; i.e., matrices regarded as tables. It models each value in the matrix as the sum of a constant plus a row effect plus a column effect. Unlike standard analysis of variance techniques, however, **twoway** uses medians or trimmed-means to calculate the effects. This makes it resistant to outliers.

The function **plotfit** provides a graphic display of the result of a twoway analysis. For example, if we decompose the **cereal** data (percent of people agreeing with various statements about popular brands of cereal) using **twoway**:

> **plotfit(twoway(cereal.attitude), rowlab=cereal.rowlab,**
+ **collab=cereal.collab)   # Figure 3**

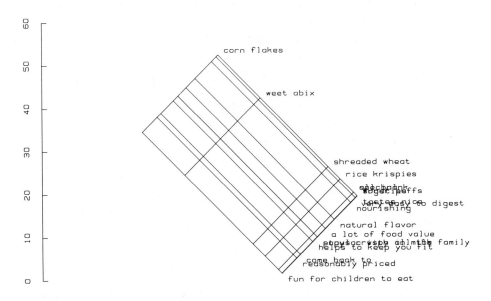

Figure 7.3. A twoway plot of fit for the cereal data.

The plot shows the row and column effects in such a way that the fitted value for each of the cells in the table can be read from the height on the vertical axis at which the line for the corresponding row label and column label meet. Only the vertical axis has meaning; the grid in the plot is only a way to relate the row and column effects. The technique is introduced in Tukey's *Exploratory* Data chapter 10. (Our example also shows the general problem of overplotting labels; at the moment, we have no corresponding general solution.)

PROBLEMS

7.4.   Carry out a least-squares regression and a robust regression on the stack loss data.

(i)   Plot the residuals from one fit against the residuals from the other.

(ii)   Plot the residuals from the robust fit against the weight produced by the robust regression.

7.5.   Generate some random **x** data and construct **y** that is linearly related to **x**. Add some Gaussian errors to the yvalues and fit a linear model using both **reg** and **rreg**. Next add some contamination (a few extreme outliers) to the y-values and fit again. Compare the results for **reg** and **rreg** in the two cases.

7.6.   Use the function **regprt** to print the results of a least-squares regression involving the stack loss data. With the assistance of a regression textbook (several are suggested in the Preface), identify the various things printed by **regprt**. Use the function **regsum** on the same regression results, and describe the data structure that is returned.

7.7.   Explain the computations in the adjusted variable plot above.

7.8.   The datasets **stack.x** and **stack.loss** provide a regression situation with five **x** variables. Construct the five different adjusted variable plots for this data.

7.9.   We can compute standardized residuals from a regression of **y** on **x** by using:

```
> r ← reg(x,y)
> h ← hat(x)
> wt.resids ← r$resid/sqrt(1−h)
```

These standardized residuals have constant variance, compensating for leverage in the design-space **x**. (See Belsley, Kuh, and Welsch *Regression Diagnostics*, Wiley, 1980, page 19). Produce a plot of standardized residual vs. fitted value for the **stack** data regression.

7.10.* Studentized residuals are defined in *Regression* Diagnostics as the standardized residuals, each divided by the residual standard error, but from the regression with the corresponding data point removed. Show that the vector of **n** the squares of the adjusted standard errors, say **vadj**, are given by the calculation

> vadj ← ((n-p)*var(resid) −
+   resid^2/(1 − h))/(n−p−1)

(see *Regression Diagnostics*, p. 20ff). Use this result to produce plots as in the previous problem, but for studentized residuals.

7.11. Do a two-way analysis of the dataset **counts** generated in section 7.1 in discussing loglinear models (it is probably easiest just to type in the **counts** data directly). Make a plot of fit for the model.

7.3 Multivariate Analysis

A number of S functions produce results useful in analyzing multivariate data. The functions **prcomp, cancor** and **discr** perform classical principal components analysis, canonical correlation analysis and discriminant analysis. For principal components

prcomp(x)

returns the principal component decomposition of the data matrix **x**. If the columns of **x** represent an original set of variables, **prcomp** finds a new set of variables (a rotation of the original) such that the new first column has the largest sample standard deviation, the second column the largest standard deviation among directions orthogonal to the first, and so on. The data structure returned by **prcomp** has three components: **x**, the data matrix of observations on the rotated variables, **sdev**, the vector of the standard deviations of the rotated variables, and **rotation**, the orthogonal matrix whose columns are the coefficients for the new variables. (Remember that component names only have to be specified sufficiently to distinguish them: the rotation can be accessed as **prcomp(x)$r**.) If **x** has **p** columns, **sdev** is a vector of length **p** and **rotation** is a **p** by **p** orthogonal matrix. For example, the famous **iris** data can be analyzed as follows. (The first line converts the three-way array into a 150 by 4 matrix.)

```
> iris.varieties ← rbind( iris[,,1], iris[,,2], iris[,,3] )
> pairs( prcomp(iris.varieties)$x )    # Figure 4
```

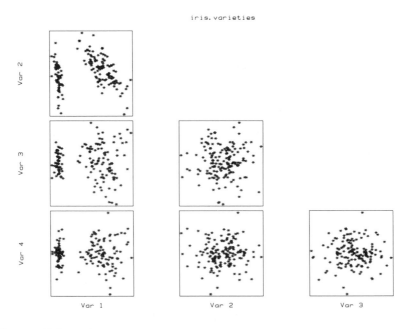

Figure 7.4. All pairwise scatter plots of the principal components of the iris data.

Principal components are sometimes done on correlation scaling; i.e., with each column divided by its standard deviation. With the iris data, the four variables have the same units, so the original scaling was appropriate. However, if correlation scaling is appropriate for a dataset, use the function **scale**:

prcomp(scale(x))

carries out principal components on scaled **x**.

Canonical correlation analysis takes two data matrices and returns a set of transformed variables for each matrix. The first pair of linear combinations (one linear combination from **x** and one from **y**) have the largest correlation among all linear combinations. The second pair have the largest correlation among variables uncorrelated with the original pair, and so forth. The S function

> cancor(x,y)

returns a structure with three components: **cor, xcoef** and **ycoef**. Suppose **x** has **p** columns and **y** has **q** columns. Let **s** be **min(p,q)**. Then **xcoef** is a **p** by **s** matrix whose columns are the linear combinations of columns in **x** producing the canonical variables. Similarly, the columns of **ycoef** give the linear combinations of columns of **y**. Both coefficient matrices are scaled so that the sum of squares of each column is one. The component **cor** is the vector of correlations between the canonical variables.

Suppose we try canonical correlation on the **iris** data:

```
> sepal ← iris.varieties[,1:2] #two sets of vars
> petal ← iris.varieties[,3:4]
> can ← cancor(sepal,petal)
> can.sepal ← sepal %* can$xcoef #transf. x data
> can.petal ← petal %* can$ycoef #transf. y data
> plot( can.sepal[,1], can.petal[,1],
+   xlab="First Sepal Canonical Variable",
+   ylab="First Petal Canonical Variable"
+   )  # Figure 5
```

Notice that matrix multiplication (the %* operator) of the original **x** data by the **xcoef** matrix gives the **x** data on the transformed scale.

The discriminant analysis function, **discr(x,k)** operates on a data matrix **x**, which should be arranged so that the first rows correspond to observations from the first group, etc. If **k** is a single number, there are **k** groups of equal size. Otherwise, **k** is the vector of group sizes. The first **k[1]** rows of **x** form the first group of observations, the next **k[2]** rows the next group, etc. If the groups are defined instead by a grouping vector, the macro version of **discr** will make the conversion. See the **discr** macro in Appendix 1.

The data structure returned by **discr** has three components: **cor, vars** and **groups**. The columns of the matrix **vars** give the coefficients (linear combinations of the columns of **x**) forming the discriminant variables. Each of these discriminates (predicts) a linear combination, specifically a contrast, of the groups. This contrast is given by the corresponding column of **groups**. The ability to discriminate the groups is measured by the correlation between the discriminant variable and the variable that defines the corresponding contrast of the groups. These correlations are the component **cor**. The linear

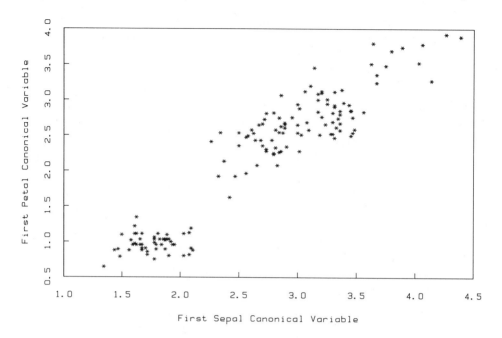

Figure 7.5. Plot of canonical variables from the iris data.

combinations in **vars** are scaled so that the discriminant variables have within-group variance equal to one. (For those familiar with discriminant analysis defined as a two-matrix eigenvalue problem, the above contains somewhat more information, and is more general and numerically well-defined. See problem 14 below.)

As an example of discriminant analysis, we can once again turn to the **iris** data. We have three varieties of iris, with 50 flowers in each group.

```
> d ← discr(iris.varieties,3)
> vars ← iris.varieties %* d$vars
> x ← vars[,1]; y ← vars[,2]
> plot(x,y,type="n",
+       xlab="First Discriminant Variable",
+       ylab="Second Discriminant Variable"
+       ) # Figure 6
> text(x,y,rep(1:3,rep(50,3)))
```

The discriminant analysis separates the first group quite well; however, so did principal components and canonical analysis (no wonder

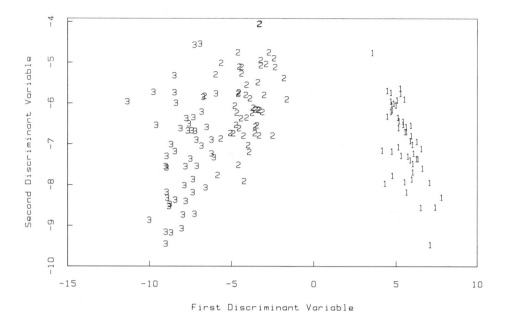

Figure 7.6. Plot of first two discriminant variables for three groups of Iris flowers. The groups are identified by the digits 1, 2, and 3.

this data set is a popular example).

PROBLEMS

7.12.   The expression

   iris.varieties ← rbind(iris[,,1], iris[,,2], iris[,,3])

   allowed us to re-arrange our 50 by 4 by 3 array **iris** into a 150 by 4 matrix. What would have happened to this approach if we had 20 groups instead of 3? Why is the following expression more general?

   iris.varieties ← matrix(aperm(iris, c(1,3,2)), ncol=4)

7.13.   Instead of a single principal component analysis of **iris**, do a separate analysis of each of the three groups and compare all of the results graphically.

7.14. Compute canonical correlations between sepal and petal variables of the **iris** data. Provide an interpretation for the first canonical variables.

7.15. What does the expression

rep(1:3,rep(50,3))

do in the discriminant analysis example? How would you have plotted letters to identify the groups?

7.16.* If **Lambda** is the vector of eigenvalues of the among cross-products with respect to the the within cross-products in the conventional definition of discriminant, and **cor** is the coefficients as defined by the function **discr**, show that **Lambda** is algebraically equivalent to

$$cor^2 \ / \ (1 - cor^2)$$

How does this indicate that the **discr** approach can be more general?

7.4 Hierarchical Clustering; Multidimensional Scaling

There are several functions which deal with *hierarchical clustering*. This statistical technique operates on a distance matrix of all pairwise distances between objects. If the objects are represented by the rows of a matrix, the function

dist(x,metric)

computes the distance matrix according to the "euclidean", "manhattan", "binary", or "maximum" metric. Then the hierarchical clustering tree is computed from the distances by

hclust(dist, method)

The **dist** argument can also be any square matrix of suitable values; for example, it could be the complement of a correlation matrix, **1−cor(x,x)**. The value of **hclust** is a *cluster tree structure*: Appendix 1 under **hclust**, gives a description of the components of the structure. However, all of the functions in this section can be used without knowledge of the internal structure of the tree.

The cluster trees produced by **hclust** can be plotted. The function

plclust(tree, label)

produces a plot called a *dendrogram* or a *cluster tree* of the structure produced by **hclust**. An argument **label** can be used to provide appropriate labels for the objects in the tree. A variety of optional arguments allow: leaves which hang a fixed distance down from the nodes or which drop to the bottom of the plot; nodes to have square or angled branches; or squashed or cut-off trees. Another function **labclust** provides more control over the labelling of dendrograms, and is described fully in Appendix 1.

The data on the system database as **auto.stats** is a matrix of statistics concerning 74 different automobiles. We can cluster it once we have scaled the variables so that distance measurements make sense:

```
> scaled.x ← scale(auto.stats)
> h ← hclust(dist(scaled.x))
> par(mar=c(10,6,4,2))   # lots of room for labels
> plclust(h, label=auto.rowlab)
```

The resulting plot is shown in Figure 7.

Other functions use the cluster tree structure to produce further information. To produce an assignment of the **n** objects into **k** clusters, use

cutree(tree,k)

It is also possible to cut according to the height, **h**, at which hierarchical clusters were merged (see Appendix 1). Given a set of objects named **leaves** i.e., a vector of integers in the range 1 to n,

subtree(tree,leaves)

returns the subtree which contains all the objects specified. For example,

subtree(tree,c(1,2,10))

returns the tree which contains observations 1, 2 and 10 in the original data. Note that the numbering in the subtree remains that used in the original tree, so that the identification of objects is correct for the original data. Subtrees can be plotted and labelled in the same way as the complete tree.

The function **cmdscale** performs classical metric multidimensional scaling; given a set of dissimilarities (distances) among some set of objects, it attempts to construct a set of observations which

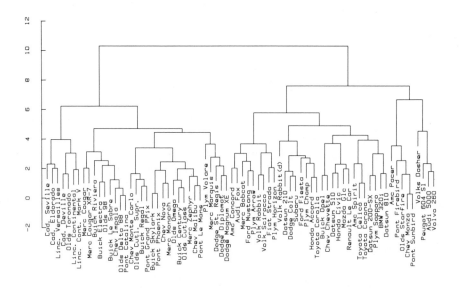

Figure 7.7. Hierarchical clustering of 74 automobile models offered for sale in the United States in 1979.

reproduce these distances. Typical applications of **cmdscale** might use perceived dissimilarities that are observed directly rather than being computed. To see how well the technique works, though, we can apply it to data of known structure. Suppose, for example, we make up some two-dimensional data, uniformly distributed in the plane.

```
> par(mfrow=c(1,2),pty="s")
> original ← matrix(runif(40),20,2)
> plot(original,original,type="n")      # set up equal axes
> text(original[,1],original[,2],1:20)
> title("Original Configuration")
```

We can use **dist** to calculate the inter-point distances and **cmdscale** to recover the positions. (Of course, the recovered configuration may differ from the original by rotation or reflection.)

```
> d ← dist(original)
> recovered ← cmdscale(d)
```

> **plot(recovered,recovered,type="n")   # equal axes**
> **text(recovered[,1],recovered[,2],1:20)**
> **title("Configuration Recovered by0mdscale")**

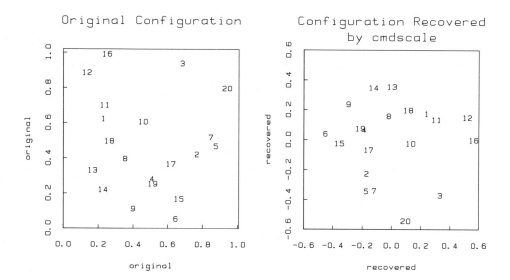

Figure 7.8. Classical multidimensional scaling. The plot on the left shows the original data configuration (points uniformly distributed in the unit square). The plot on the right shows the two-dimensional configuration recovered by **cmdscale** using only the inter-point distances of the data on the left.

Of course, with original data of higher dimensionality, we would not in general be able to construct a two-dimensional solution to display the distances without error.

PROBLEMS

7.17. Perform a hierarchical clustering on the **iris.varieties** data. Produce a cluster-tree plot labelled with three symbols or colors for the three species. How does the hierarchical clustering correspond to the three pre-defined groups?

7.18. Suppose you have a clustering of many objects so that the cluster tree is quite large (for example, the 150 irises). In this case, the labels plotted by **plclust** will hopelessly overwrite one another. The approach is to use **label=FALSE** to suppress labelling, and then use **identify** (along with coordinates obtained from **plclust(..., plot=F)** to insert selected labels on the plot. Try this out. Use the value returned by **identify** to extract and plot a subtree.

7.19. Apply **cmdscale** to distances of the rows of **scale(auto.stats)**. to construct a two-dimensional approximation. Compare the reconstruction to the hierarchical clustering of this data above.

7.20. The next three problems refer to a set of data of your own choosing, or some random data you construct. Carry out a test of the **cmdscale** function on your data. Compute the distances between the points in the output. Show a stem-and-leaf display of the differences between the original and the recovered distances. Plot the recovered vs. original distances.

7.21. The macro **mdsplot** produces a plot to display the results of multi-dimensional scaling. Use **mprint** to print the definition of **mdsplot**. Use the macro on the output of **cmdscale** generated in the previous problem.

7.22. Use hierarchical clustering to cluster your data. How might you use multi-dimensional scaling and hierarchical clustering together to display data.

7.23.* Use the function **cutree**, to cut your hierarchical clustering tree into two separate clusters. Could you use the function **chull** to construct lines on a scatter plot around the areas which represent those two clusters?

7.5 Time-Series Analysis

Time-series in S are differentiated from vectors by the information concerning their starting and ending dates, and the number of observations per period. The functions **lag**, **diff**, and **window** are often used together with **tsmatrix** for carrying out regression problems involving time-series. For example, to model housing starts by manufacturing shipments and a lagged version of shipments,

```
> y.and.x ← tsmatrix(hstart,ship,lag(ship))
> r ← reg(y.and.x[,−1], y.and.x[,1])
```

The important thing here is that, even though **hstart** and **ship** cover different time periods, **tsmatrix** aligns them so that all the observations in each row of the **y.and.x** matrix come from the same month.

PROBLEMS

7.24. What advantage is there to putting the **y** variable first in the **y.and.x** matrix?

7.25. How would you include differences in **ship** as an additional **x** variable? Do some plots or other displays help you to decide whether the third variable is useful in the regression?

7.26. Do the regression above using only data from July, 1970 to June, 1974.

7.27.* Make a macro version of **tsmatrix** that applies a window (supplied as an argument to the macro) to all the time-series.

The function **smooth** provides a robust smooth for equally-spaced data, hence is quite useful for smoothing time-series. The documentation in Appendix 1 gives some options to control the smoothing algorithm used, but just saying **smooth(x)** gives a reliable, robustly smoothed version of **x** in almost all cases.

The **sabl** function is a modern statistical technique for carrying out seasonal decomposition of time-series. It is robust to outliers, helps to choose a transformation for the data that assists in the decomposition, and can adjust for calendar effects. **Sabl** decomposes a time-series into components named **trend**, **seasonal**, and **irregular**, so that the sum of the three components is equal to the original series. The **trend** component is a smooth series that varies only slowly through time; **seasonal** follows a similar pattern year to year; **irregular** is the residual that is not explained by either **trend** nor **seasonal**.

One of the strengths of **sabl** is the collection of graphical methods developed with it for displaying results. The primary display of the decomposition is done by the **sablplot** macro.

> > s ← sabl(hstart)
> > ?sablplot(s,"US Housing Starts")   # Figure 9

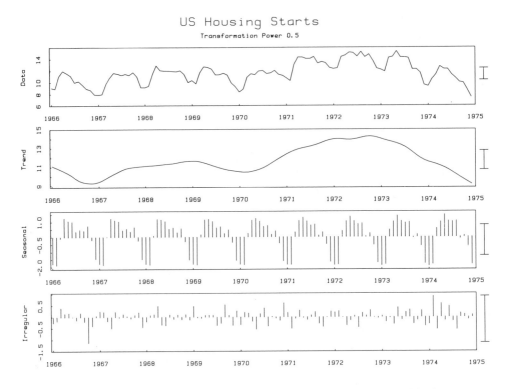

Figure 7.9. Four plots show a **sabl** decomposition of monthly housing start series into components: trend, seasonal effect, and irregular (residuals). The data is on a transformed scale, selected by the **sabl** algorithm.

Since **sabl** allows the seasonal pattern to evolve over time, it is important to be able to see this variation; **monthplot** is often used for this.

> > **monthplot(s$seasonal) #Figure 10**
> > **title("Seasonal Component of US Housing Starts")**

In this plot, horizontal lines are drawn at the midmean of the seasonal effect for each month, giving a general picture of the seasonal effect. The vertical lines emanating from these midmeans show the actual seasonal effect computed for the various years of data.

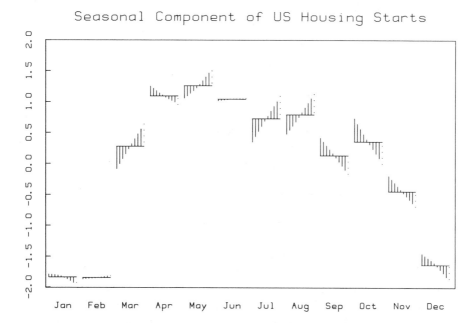

Figure 7.10. Plot of seasonal component of housing starts data. Horizontal lines are at the midmean of the seasonal component for each month; vertical lines show year-to-year evolution of the seasonal component.

Calendar effects in monthly series can be estimated by giving the **calendar=T** argument to **sabl**. The calendar effect should be estimated only for aggregate series (those where the monthly number represents the sum of daily numbers). It adjusts the series for the number of days in a month, as well as the number of different days-of-the-week occurring in the month. If calendar effects are estimated, **sabl** will return another component, named **calendar**, and **sablplot** will plot this component along with the others.

PROBLEMS

7.28.   Use the functions **smooth, tapply,** and **cycle** to form a primitive seasonal decomposition macro. Try it on the housing starts data **hstart**. (*)When would you expect **sabl** to be superior to this technique?

7.6 Statistical Tests and Estimates

Many classical statistical techniques are very easy to carry out in S, even though there is often no specific function for them. For example, consider the *paired t-test*. Suppose two vectors **x** and **y** give paired measurements, for example, blood pressure before and after administration of a drug. We are interested in determining whether the treatment has any effect on the measured value. The paired t-test constructs the vector of changes, **y−x**, and looks at the ratio of the mean change to standard deviation of the mean change.

```
> change ← y − x
> stddev ← sqrt( var(change)/len(change) )
> t.stat ← mean(change) / stddev
```

Once a t-statistic is computed, like **t.stat** here, the probability of getting such a value from a t-distribution on **len(x)−1** degrees of freedom is determined from a) looking at a table or b) using the function **pt** to determine the probability level. If the computed value is far out in the tails of the t-distribution, then there is evidence that the mean change really was non-zero. While **pt(t,df)** returns the probability that a value from the specified t-distribution is less than **t**. we usually want to know the probability that a value from the specified t-distribution is larger in magnitude than **t**. Because the t-distribution is symmetric, this is computed as

```
> prob ← 2 * pt( −abs(t.stat), len(x)−1)
```

That's it—computation of a paired-t test.

We could turn this expression into a macro, here shown obeying our guidelines for good macro design (see chapter 6):

```
MACRO paired.t(x,y)
({
        ?T(change) ← y − x
        ?T(obs) ← len(?T(change))
        ?T(stddev) ← sqrt( var(?T(change))/?T(obs) )
        ?T(t.stat) ← mean(?T(change)) / ?T(stddev)
        rm(?T(change),?T(obs),?T(stddev),?T(t.stat),
            value= cstr(
            t.stat = ?T(t.stat),
            df = ?T(obs)−1,
            prob = 2 * pt( −abs(?T(t.stat)),
              ?T(obs)−1))
        )
})
END
```

Other classical statistical tests are often as simple.

PROBLEMS

7.29. Why not use **2*(1−pt(abs(t.stat),len(x)−1))** in the paired t-test?

7.30. Should the **paired.t** macro print the results or return a data structure?

7.31. Modify the macro so it does a set of paired t-tests at one time, comparing columns of one matrix to corresponding columns of another matrix.

7.32. Construct a macro to carry out an unpaired t-test.

7.33. Write *and document* a macro that carries out a statistical test of your choosing. Provide evidence that your macro works, for example by using a textbook problem.

To make the process clearer, we will work through another standard statistical test, the *chi-square* test. One use of this test is to assess distributional assumptions, for example, do we have reason to believe that the values of lottery winning numbers introduced in Chapter 1 are not uniformly distributed? The basic idea is to count the number of observations that occur in each of a number of bins, and compare the actual count with the expected count. With the lottery winning numbers, we can use the **cut** and **table** functions to compute the number of observations occurring between 0 and 100, 100 and 200, etc.

```
> observed ← table( cut( lottery.number, seq(0,1000,100)−1 ) )
> tprint(observed)
observed
```

	−1+ thru 99	99+ thru 199	199+ thru 299
:			
:			
	26	31	32

```
:    299+ thru 399  399+ thru 499  499+ thru 599
:
         24              29              21

:    599+ thru 699  699+ thru 799  799+ thru 899
:
         23              20              21

:    899+ thru 999
:
         27
```

> print(expected ← sum(observed) / len(observed))
    25.4
> print(chi.stat ← sum((observed − expected)^2 / expected))
    6.55118
> print(prob ← pchisq(chi.stat, len(observed)−1))
    0.3162653

Thus, in this case, 32% of chi-square values on 9 degrees of freedom would be smaller than the one we computed, 68% would be larger. From this test, we have no evidence that **chi.stat** did not come from the specified chi-square distribution, hence no reason to question the uniformity of the winning numbers. Of course, this does not assure us of uniformity. We tested the first digit of the winning number for uniformity; similar tests could be performed on second or third digits, etc.

PROBLEMS

7.34.   Carry out a chi-square test on the second and third digits of the lottery winning number.

7.35.   What does the expression

        seq(0,1000,100)−1

  do?

7.36.   Create a chi-square test macro.

7.37. The Durbin-Watson statistic tests for serial correlation in the residuals of a time-series regression. It is defined as

$$d = \frac{\sum (z_t - z_{t-1})^2}{\sum z_t^2}$$

Write a macro to compute d and apply it to the regression at the beginning of section 7.5.

Distribution-free or *nonparametric* statistics are often easy to do using the **rank** function. Nonparametric statistics are often based only on the ordering of a set of data, not on the numerical values. For example, the *two-sample rank test* uses a test statistic consisting of the sum of the ranks for one set of data, say **x**, when both **x** and **y** are ranked together. The expression **rank(c(x,y))** gives a vector containing the ranks taken together; the first **len(x)** elements of this are the ranks associated with **x**. A single expression to give the desired statistic is then:

sum(rank(c(x,y))[1:len(x)])

This is precisely the expression that the system macro **ranktest** evaluates.

A *one-way analysis of variance* is also quite easy to carry out, although it is not as compact as it might be. We want to take a data vector and a category giving group information, and produce a standard ANOVA table. The following might be a first sketch for a macro to carry out the analysis, without applying the good-design principles of Chapter 6 (see Problem 36 below).

```
> mprint(mac.one.way)
MACRO one.way(data,group)
print(quote=FALSE,
    "One-Way Analysis of Variance of data",
    "grouped by group")
grand ← mean(data)
n ← table(group)$Data
i ← index(group)
group.mean ← tapply(data,i,"mean")
within ← data − group.mean[i]
SSbetween ← sum(n*(group.mean-grand)^2)
SSwithin ← sum(within^2)
df1 ← len(group.mean)−1
df2 ← len(data) − len(group.mean)
```

```
m ← matrix(c(
    SSbetween, df1, SSbetween/df1,
    SSwithin, df2, SSwithin/df2,
    sum((data-grand)^2), df1+df2, NA),
    byrow=T,ncol=3)
print(m,head=F,
    rowlab=c("Between Groups","Within Groups","Total"),
    collab=c("SS","df","MS"))
f ← (SSbetween/df1)/(SSwithin/df2)
print(quote=F, encode("F ratio is",round(f,2)),
    encode("Probability of exceeding this is",
        1-pf(f,df1,df2)) )
END
```

We can execute this macro on the lottery payoff data as well:

> **?one.way(lottery.payoff, code(lottery.number%/100))**

```
One-Way Analysis of Variance of lottery.payoff
grouped by code(lottery.number%/100)
```

	SS	df	MS
Between Groups	1014119	9	112679.9
Within Groups	3188771	244	13068.73
Total	4202890	253	NA

```
F ratio is 8.62
Probability of exceeding this is  0
```

PROBLEMS

7.38. As a macro for general use, there are several bad features in
       one.way: intermediate datasets are not removed, data is
       assigned to commonly occurring names like **n** and **i**, and the
       results of the analysis are printed, rather than being returned.
       Apply some of the principles of Chapter 6 to produce a version
       of **one.way** that has a printing option, but returns a data struc-
       ture representing the analysis, as well.

7.39. The probability quoted as "0" in the last line is, actually, just very small. The problem is that the value of **pf(f,df1,df2)** is exactly 1, to the accuracy of numbers in S, with the result that the subtraction returns 0. Using the relation that **pf(f,df1,df2)** equals **1−pf(1/f,df2,df1)**, modify **one.way** to avoid the loss of accuracy.

7.7 Probability and Quantile Functions

Section 3.5.1 described random number generators for various probability distributions. Corresponding to these are probability and quantile functions. The probability functions take a vector of data values, plus parameters for the distribution, if any, and return a set of probability values (i.e., values of the cumulative distribution function) corresponding to the data values. Each of the probability functions has a name consisting of "p" followed by a code for the distribution family, such as "norm" for the normal. See the list of code names for distributions in Section 3.5.1.

The quantile functions perform the inverse calculation. Given a vector of probability levels, they return corresponding quantile values. Quantile functions are named "q", followed by the name for the distribution. One application of quantile functions is to generate probability plots from arbitrary distributions (see Appendix 1 for the macro **qqplot**).

In analyzing data, we are often interested in empirical quantiles from a sample of data. For example, suppose we have a set of data **x**, and wish to get the 10, 25, 50, 75, and 90 percentiles of the data.

quants ← approx(ppoints(x), sort(x),c(.1, .25, .5, .75, .9))\$y

This works because of the two functions **ppoints** and **approx**. The **ppoints** function generates the values of *plotting points*, **(i−.5)/n** for an argument of length **n**. The **approx** function carries out linear interpolation: we give it points on the cumulative distribution function (the plotting points and the sorted data), as well as points at which we would like to have fitted values.

PROBLEMS

7.40. The S quantile routines are equivalent to the tables of distributions that are in ordinary statistical texts. Try the following:

> ```
> > df ← c(1,2,3,5,10,100)
> > p.val ← c(.2,.1,.05,.02,.01,.001)
> > tab ← −qt( matrix(p.val,6,6,byrow=T),
> +    matrix(df,6,6) )
> ```

Print **tab** with meaningful row and column labels. Compare it
to a table of one-sided percentage points of the t-distribution.
How would you modify this to get two-sided percentage
points?

7.41. In the previous problem we computed $-qt(p,df)$. Why negate
the quantiles? What would happen if we had used

> ```
> > p.val ← c(.8,.9,.95,.98,.99,.999)
> ```

instead? Which method should we prefer?

7.8 Matrix Methods; Linear Algebra

For users familiar with numerical matrix methods and numerical linear algebra, S provides a number of functions which carry out
major operations in this area. These S functions can be combined (frequently through macros) to generate customized numerical or statistical procedures.

Matrix products and cross products are produced by the operators:

```
x %* y
x %c y
```

The **%c** operator is equivalent to **t(x) %* y**. Cross products can also be
produced through the functional form:

```
crossprod(x,y)
crossprod(x)
```

where the second form is equivalent to **x %c x**.

The %o operator forms outer products. For example, if **x** and **y** are vectors, then **x** %o **y** is a matrix whose **[i,j]** element is **x[i]** * **y[j]**. It is possible to use %o with arrays and to use the functional form, **outer**, to combine elements with an operation other than "*" (see **outer** in Appendix 1).

Inversion of matrices and the solution of systems of linear equations are handled by the function **solve**. With one argument, **solve(x)** inverts the square matrix **x**. With two arguments, **y←solve(A,b)** solves the system of linear equations A %* y = b.

Matrix decompositions are available: **eigen(x)** and **svd(x)** return structures representing the (symmetric) eigenvalue and the singular-value decompositions respectively. The function **gs(x)** returns the explicit orthogonal decomposition of a rectangular matrix **x** using the iterated Gram-Schmidt algorithm (i.e., a matrix of orthonormal columns, **q**, and an upper-triangular matrix, **r**). The function **chol(ss)** returns the upper-triangular factor for the Choleski decomposition of the (square, positive-definite) matrix **ss**. With either of these decompositions, one often needs subsequently to solve triangular systems of equations: this is done by the function **backsolve(R,b)**, which returns the solution of the upper-triangular system of equations **R %*y = b** (the optional argument **lower=T** solves lower-triangular systems). Notice that triangular matrices are not represented by a special data structure; therefore, they can be used wherever a standard matrix can.

The linear-algebra computations are useful building blocks for simulation and for statistical summaries. In working out accurate and efficient computations, some familiarity with numerical methods for linear models and multivariate analysis is needed; see, for example, chapter 5 of *Computational Methods for Data Analysis*.

For example, suppose we would like to generate a sample of 100 points from a multivariate normal distribution, with a specified correlation matrix, **rho**, a vector of standard-deviations for the variables, **sd**, and a vector of means, **means**. A good way to accomplish this starts with the covariance matrix, formed by multiplying the elements of the correlation matrix by the product of the row and column standard deviations. The sample itself is generated first as 100 samples from the *standard* multivariate normal. Then we multiply by the Choleski decomposition of the covariance matrix and add on the means. (See *Computational Methods for Data Analysis*, page 185, for the computational background.)

```
> var.cov ← rho * outer(sd,sd)
> x ← matrix(rnorm(100*len(sd)),100,len(sd))
> samp ← x %* chol(var.cov) +
+    matrix(means,100,len(sd),byrow=T)
```

We used an expression for the number of variables; replacing 100 by a variable name, using some intermediate datasets and setting the computation up as a compound expression, would enable us to define a macro for multivariate normal simulation.

As a second case study, suppose we would like to determine confidence limits for the predicted value in a linear regression situation; that is, for a particular **x** value, we would like a confidence region for the corresponding predicted value, **yhat**. However, rather than construct these intervals for individual **x** values, we will construct a sequence of values over the range of the data in the original regression, and then draw lines showing the corresponding upper and lower limits on the scatter plot of **(x,y)**.

The main argument is as follows (see also Problem 44). The vector of upper confidence limits has the form

> **yhat + qt(.975,df) * sqrt(diag(vhat))**

where **vhat** is the (estimated) covariance matrix of **yhat**, and .975 comes from assuming we want 95 percent confidence bars. The predicted values are, in S notation, **xmat %* coef**, where **coef** are the least-squares coefficients, and **xmat** is a matrix with ones in the first column and the desired sequence of **x** values in the second. Therefore, the covariance matrix, **vhat**, is equal to

> **xmat %* cov %* t(xmat)**

where **cov** is the covariance matrix of the coefficients. The regression summary function, **regsum**, computes **cov**, so we could evaluate this expression easily. Two improvements, however, come from thinking about the algebra. First, it is usually better to work with the triangular decomposition of a covariance matrix in evaluating quadratic forms; the expression above is just the cross-product matrix of the transformed data

> **x1 ← xmat %* t(chol(cov))**

Second, we really would rather not form the large covariance matrix for the predicted values, when we will only use the diagonal. The diagonal elements are the row sum-of-squares of **x1**.

Here are the computations, applied to the **auto.stats** data for automobile length and weight.

> > **weight ← auto.stats[,8]**
> > **length ← auto.stats[,9]**

First, we carry out a standard linear regression of **weight** on **length**.

```
> z ← reg( length, weight )
```

Next, we need to compute fitted values for the sequence of **x0** values, and to construct the confidence intervals.

```
> x0 ← seq( min(length), max(length), len=50 )
> xmat ← cbind(1,x0)
> yhat ← xmat %* z$coef
>
> triang ← chol( regsum(z)$cov )
> x1 ← xmat %* t(triang)
> lims ← qt(.975,len(weight)−2) *sqrt( ?row( x1^2, sum) )
```

The confidence bars should be plotted at **x0** and **yhat ± lims**.

```
> plot(length, weight,   # Figure 11
+    ylim=range(weight,yhat+lims,yhat−lims) )
> abline(z)
> lines( x0, yhat+lims, lty=2 )
> lines( x0, yhat−lims, lty=2 )
> title("Automobile Data - Linear Least-Squares Fit",
+ sub="For any particular x-value, the lines give
+ a 95% confidence interval for true mean")
```

We used a **ylim=** argument to **plot** to ensure that the points and confidence intervals will both fit on the plot, and used the **lty=** parameter to **lines** to give the line segments joining the confidence limits a non-solid line type.

PROBLEMS

7.42.   Turn the expressions for computing regression confidence intervals into a macro, **regconf** which takes as input: 1) a regression structure, 2) a vector of x-values for which confidence intervals are to be constructed, and 3) a confidence level, and returns a matrix with 2 columns containing the lower and upper confidence limits. You could use the macro like this:

```
> plot(x,y)
> z ← reg(x,y)
> abline(z)
```

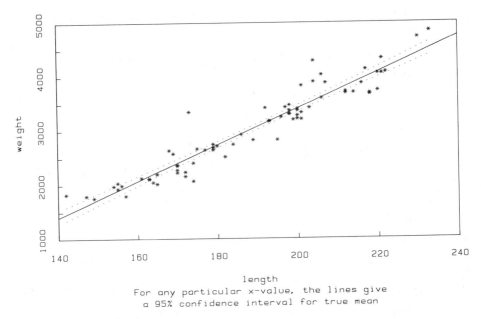

Figure 7.11. Automobile data with least-squares regression line and 95% confidence bars for the predicted values.

```
> newx ← seq(min(x),max(x),len=100)
> intervals ← ?regconf(z, newx, .95)
> matlines(newx, intervals, lty=2)
```

7.43.   Should the macro in the previous problem have default values for its arguments? Should it produce labels? Would it be better to include the plotting in the macro?

7.44.*   Fill in the sketch above that justifies the computation of the confidence bars. If necessary, review the definition and use of Choleski decompositions; for example, in *Computational Methods for Data Analysis*, Chapter 5.

Although not formally a technique of linear algebra, the S function **napsack**, which solves knapsack problems, is related to the functions we have discussed in this section. We can use it in conjunction with many of the techniques discussed before to answer some interesting questions.

In a 0-1 knapsack problem, we are given a number of pieces of varying sizes, and we want to pack them into a knapsack of a specified size. The problem is to find the subsets of the pieces which most nearly fit the knapsack. General knapsack problems are difficult to solve, since they appear to require algorithms whose running time is exponential in the problem size. For this reason, knapsacks have been used in public key cryptosystems. The knapsack algorithm used in S can reasonably accommodate problems with approximately 40 pieces.

In any case, let us construct a (useful) knapsack problem of our own. Suppose we wish to design a plot to investigate perception of areas in bar graphs. The plot will show a number of bars which are divided into, say, 20 parts. The bars will each be divided in the same way, and the pieces of each bar will be partly shaded-in, with the shading pattern varying from one bar to the next, but with the total shaded area being nearly the same in all bars, so we can test whether shade pattern affects perception of area.

First, we construct 20 random-sized bar divisions. Let us also sort them and force their total to be 1000.

```
> x ← rev(sort(runif(20,0,100)))
> x ← x*1000/sum(x)  # force x to sum to 1000
```

The **napsack** function can find the 10 subsets of x whose sums most nearly match 500. (The idea will be to make a barplot out of these 20 divisions, with 10 identical divided bars, each shaded to correspond to one of the **napsack** subsets.)

```
> nap ← napsack(x, 500)
> x %* nap   # print fitted values
      500.0000 500.0000 499.9998 500.0001 500.0001
[ 6]  499.9996 500.0003 500.0004 499.9993 499.9992
```

As we can see, each subset defines a divided bar with exactly half its area shaded, to the accuracy the eye can perceive.

As a final nicety, lets try to order the 10 bars so that similar shading patterns tend to be nearby. An approximate way to do this is to cluster the columns of the **nap** matrix, using the **hclust** function. The component **order** of **hclust** (see Appendix 1) gives the re-ordering of the objects. This should, on the whole, put similar objets nearby.

```
> order ← hclust(dist(t(nap)),"binary") $ order
```

Now we are ready to do the barplot

```
> barplot(
+     matrix(x,20,10),
+     density = 20*nap[,order],
+     main="Result of Knapsack Algorithm",
+     sub="Each Bar has Half of its Area Shaded"
+     )
```

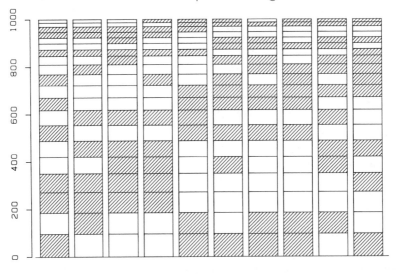

Figure 7.12. A barplot showing the result of the knapsack algorithm. Each bar has height 1000, with half its area shaded, and half unshaded.

Of course in our perception experiment, we would probably leave the pieces and bars in a random order when presenting them to subjects.

PROBLEMS

7.45. Using the technique in this section, generate a bivariate normal sample of length 100, with standard deviations 1, means 0, and correlation .7. Plot the data. Compute the means, standard deviations, and correlation. Should the results be exactly what we specified? (Hint: is **mean(rnorm(100))** exactly zero?)

7.46.* How could you construct a random sample with exactly a specified mean vector and correlation matrix.

7.47. Re-create the knapsack example given in the text, only do not sort the areas within the bars or cluster the bars. (For the experiment, we probably do not want the method of construction to be obvious.) What would be the effect of interchanging which areas are shaded and which are non-shaded? How could you use this to help determine whether people perceive shaded or non-shaded areas as larger? Try the experiment on several people.

This problem and example was inspired by Cleveland and McGill, "A Color-Caused Optical Illusion on a Statistical Graph", *The American Statistician*, **37**, No. 2, pp. 101-105, 1983.

7.9 Simulation; Monte-Carlo Methods; Internal Sampling

When we apply even a moderately complicated model-fitting or other estimation procedure, the statistical properties of the procedure are important to an understanding of our results. If the procedure depends on some theoretical probability model, we need to know what the distribution of the results would be like if this model were true. In a careful analysis, we should also check how much the distribution would be affected by possible failures in the model assumptions. The techniques presented in this section will help to estimate such distributions by *Monte-Carlo* methods; that is, by generating pseudo-random datasets from the model and applying the procedure to each of these. Having, say, M such estimates gives us an empirical distribution to use as an approximation to the (generally, unknown) theoretical distribution of the estimation procedure. We may also want to understand how the results depend on the different observations within the current data. We will show later in the section various ways to construct internal samples (both random and systematic) from a single dataset.

Let's look at the stack loss regression example. If we assumed the usual statistical regression model with normally distributed errors, the distribution of the least-squares estimates is also known. But if either the model or the estimates are not the standard, it will usually be impossible to describe the distribution simply. In particular, suppose we compute the regression by minimizing absolute residuals (the function **l1fit**). Simulation then becomes the most practical approach to estimating the distribution of the coefficients. The model being simulated says that the data **y** is of the form

   X %* beta + error

where **X** is the (fixed) matrix of independent variables, extended by a column of 1 for the intercept term, **beta** is some unknown population regression coefficient vector and **error** is a vector of independent and identical normal variables with zero mean and unknown standard deviation, **sigma**. We must choose values for the unknown population values; frequently, we start by setting them to the estimates generated from the original dataset. We also have to decide how many Monte-Carlo replications of the estimation to do, balancing off the cost of the computations against the greater accuracy obtained from more replications. In S, as we will see, it is easy to do some replications and then decide whether more are needed. Suppose we decide to do 20 replications to start with.

```
> l2reg ← reg(stack.x,stack.loss)
> yhat ← stack.loss − l2reg$resid
> sigma ← sqrt(var(l2reg$resid))
> n ← len(stack.loss)
>
> y ← yhat +
+      sigma*matrix(rnorm(20*n),ncol=20)
> coefs ← NULL
> for( j in 1:20){
+     coefs ← rbind( coefs,
+         l1fit(stack.x,y[, j])$coef)
+     }
```

The matrix **coefs** has 20 rows corresponding to replications of the simulation, and 4 columns corresponding to the coefficients in the regression model. We can look at the columns to approximate quantities of interest in the probability distribution of the estimates; for example,

```
> mc.var ← var(coefs)
> sqrt(diag(mc.var))
    16.42091      0.1778683     0.4213854     0.2184393
```

to estimate the variance-covariance matrix of the coefficients and print the approximate standard deviations of the coefficient estimates.

PROBLEMS

7.48.   Even a relatively simple Monte-Carlo like the example above is usually too much to have to type each time. Turn our llfit simulation into a macro. You should be able to apply the macro easily to different data, population estimates and number of simulations. Try to follow the guidelines in Chapter 6 on good macro design.

7.49.   Explain in words the steps in the example. Why is the computation in the second line of **yhat** equivalent to X%* **beta**?

7.50.   Set up a similar Monte-Carlo for least-squares regression. Notice that because **reg** accepts a matrix for **Y**, you can avoid the **for** loop.

The general steps in the design of the regression simulation are typical of other simulations in model-fitting and estimation. We rearrange the problem, if possible, so that the quantities to be simulated are a simple function of some population values (**yhat** and **sigma** in this case) and some basic random variables that can be generated in S. We then repeatedly do the simulation, apply the estimation to the results, and bind in the new estimates with the previous values. The result is one or more data structures representing the complete simulation. In S, we don't usually have to reduce the simulation to just a few numbers. If the simulation does not generate an enormous volume of data, we can and should keep the data structures around to study the simulated distributions in a variety of ways.

A second example will illustrate the generality of the process. Suppose we want to study the distribution of the estimates in principal component analysis, particularly the standard deviations of the principal directions (the component **sdev**). Given a dataset, we will take the estimated variances to define the population variance (if we assume a normal distribution, the population means are irrelevant). Then we can construct successive simulations and collect the estimates as follows, illustrated for the first species of iris data:

```
> setosa ←iris[,,1]
> theta ← t(chol(var(setosa)))
> prcoefs ← NULL
> for( i in 1:20) {
+     y ← matrix(rnorm(200),50,4) %* theta
+     prcoefs ← rbind(prcoefs,prcomp(y)$sdev)
+     }
```

Again, we have a matrix of the coefficients from which to compute
any summaries of the approximate distribution.

PROBLEMS

7.51.* (This and the next problem require some knowledge of matrix
       algebra and multivariate analysis.) Why does the calculation
       with **chol** produce multivariate normal data with the correct
       population variance?

7.52.* Set up a similar calculation for a discriminant analysis problem,
       based on the whole iris data.

     In carrying out Monte-Carlo estimation to support an analysis,
we should make the results reproducible. Two steps are needed: to
record accurately the method used and to enable someone else to generate the same pseudo-random values. The first part is best handled
by the **diary** function in S (see section 5.3.2). Whenever we do a non-
trivial simulation, we should record a diary file and keep this as evidence of the method used. To make the sequence of pseudo-random
values accessible, it is necessary to save the *seed* used by the random-
number generators in S before the simulation starts. In S the random
number generators always look for a dataset called **Random.seed** to
use as a seed, and always return values to that dataset that will act as a
generator for the next set of pseudo-random values needed. To record
the starting point of a particular simulation, then could be done as follows.

```
> diary("simul.case3")
> print( seed.case3 ← Random.seed )
> ... now do the simulation ...
```

The second line keeps the current seed, as **seed.case3**. We also printed the seed value on the diary file as a precaution. So long as either the file or the dataset **seed.case3** are not both lost, we can reset the generator. Specifically, to regenerate the same simulation, one would edit the assignment in the second line to

> > **Random.seed ← seed.case3**

and just run the rest of the expressions in the diary without change.

PROBLEMS

7.53. Write a macro that generates the saved seed and the diary file invocation, given a name for the simulation, and another that reruns the simulation. (*)Make the macro generate a unique name for the simulation automatically.

7.54. The dataset **Random.seed** is actually a vector of 12 6-bit integers. (See the documentation in Appendix 2.) Show that a pseudo-random seed is generated by

> > **Random.seed ← runif(12,0,64) %m INT**

Why would you want to do this?

The various random number generators simulate continuous distributions for the most part. The function **sample** simulates samples from finite populations, either from an explicit vector of values, or implicitly from the population **1:n** for specified **n**. So, for example, if we wanted to run our own version of the lottery discussed in Chapter 1, we could generate 10 random numbers from the set of 3 digit numbers by

> > **pickit ← sample(1000,10) − 1**

Closer to the topic of this section, suppose we want to divide the rows of a matrix into two, perhaps so we can do an analysis on one half of the data and then test it on the other half.

> > **n ← nrow(x)**
> > **set1 ← sample(n,n/2)**
> > **x1 ← x[set1,]; x2 ← x[−set1,]**

A neat, general way to break a set of **n** items into **k** mutually exclusive sets is:

```
> sets ← matrix(sample(n), ncol=k)
```

This generates a random permutation of **1:n** and then splits this up into a matrix whose columns give the various subsets.

Finally, there are a number of statistical techniques that compute estimates on *systematic* samples from one data set; for example, on each of the samples obtained by dropping one observation. Suppose observations are represented by rows of the matrix **x**. Then suppose we want to do the **l1fit** computation mentioned earlier on all the datasets with one observation deleted. Using negative subscripts to delete observations, and observing that **len(stack.loss)** is 21:

```
> c.minus1 ← NULL
> for(i in −(1:21)){
+     x ← stack.x[i,]
+     y ← stack.loss[i]
+     c.minus1 ← rbind(c.minus1,
+     l1fit(x,y)$coef)
+     }
```

(Clearly, doing all subset calculations on a large dataset for a substantial analysis will take quite a while.) The technique known as the *jackknife*, due to J. W. Tukey, computes estimates called *pseudo-values* as **n** times the estimate based on all the data minus **(n−1)** times the estimate with one observation removed. The average of the pseudo-values is considered as a single estimate for all the data, and the sample variance of the pseudo-values estimates the population variance of that single estimate. For our example above, the pseudo-values can be computed as follows:

```
> c.all ← l1fit(stack.x, stack.loss)$coef
> c.pseudo ← 21*matrix(c.all,21,4,byrow=T)
+     − 20*c.minus1
```

PROBLEMS

7.55. Complete the jackknife calculations for the example above. Also, look at the pseudo-values. Are there unusual features?

7.56. Comment on the numerical problems, if any, with the

jackknife.

7.57.   How would you modify the computations of the "minus 1" estimates to drop out blocks of, say, 10 observations at a time?
Arrange that the observations are assigned at random to the
blocks of 10.

7.10 A Numerical Experiment

The purpose of this section is to show how questions in statistical methodology can be attacked with the help of S. Consider the
question "what would happen to a paired t-test if one of the X or Y
values was grossly out of line?" Many people think, since the t-test is
composed of non-resistant parts (mean and variance), it would tend to
give false evidence that the difference was non-zero.

There are several ways of conducting a numerical experiment
to investigate the resistance of a t-statistic to outliers. One method is
to construct sample data, initially satisfying the assumptions of nor-
mality made by the t-test, and then perturb the data by adding an
outlier. A second method is to take a fixed sample and determine
what will happen to the t-value as an additional observation is added
with some arbitrary value. We will pursue both of these methods.

To do the numerical experiment, let us first choose a sample
size for study. As sample size increases, the influence of a single
outlier will decrease, so we settle on a moderate-size sample, say of 10
points. The basis for a paired-t test is that the t-statistic will follow
the t-distribution if the differences follow a Gaussian distribution
with mean zero. To test this, lets construct 500 samples of size 10,
compute t-statistics, and see that the background theory holds true.

```
> samples ← matrix(rnorm(5000),nrow=10)
> t.stats ← apply(samples,2,"mean")/
+       sqrt(apply(samples,2,"var")/10)
> # now test to see if these t.stats look like a t on 9 d.f.
> ?qqplot(t.stats,t,9)
```

The plot is shown in Figure 13, and generally follows a straight line.

We perturb the data by adding an outlier to each sample. We
can do this by changing the first row of the matrix so that the values
now come from a Gaussian distribution with a much larger variance.

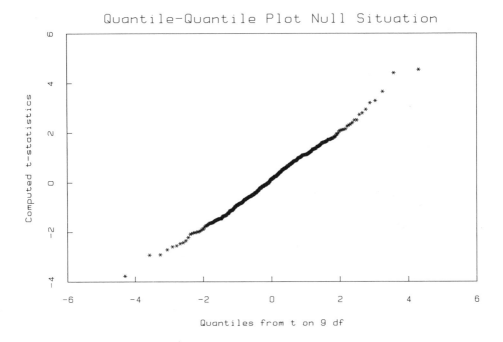

Figure 7.13. Theoretical Quantile-Quantile plot of t-statistics computed from 500 samples of size 10 from a standard Gaussian distribution.

> **perturbed ← samples**
> **# add an observation with 5-times the std dev**
> **perturbed[1,] ← rnorm(500,0,5)**

Now we carry out the t-test and Q-Q plot again, giving Figure 14. This time our values are not so well t-distributed. Generally, the computed values have somewhat shorter tails than the t on 9 degrees of freedom. We could, of course, try adding much larger outliers than we did here. The short tails suggest that the test is being conservative; i.e., that it will fail to detect significant results as often as it would under the model assumptions.

As a second test, we want to see the effect on the plot of sampling from a normal distribution but with non-zero mean.

> **non.zero ← samples + 2   # now the mean is 2**

We carry out the t-computation and plot as before and obtain Figure 15. Now the striking thing is not the straightness of the line, but simply that the new statistics are very different than one would expect from a t on 9 d.f.. The smallest is about 2.5, which corresponds to a

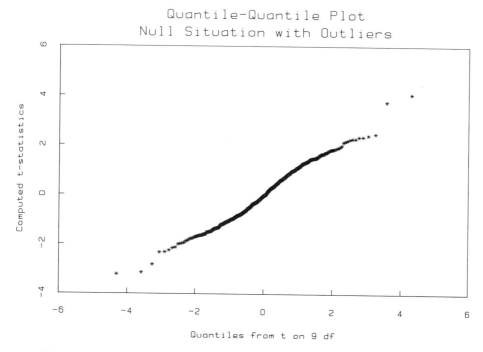

Figure 7.14. Theoretical Quantile-Quantile plot of t-statistics computed from 500 samples of size 10. Nine of each sample come from a standard Gaussian distribution, the other one from a Gaussian with standard deviation equal to 5.

two-sided 3% probability level.

PROBLEMS

7.58. Carry out the analysis in the example above (changing the size of the sample, if you like). Study explicitly the suggestion that the test is conservative.

7.59. A different kind of perturbation generates all the data from a distribution other than the assumed normal. One family of such distributions is the *t* family itself, which tends to the normal as the degrees-of-freedom parameter gets large. Redo the analysis above, but generate the data from a *t* distribution with 5 degrees of freedom, rather than from the normal.

7.60. Repeat the previous problem with each of the degrees of

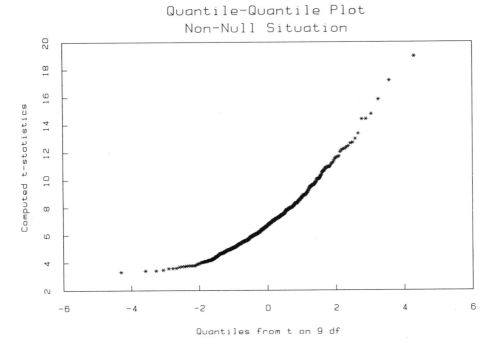

Figure 7.15. Theoretical Quantile-Quantile plot of t-statistics computed from 500 samples of size 10 from a Gaussian distribution with mean 2 and standard deviation 1.

freedom **c(2,3,10,15,50)**. Try not to use explicit iteration in the calculation of the t statistics. (Hint: consider the general form of **apply** for a multi-way array.)

For a third test, we add 2 to the **perturbed** dataset, giving a sample centered at 2 but with one outlier, and recompute. The Q-Q plot is shown in Figure 16. Once again, shorter tails for the simulated distribution are indicated. The t-test is *conservative* in the presence of outliers since it does not *incorrectly* flag zero-origin populations as non-zero. However, in the presence of outliers, it less frequently recognizes a sample from a distribution with non-zero location. If you were testing drugs using experiments that had occasional outliers, the t-test might cause you to pass-over effective drugs.

Another way of looking at the whole thing is to produce *empirical influence curves*. Suppose we generate a sample of size 9 from a standard Gaussian. If we then add one new point, say at **x**, we can compute a t-statistic. We can do this for a collection of x-values to

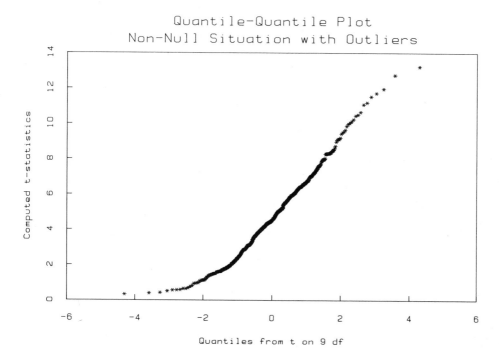

Figure 7.16. Theoretical Quantile-Quantile plot of t-statistics computed from 500 samples of size 10. Nine of each sample come from a Gaussian distribution with mean 2 and standard deviation 1, the other one from a Gaussian with mean 2 and standard deviation equal to 5.

define a curve **t(x)**, the t-statistic based on the value of the 10th point. If we plot **t(x)** vs. **x**, we can see how various values of **x** affect the t-statistic.

```
> single.sample ← rnorm(9)
> x.values ← seq(−100,100,1)
> mat ← rbind( x.values, matrix(single.sample,9,len(x.values)) )
> t.stats ← apply(mat,2,"mean")/sqrt(apply(mat,2,"var")/10)
> plot(x.values, t.stats, type="l",
+      main="Influence Curve for t on 9 df",
+      sub="9 points centered at 0",
+      xlab="value of 10th point")   # Figure 17
```

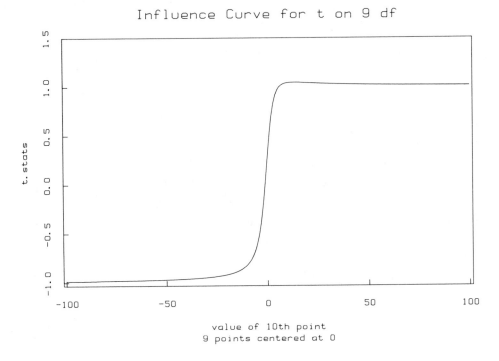

Figure 7.17. Influence curve for computed t-statistic. Given nine points from a standard Gaussian distribution, the curve shows the t-statistic that would be computed if a tenth point were added at different **x** values.

```
> mat[−1,] ← mat[−1,]+3   # now give the sample a mean of 3
> t.stats ← apply(mat,2,"mean")/sqrt(apply(mat,2,"var")/10)
> plot(x.values, t.stats, type="l",
+      main="Influence Curve for t on 9 df",
+      sub="9 points centered at 3",
+      xlab="value of 10th point")   # Figure 18
```

The two figures show clearly that, as the outlier gets larger, the t-statistic approaches ±1.

The conclusion: numerical experiments are easy to do using S, and often can provide a quick look at theoretically complex problems.

PROBLEMS

7.61.   Conduct a numerical experiment to determine the distribution of the order statistics of a sample of size 20.

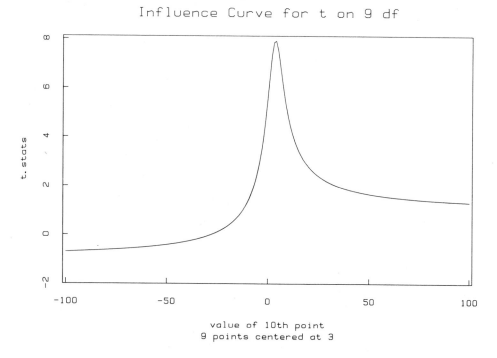

Figure 7.18. Influence curve for computed t-statistic. Given nine points from a Gaussian distribution with mean 3 and standard deviation 1, the curve shows the t-statistic that would be computed if a tenth point were added at different **x** values.

7.62. Carry out the simulation described in this section with a different sample size. How similar are your results to those produced here?

8

Presentation:
Graphics and Reports

The previous chapters were primarily concerned with how to use S to carry out data analysis. However, the analysis is not normally the final operation; a data analyst must communicate the results of the analysis to others. The success of such communication is, of course, very important. Simplicity, attractiveness, and familiarity in the presentation are important to good communication, particularly to non-specialist audiences. While S is primarily intended as an *analytic* system, this chapter shows how to combine S and the facilities of the UNIX system to present graphics and reports that are attractive and effective. Section 8.1 discusses graphical displays; section 8.2 describes report generation.

8.1 Graphical Presentation

Many of the analytic plots described earlier in the book are useful for presentation as well. Naturally, for presentation you may want to choose a graphics device that generates high-quality graphics, with color if you can use that. For example, pen plotters with drafting-quality pens drawing on good paper or on special transparency material can give good results. A high-resolution raster terminal with a good hardcopy device may also be useful, particularly if it can generate 35mm slides directly. Note also that a phototypesetter can be used for graphics output via the *pic* interface (see section 8.2).

Getting attractive plots for presentation may also involve you in special choices of graphic parameters; for example, to choose character sizes, line textures, colors or plotting symbols that carry your message vividly. See **par** in Appendix 1. Such choices depend on local conventions, your own aesthetic judgements and the characteristics of the graphics device being used: we suggest that you do some experimenting and then keep a record (say, in a set of macros or source files) of the plotting styles you found pleasing.

8.1.1 *Bar Plots; Pie Charts*

The *bar plot* or *bar graph* is perhaps the most common form of presentation graph. A bar graph represents a measurement or count by the length of a bar; many bars can be displayed on a single plot to represent different groups of data, and bars can be sub-divided to reflect sub-groups. The function

barplot(height)

is given vectors of bar heights. The argument **height** can also be a matrix, each column corresponding to one bar and the values in the column representing the sizes of the pieces into which the bar is sub-divided.

The **barplot** function also takes a number of optional arguments. Giving **names** as a character vector argument labels the bars with the elements of the character vector. Other arguments, **density**, **angle**, and **col** combine to describe the way in which bars (or bar segments) should be shaded. If either **angle** or **density** are specified, the bars are shaded with lines at the specified **angle** and with **density** lines per inch. If **col** is specified in conjunction with **angle** or **density**, it controls the color of the shading lines: if **col** is specified alone, and if the graphic device has area-filling capabilities, the bars are filled with solid colors. (To get solidly filled bars on a pen-plotter, however, use a large value of density and be prepared to wait a long time!) Other arguments to **barplot** control details such as bar spacing; see Appendix 1.

Often, a bar graph encodes grouping information in the colors or shading of bars or bar segments. The argument **legend** allows the user to specify a character vector of names corresponding to the various levels of **angle**, **density**, and **col**. The legend is placed in the upper right-hand corner of the bar graph. When more control over legend placement is desired, use the function **legend**. Not only can **legend** provide legends for bar graphs, but since it also allows specifications of line styles (parameter **lty**), and plotting characters (**pch**), it is useful with scatter plots, time-series plots, and other displays.

The following example uses some fictitious data about the per-cent expenditures in three communities for five charitable services. The 15 data items are typed to a file, "perc.data":

```
5.4   3.1   3.5
5.7   8.6   25.0
20.4  26.0  22.0
36.3  34.1  28.0
14.4  11.4  4.5
```

and then are read in. We create the character vectors for community and services directly:

```
> percent ← matrix( read("perc.data"),
+     ncol=3, byrow=T )
  15 items read
> community ← c("Old Suburb", "Coast County", "New Suburb")
> service ← c("Child Care", "Health Services", "Comm'y Centers",
+     "Family & Youth", "Other")
```

In Figure 1 we create a vector of angles for shading the bars, going from 45 degrees to 135 degrees, and plot divided bars for the three communities. Having set up 4 extra lines of margin on the top to ensure enough room, before doing the barplot, we can use the **legend** function (see Appendix 1) to draw a legend box defining the angles in terms of the services.

```
> angles ← seq(45, 135, len=5)
> par(mar=c(4,4,8,1))
> barplot(percent, names=commy, angle=angles)
> legend(rdpen(2), service, angle=angles)
```

Notice that we used the ability to read two points interactively in order to position the legend in an aesthetically pleasing spot on the graph.

When dealing with a category (Section 5.1.3), the macro **bart-able** can be useful. It will produce a bar graph of the counts corresponding to the different levels of the category, and will produce the appropriate bar labels. For example, the following computations produce a category of national income by thousands of dollars, from the system dataset **saving.x** and then plots a bar graph of the corresponding counts.

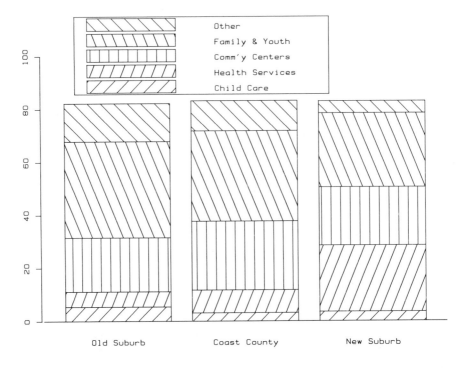

Figure 8.1. Barplot showing charitable contributions by percent in three communities to five services.

> income ← saving.x[,3] #national income data
> catgy ← cut(income/1000,0:5)
> ?bartable(catgy)
> title("National income, in thousands of dollars")

produces Figure 2.

A common, although not analytically very effective, plot is the *pie chart*, a disc divided up into segments whose angles represent portions of a total given to different components. S does provide a function for creating pie charts:

pie(x, label)

draws a circle and divides it into labelled parts with the **x** vector specifying relative areas, i.e., the *i*th slice takes up a fraction **x[i]/sum(x)** of the pie. By the way, arguments **density, angle**, and **col** also work as

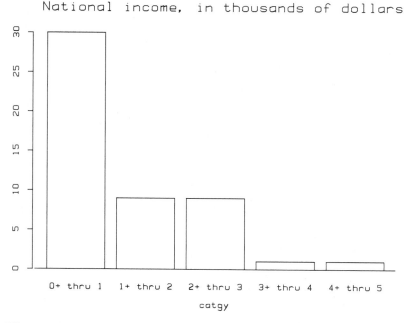

Figure 8.2. Bar graph of counts of national income in the **saving** data, using the **bartable** macro.

they do with **barplot** to control hatching or filling of pie segments. Using the charity data introduced in Figure 1,

```
> dens ← c(40,15,30,20,10)
> pie(percent[,1],main=community[1],
+     density=dens,angle=angles,explode=1)
> legend(rdpen(2), service,
+     density=dens,angle=angles)
```

puts up a pie chart with an "exploded" segment and a legend, shown in Figure 3.

8.1.2 *Text on Graphic Devices*

Section 8.2 describes how to put S results into reports generated from a printer or phototypesetter. However, it is often convenient to generate simple text slides or transparencies on a graphics device, particularly if the device plots in color. The S function **vu** generates a plot of text information on the current graphics device, choosing appropriate character sizes, providing layouts for lists, changing colors and fonts, etc.

Old Suburb

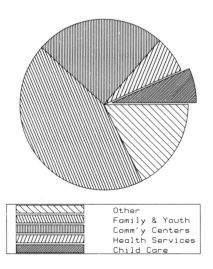

	Other
	Family & Youth
	Comm'y Centers
	Health Services
	Child Care

Figure 8.3. Pie chart for fraction of contribution to various services.

The function

vu(text)

takes a character vector and plots the individual character strings as lines of text. Character strings that start with a period ("."), are interpreted as layout commands, and control aspects such as character-size (.S), color (.C), and font (.F) in a style analogous to **nroff/troff** macro systems.

When the material to be presented by **vu** is in the form of lists of items, the function **quickvu** generates the character vector suitable as the argument to **vu** without the user having to know any of the layout commands. **quickvu** prompts the user for lines for a title. An empty line signals the end of the title; then, the user is prompted for list items (by default, one line per item). Again, an empty line signals the end of the input, upon which **quickvu** returns a character vector complete with layout commands.

```
> my.text ← quickvu()
Title line: Computing for
Title line: Good Data Analysis
Title line:
```

```
List item: Interaction, Graphics
List item: Flexibility
List item: Extensibility
List item:
> vu(my.text,font="sr")
```

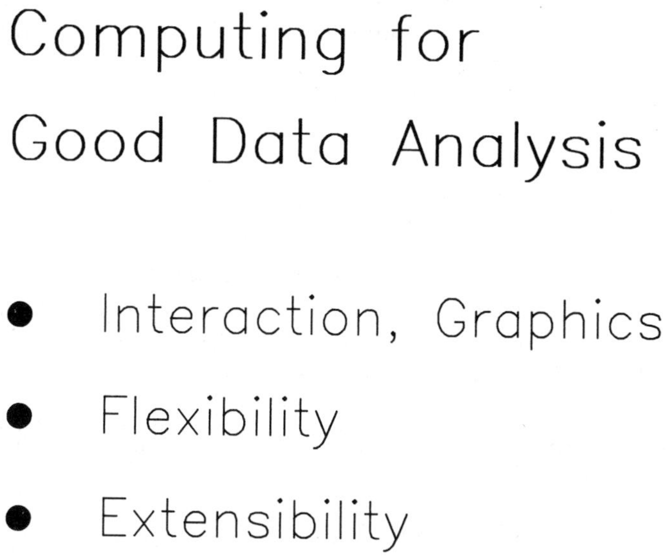

Computing for
Good Data Analysis

- ● Interaction, Graphics

- ● Flexibility

- ● Extensibility

Figure 8.4. Example of output created by the **vu** function.

If you want to have full control over formatting, you can use the **edit** function to modify the details of **my.text**. For example, we could use

> **edit(my.text)**

to produce the following:

```
.S 2
.CE
Computing for
Good Data Analysis
.S 1
.BL
.C 2
.LI
Interaction, Graphics
.LI
Flexibility
```

```
.LI
Extensibility
.LE
```

If this modified version of **my.text** were given to **vu**, it would produce a centered title and a bullet list (.BL) of three items. The title would be twice the size of the list items and in the standard color (color 1). The list items would be in color 2.

The actual size of the characters is chosen by **vu** to fill the plotting surface to the width and height desired (7 inches square by default).

PROBLEMS

8.1.   Given the data in the first 5 rows of dataset **telsam.response**, display the percentage of respondents giving each of the four answers for each of the 5 interviewers. Try both pie charts and bar graphs. Which do you prefer? Why?

8.2.   Suppose the three communities in Figure 1 had total charitable collections of

> **total ← c(1104, 626, 2531)**

in thousands of dollars. Redo the barplot in Figure 1, but in terms of money rather than percent.

8.3.   With the data above, draw a pie chart for the total expenditures for all three communities, including the five services and also the remaining money (representing overhead).

8.4.   Experiment with **quickvu** and **vu** to create vugraphs with various fonts. You can find the names of fonts by looking under **font** in Appendix 2.

8.2 Reports

S is not intended to produce presentation-quality reports internally. Instead there are several facilities in S that make it easy to interact with the word-processing tools which are part of the UNIX system environment.

Let's first address the need for tabular output in a report. S can easily produce the numbers that should appear in a table, but it has no provision for the special layouts that may appear in a typeset table. Fortunately, the UNIX system command *tbl* provides a very flexible way for formatting tables. The S function **tbl** writes a file that is precisely what the command *tbl* wants. The best way to think of **tbl** is as a version of printing in S that provides the additional information needed to create a nice table in a document to be processed by *nroff/troff*. Simply invoking **tbl(x)** where **x** is a matrix, produces a file "tbl.out" ready for processing. Optional arguments let you supply a heading, row or column labels for the table. You can also specify the file for output. For fine-tuning of the results, it is best to generate the file and then to edit it. On the other hand, to break up large amounts of data into manageable tables, it is better to use S to generate moderate-sized matrices for *tbl*.

As an example, here is a table produced from two columns of regression coefficients, which we assume have already been computed on the **stack** system dataset:

```
> coefs ← cbind(l2reg$coef, l1reg$coef)
> tbl( round(coefs,3),
+     row = c("Intercept",stack.collab),
+     col = c("Least-Squares", "L1 Norm"),
+     head = "Stack Loss: Coefficients") )
```

Running this through tbl and troff shows that the labels in **stack.collab** could be more uniform in length. With a little editing of the file "tbl.out", however, we get the following table.

Stack Loss: Coefficients		
	Least-Squares	L1 Norm
Intercept	−39.920	−39.690
Air Flow	0.716	0.832
Water Temp.	1.295	0.574
Acid Conc.	−0.152	−0.061

A similar situation is available for graphics: the UNIX system provides the command **pic** to allow the user to specify pictures that should appear in typeset documents. S interfaces to this by means of a device driver, named **pic** which produces a file named "pic.out" which is what the **pic** command wants. As a result, any plot can be produced on a troff output device (e.g., a phototypesetter), by replacing your usual device call with **pic**. See the case study in section 8.3

for an example.

Finally, there are often times in a report when the result of an S computation is desired, not as a figure or table, but as an integral part of the text. For this situation, we provide the utility function

$ S REPORT input output

which scans the file **input** for S expressions surrounded by braces "{ ... }", evaluates them, and replaces them in the output by the output of the S expressions. As a very simple example, suppose the input file contained the lines:

```
The lottery payoffs had median
value {median(lottery.payoff)}, and
standard deviation {sqrt(var(lottery.payoff))}.
```

Then the output file would contain the corresponding lines:

```
The lottery payoffs had median value 270.25, and
standard deviation 128.8884.
```

Again, you may want to polish up either the S expressions or the output file to improve appearance. The advantage of incorporating the S expressions directly into the text of the report is that the same report may be generated easily at different times or for different data. Also, errors of transcription from the analytical results to the report are eliminated.

PROBLEMS

8.5.   From the charitable donation data used for Figure 1, create a table of **percent**, with labels based on **service** and **community**. Do whatever editing of the **tbl** output file is needed to make a nice table when processed by *tbl* and *nroff* or *troff*.

8.6.   Use the **pic** device driver to generate a version of one of the pie charts suitable for processing with *pic* and *troff*.

8.3 A Case Study

The best way to illustrate all of these facilities working together is through a single, comprehensive example. Suppose we want a report that describes our analysis of the randomness of the winning numbers drawn in the New Jersey Pick-It Lottery. We might want something like Figure 5.

Investigation of New Jersey Lottery Winning Numbers

We recently investigated the randomness of the digits of the winning numbers in the NJ Pick-It Lottery. There were 254, 254, and 252 numbers in each of three sets of data.

We can construct a chi-square test on the first digit of the winning number. Remember that the test evaluates

$$\chi^2 = \sum \frac{(\text{observed} - \text{expected})^2}{\text{expected}}$$

The result is a value of 9.13158 on 9 degrees of freedom. Since the probability that a value from a true chi-square distribution is at least this large is 0.574782, we find no evidence of non-uniformity in the leading digit.

A table of the data might make this clearer:

Leading Digit of Winning Number in NJ Lottery		
	Group 1	Group 2	Group 3
0	26	23	19
1	31	36	29
2	32	23	26
3	24	17	29
4	29	22	23
5	21	23	28
6	23	26	21
7	20	26	26
8	21	30	21
9	27	28	30

We can also look at it graphically:

Figure 8.5. Reduced version of "Investigation of New Jersey Lottery Winning Numbers".

To understand how this report was produced, let's examine some fragments of the corresponding input to REPORT.

```
.TL
Investigation of New Jersey Lottery Winning Numbers
.PP
We recently investigated the randomness of the
```

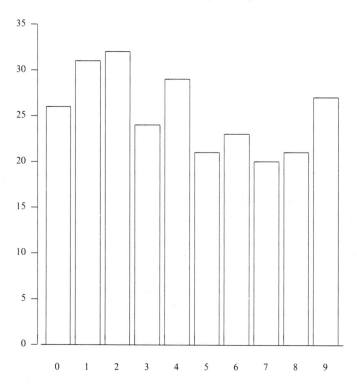

As you can see, the ability to combine graphical displays, tabular output, and general S computations provides a very powerful environment for report writing.

Figure 8.5. (continued) Reduced version of "Investigation of New Jersey Lottery Winning Numbers".

```
digits of the winning
numbers in the NJ Pick-It Lottery.
There were
\{len(lottery.number)},
\{len(lottery2.number)}, and
\{len(lottery3.number)}
numbers in each of three sets of data.
```

This segment shows the use of a standard macro package for *troff* (the —**ms** macros, in this case) and a very simple use of S expressions inside the text.

You may have wonder why we have "\" characters before the braces. The next segment of the input file uses *eqn* to produce an in-line equation. We must make special arrangements to use both *eqn* and **REPORT** on the same file, since both recognize "{" and "}" as delimiters. Two steps are needed: we invoke **REPORT** in the form

    S REPORT −e

and then use a back-slash escape before each brace or double brace that introduces S computations.

```
.PP
We can construct a chi-square test on the first
digit of the winning number.
Remember that the test evaluates
.EQ
chi sup 2  =  sum {
   ( roman observed - roman expected ) sup 2 }
   over { roman expected }
.EN
The result is a value of \{
o <- table(code(
   c(lottery.number, lottery2.number, lottery3.number)%/100
   ))
e <- sum(o)/len(o)
chi <- sum((o-e)^2/e)
df <- len(o)-1
chi
} on \{ df } degrees of freedom.
Since the probability that a value
from a true chi-square distribution is at least this
large is \{ pchisq(chi,df) },
we find no evidence
of non-uniformity in the leading digit.
```

Next, we would like to produce a table. The only trick here is to note that the UNIX system command *tbl* does not follow files inside documents that are introduced by ".so" (source) commands in *troff*. We create the file "tbl.out", and then invoke *tbl* on it to produce pure *troff* commands.

```
.PP
A table of the data might make this clearer:
\{
c1 <- table(code(lottery.number%/100))
c2 <- table(code(lottery2.number%/100))
c3 <- table(code(lottery3.number%/100))
tbl( cbind(c1,c2,c3), rowlab=encode(0:9),
   collab=encode("Group",1:3),
   head="Leading Digit of Winning Number in NJ Lottery",
   file="tbl.out")
```

```
sys("tbl tbl.out >table.output")
}
.TS
.so table.output
.TE
```

Next, we want to produce a plot, using the S device driver **pic** and then processing its output by the UNIX system *pic* command. The ".PS <pic.out" command directs *pic* to produce a picture from the named file.

```
We can also look at it graphically:
\{
pic
?bartable(code(lottery.number%/100))
title("New Jersey Lottery Winning Numbers")
printer
}
.PS <pic.out
.PP
As you can see, the ability to combine graphical displays,
tabular output, and general S computations
provides a very powerful environment for report writing.
```

We have now shown all of the fragments of the input to **REPORT**. Assuming that the fragments are together in one file named "lottery.in", we can produce the report in Figure 5 by the following commands:

> $ **S REPORT −e lottery.in lottery.out**
> $ **pic lottery.out | eqn | troff −ms**

The **REPORT** utility provides a simple way of tying S computations to the powerful word-processing capabilities of the UNIX system.

Appendix 1
Function and Macro Documentation

This documentation reflects S as of the time this book was written. Your installation may have a version older or newer than that described in this book. Therefore, if a function behaves in a manner inconsistent with this documentation, consult the on-line documentation, available via the **help** function. It contains the definitive documentation for the version of S that you are running.

---

abline	Plot Line in Intercept-Slope form	**abline**

USAGE:

> **abline(a, b)**
> **abline(reg)**
> **abline(coef)**
>
> **abline(h=)**
> **abline(v=)**

ARGUMENTS:

reg: a regression structure; i.e., **reg** is a structure that contains a component **coef** giving coefficients of a fitted line (with intercept term first). If the **coef** component contains a single number, it is treated as the slope of a fit through the origin.

coef: vector containing the intercept **a** and slope **b** of the line y=a+b*x.

a,b: intercept and slope as above.

h: vector of y-coordinates for horizontal lines across plot.

v: vector of x-coordinates for vertical lines across plot.

> Graphical parameters may also be supplied as arguments to this function (see **par**).

> The effect of **abline** is that the line y=a+b*x or the specified horizontal and vertical lines are drawn across the current plot. A warning is given if an intercept/slope line does not intersect the plot.

EXAMPLES:

> **abline(0,1)**   #draws line with 0 intercept and slope 1
> **abline(reg(x,y))**   #line produced by least-squares fit to x and y
> **abline(v=c(0,10),lty=2)**   # dotted vertical lines at x==0 and 10

abs	Absolute Value	**abs**

USAGE:

   abs(x)

ARGUMENTS:

x: numeric structure. Missing values (NAs) are allowed.

VALUE:

   numeric structure like x, with absolute value taken of each data
   value.

acos	Inverse Trigonometric Functions	**acos**

USAGE:

   acos(x)
   asin(x)

ARGUMENTS:

x: numeric structure. Missing values (NAs) are allowed.

VALUE:

   structure with angles in radians. For values of **x** outside the interval [−1,1], NA is returned and a warning is given. The range
   of the result is **0 <= acos(x) <= pi** and **−pi/2 <= asin(x) <=
   pi/2**.

AED	see aed512	**AED**

aed512	AED 512 Color Graphics Terminal	aed512

USAGE:

> **aed512(ask)**
> **!S AED** # UNIX command

ARGUMENTS:

ask: logical, should device driver ask before clearing graphic display for a new figure? Default TRUE.

COMMENTS:

The function **aed512** is the device driver for the Advanced Electronics Design 512 color graphics terminal.

The **S AED** utility should be used immediately after logging on; it provides initial values for the color table and the function buttons.

Color 0 (the background) is dark blue and colors 1 through 8 are cyan, green, yellow, blue, magenta, red, white, and black. Colors 20 through 35 form a gray scale from black to white, and colors 50 through 89 are a spectrum of colors.

A feature of this terminal is that alphanumeric information is displayed in an overlay separate from graphical information (text in titles, axis labels, etc. is considered graphical). Function keys 0 to 2, should be re-labelled "view alpha", "view graph", and "view both", to cause either the alphanumeric information, graphics, or both to be displayed. Function key 4, marked "erase", clears the alpha overlay, and successive pushes of the buttons "full..." (key 3) and "erase" completely clear the screen. When the bottom of the screen is reached, alpha information scrolls off the top. The graphical display also scrolls, but wraps around. Erasing the alpha display will re-center the graphical information.

Buttons marked "zoom in", "zoom out" and "pan" control zooming and panning operations which allow portions of the overall display to be viewed.

If **ask** is TRUE, whenever a new plot is about to be produced the message "GO?" appears. This allows further viewing, making a hard copy, etc., before the screen is erased. When ready for plotting to proceed, simply hit carriage return.

When graphic input (**identify, rdpen**) is requested from this terminal, an "X"-shaped cursor appears on the screen. The joystick should be moved to position the cursor at the desired point, and then any single character should be typed. A carriage return will terminate graphic input without transmitting the point.

The terminal can change character size and line type. Character sizes are .75, 1, 1.5, and 2. The characters of size .75 and 1.5 are from a lower-resolution font. There are many line types, 1 through 7 being solid, dot, dash, long dash, dash-dot, long dash with short space, and dash-dash-dot. Line types above 7 are determined by the left-to-right bit-representation of the integer **lty+1**. Characters cannot be rotated on this device.

A device must be specified before any graphics functions can be used.

SEE ALSO:

> **color** and **ctable**.

---

again	see **edit**	**again**

---

all	Logical Sum and Product	**all**

USAGE:

> **all(x)**
> **any(x)**

ARGUMENTS:

x: logical expression. Missing values (NAs) are allowed.

VALUE:

> either TRUE or FALSE. **all** evaluates to TRUE if all the elements of the argument are TRUE; FALSE if there are any NAs or FALSEs. **any** evaluates to TRUE if any of the elements of the ar-

gument are TRUE; FALSE otherwise.

EXAMPLES:
> **if(all(x>0))x ← sqrt(x)**

---

any	see **all**	**any**

---

aperm	Array Permutations	**aperm**

USAGE:
> **aperm(a, perm)**

ARGUMENTS:

a: array to be permuted. Missing values (NAs) are allowed.

perm: vector containing a permutation of the integers 1:n where n is the number of dimensions in the array **a**. The old dimension given by **perm[j]** becomes the new **j**-th dimension.

VALUE:

array like **a**, but with the observations permuted according to **perm**, e.g., if **perm** is **c(2,1,3)**, the result will be an array in which the old second dimension is the new first dimension, etc.

If **a** contains a component **Label** (as in tables constructed with the **table** function), it will also be permuted.

EXAMPLES:

> **myiris ← aperm(iris,c(1,3,2))**   # turns 50 x 4 x 3 to 50 x 3 x 4

> **myiris ← matrix(aperm(iris,c(1,3,2)),150,4)** # make 150 x 4 matrix

---

| **append** | Data Merging | **append** |

USAGE:

> **append(x, values, after)**
> **replace(x, list, values)**

ARGUMENTS:

x: vector of data to be edited. Missing values (NAs) are allowed.
list: indices of the elements in **x** to be replaced.
values: vector of values to replace the list of elements, or to be appended
                after the element given. If **values** is shorter than **list**, it is reused
                cyclically. Missing values (NAs) are allowed.
after: index in **x** after which **values** are appended. **after=0** puts **values**
                at the beginning.

VALUE:

> the edited vector, to be assigned back to **x**, or used in any other
> way. Remember, unless the result is assigned back to **x**, **x** will
> not be changed.

EXAMPLES:

> **x ← replace(x,3,1.5)**   #Replace x[3] with 1.5
> **x[3]←1.5**   # Alternative: replaces x[3] with 1.5
>
> **x ← append(x,c(3.4,5.7),6)**   #Append two values in x[7],x[8]
>
> **y ← replace(x,c(3,7,8,9),0)**   #Replace the four elements with 0

| **apply** | Apply a Function to Sections of an Array | **apply** |

USAGE:

> **apply(a, margin, fun, arg1, arg2, ...)**

ARGUMENTS:

a: array. Missing values (NAs) are allowed if **fun** accepts them.
margin: the subscripts over which the function is to be applied. For ex-
                ample, if **a** is a matrix, 1 indicates rows, and 2 indicates columns.
                For a more complex example of the use of margin, see the last ex-

ample below. In general, a subarray is extracted from **a** for each combination of the levels of the subscripts named in **margin**. The function **fun** is invoked for each of these subarrays, and the results, if any, concatenated into a new array. Note that **margin** is also the dimensions of **a** which are retained in the result.

fun: character string giving the name of the function to be applied to the specified array sections.

argi: optional, any arguments to **fun**; they are passed unchanged.

VALUE:

array whose dimensions are defined by **margin**. If the result of each application of **fun** has a length that is greater than 1, this length will be the new first dimension of the result. If **margin** only specifies one dimension to be saved, and **fun** returns results of length 1, a vector will be returned.

The system macros **?row(a,fun,...)** and **?col(a,fun,...)** apply a function over rows and columns of a matrix.

EXAMPLES:

    apply(x,2,"mean",trim=.25)   # 25% trimmed column means
         # The result is a vector of length **ncol(x)**

    apply(x,2,"sort")   # sort columns of x

    t(apply(x,1,"sort"))   # transpose result of row sort

    apply(z,c(1,3),"sum")

The sorting examples show the difference between row and column operations when the results returned by **fun** are vectors. The returned value becomes the FIRST dimension of the result, hence the transpose is necessary with row sorts.

In the last example, **z** is a a 4-way array with dimension vector (2,3,4,5). The expression computes the 2 by 4 matrix obtained by summing over the second and fourth extents of **z** (i.e., **sum** is called 8 times, each time on 15 values).

Each section of the input array is passed as the first argument to an invocation of **fun**. It is passed without a keyword modifier, so, by keywords attached to **argi**, it should be possible to make the array section correspond to any argument to **fun**.

The function must return the same number of values (possibly 0) each time it is invoked on an array section.

System infix operators such as "+" can also be passed as functions.

Function **sapply** can be used to apply over all components of structures. Function **tapply** is useful for ragged arrays. The language looping construct **for** is more general than **apply**, but is less efficient computationally.

?apply	Apply a Macro to a Number of Arguments	**?apply**

USAGE:

> **?apply(macro, arg1, arg2, ...)**

ARGUMENTS:

macro: the name of a macro.

argi: arbitrary list of arguments that are to be given in turn to **macro**.

EXAMPLES:

> If the macro **analyze** performs an analysis of a dataset, then
> > **?apply(analyze,ds1,ds2,ds3,ds4,ds5)**
> is equivalent to typing
> > **?analyze(ds1)**
> > **?analyze(ds2)**
> > ...

---

approx	Approximate Function from Discrete Values	**approx**

USAGE:

> **approx(x, y, xout, method, n, rule)**

ARGUMENTS:

x,y: pairs of points giving $(x, f(x))$. Can be a time-series or a structure containing components named **x** and **y**.

xout: optional set of x values for which function values are desired. If

                 xout is not specified, the function will be approximated at **n** equally spaced data points, spanning the range of **x**.

method: character describing the method to be used in approximating the function. Possible values are "linear" for linear interpolation (later may have "spline", etc).

n: optional integer giving the number of points evenly spaced between the minimum and maximum values in **x** to be used in forming **xout**. Default 50.

rule: integer describing the rule to be used for values of **xout** that are outside the range of **x**. If **rule** is 1 (the default), NAs will be supplied for any such points. If **rule** is 2, the y values corresponding to the extreme **x** values will be used.

VALUE:

      structure with components named **x** and **y**.

EXAMPLES:

      z ← approx(x,y,newx)   # linear interpolation at newx
      quants ← approx(ppoints(x),sort(x),c(.1,.25,.5,.75,.9))$y
      # get the 10, 25, 50, 75 and 90 percentiles of x

---

arith	Arithmetic Operators	**arith**

USAGE:

      expr1 op expr2

ARGUMENTS:

expr: numeric structure. Missing values (NAs) are allowed.

VALUE:

      numeric result, mode is real if either **expr** is real. The exception is integer divide (%/), which truncates its result, returning an integer.

      The operator %% is the modulus function defined as **expr1−floor(expr1/expr2)** for **expr1!=0** and **expr1** if **expr2==0** (see Knuth); "^" and "**" are synonyms for exponentiation.

REFERENCE:

D. E. Knuth, *The Art of Computer Programming, Fundamental Algorithms* Vol. 1, Section 1.2.4., Addison Wesley, 1968,

SEE ALSO:

precedence.

EXAMPLES:

x—mean(x)
(1+(5:8)/1200)ˆ12   # compound interest, 5:8 per annum monthly

---

array	Create an Array	**array**

USAGE:

array(data, dim)

ARGUMENTS:

data: numeric vector containing the data values for the array in the normal array order: the first subscript varies most rapidly. Missing values (NAs) are allowed.

dim: vector giving the extent for each dimension. Default is **len(data)**.

VALUE:

an array with the same mode as **data**, and dimensionality described by **dim**. If **data** does not completely fill the array, it is repeated until the array is filled (giving a warning if it does not take an integral number of repetitions of **data**).

EXAMPLES:

myiris ← array(read("irisfile"),c(4,3,50))
        # creates a 4 by 3 by 50 array

---

arrows	see **segments**	**arrows**

asin	see **acos**	asin

assign	Assignment	assign

USAGE:

> name ← expression
> name _ expression
> expression → name

The assignment operator, in any of its forms, causes the value of **expression** to be saved on the working storage file under **name**. Note that the arrow always points toward **name**. The left-arrow is made up of the two characters less-than and minus. An underscore can also be used for left-arrow. Right-arrow is the two characters minus greater-than.

Once an assignment is made, the data can be retrieved by mentioning the name used in the assignment. The value of the assignment (when used as an expression) is the **expression** part of the assignment. Missing values (NAs) are allowed in **expression**.

Assignment can also be done with the function **assign**.

USAGE:

> **assign(cname, expression, pos=)**

ARGUMENTS:

cname: character string giving the name of a dataset to create.

pos=:   the position in the current search list of the data directory on which **expression** is to be stored. Default 1. See **attach** and **save**. **save** is the usual way to assign, other than to the working data.

EXAMPLES:

> y ← sqrt(x←runif(100))   # 100 uniforms saved as x
>   # square root of sample saved under the name y
>
> assign("x", runif(100))   # equivalent to x ← runif(100)

```
for(i in seq(ncol(x)))
   assign(colname[i],x[,i])
   # saves each column of matrix x under different name
```

---

atan	Inverse Tangent	**atan**

USAGE:

atan(a)
atan(a, b)

ARGUMENTS:

a: numeric structure giving tangents of desired angles. Missing
          values (NAs) are allowed.

b: numeric structure which along with **a**, determines the tangent of
          the desired angle as **a/b**. The two-argument version should be
          used if **b** may be near zero. Missing values (NAs) are allowed.

VALUE:

the angles in radians. Where both **a** and **b** are zero, the value is
NA. The range of the value is $-pi/2 <= atan(a) <= pi/2$ and
$-pi < atan(a,b) <= pi$.

---

attach	Attach a New Data Directory	**attach**

USAGE:

attach(file, pos)

ARGUMENTS:

file: character string giving the name of a new data directory to be ac-
          cessed. The name must be a correct UNIX file-system pathname,
          such as "/usr/abc/swork".

pos: position in data directory search list that **file** should occupy. The
          data directories originally in position **pos** or beyond are moved
          down the list after **file**. Default, after all current data directories.
          If pos=1, **file** will be attached as the working directory (you must
          have write permission).

When a user logs on, three data directories are attached: working, save, and shared. See **detach** for removing a data directory from search list, and **search** for the current list of attached data directories. Data directories are attached only for the current invocation of S. Only directories containing S datasets can be attached as S data directories; ordinary ASCII data files should be read by means of **read**.

EXAMPLES:

attach("/usr/joe/sdata") # attach save directory belonging to joe

axes	see **title**	**axes**

axis	Add an Axis to the Current Plot	**axis**

USAGE:

axis(side, at, labels, ticks, distn, line, pos, outer)

ARGUMENTS:

side: side of plot for axis (1,2,3,4 for bottom, left, top, right).

at: optional vector of positions at which the ticks and tick labels will be plotted. If **side** is 1 or 3, **at** represents x-coordinates. If **side** is 2 or 4, **at** represents y-coordinates. If **at** is omitted, a the current axis (as specified by the **xaxp** or **yaxp** parameters, see **par**) will be plotted.

labels: optional. If **labels** is logical, it specifies whether or not to plot tick labels. Otherwise, **labels** must be the same length as **at**, and **label[i]** is plotted at coordinate **at[i]**. Default TRUE.

ticks: logical, should tick marks and axis line be plotted? Default TRUE.

distn: optional character string describing the distribution used for transforming the axis labels. If **dist** is "normal", then values of **at** are assumed to be probability levels, and the labels are actually plotted at **qnorm(at)**. This also implies a reasonable default set of values for argument **at**.

line: line (measured out from the plot in units of standard-sized character heights) at which the axis line will be plotted. Tick labels will be plotted relative to this position (as much as the tick labels and axis line differ in graphical parameter **mgp**). Default 0, on the edge of the plot.

pos: x or y coordinate position at which the axis line should be plotted. Labels will be on the side of the axis specified by **side**. If **pos** is omitted, argument **line** controls positioning of the axis.

outer: if TRUE, the axis will be drawn in the outer margin rather than the standard plot margin. Default FALSE.

Graphical parameters may also be supplied as arguments to this function (see **par**).

EXAMPLES:

```
axis(3)  # add axis on top
axis(4,label=F)  # tick marks only on right

qqnorm(data)
axis(3,distn="normal")  # add normal probability axis at top

qqnorm(data,xaxt="n")   # normal prob plot, no x axis labels
probs ← c(.01, .05, .1, .9, .95, .99)
axis(1,dist="norm",at=probs,
   lab=encode(probs*100,"%"))  # add user-defined probability axis

plot(x,y,axes=F)  # scatter plot with no box or axes
axis(1,pos=0); axis(2,pos=0)  # coordinate axes through origin

plot(fahrenheit)  # time-series record of temperatures
celsius ← pretty((range(fahrenheit)-32)*5/9)
axis(side=4, at=celsius*9/5+32,lab=celsius)  # celsius at right
mtext(side=4,line=4,"Celsius")
```

---

backsolve	Backsolve Upper- or Lower-triangular Equations	**backsolve**

USAGE:

    backsolve(r, x, lower)

ARGUMENTS:

r: upper or lower triangular matrix.
x: right-hand sides to equations.
lower: logical flag to indicate lower-triangular **r**. Default FALSE.

VALUE:

    matrix like **x** of the solutions **y** to the equations **r y=x**.

---

barplot	Bar Graph	**barplot**

USAGE:

    barplot(height) #simple form
    barplot(height, width, names, space, inside, beside, horiz,
         legend, angle, density, col, blocks)

ARGUMENTS:

height: matrix or vector giving the heights (positive or negative) of the bars. If **height** is a matrix, each column represents one bar; the values in the columns are treated as heights of blocks. Blocks of positive height are stacked above the zero line and those with negative height are stacked below the line. Matrix values in **height** can also be treated as the positions of the dividing lines; see argument **blocks**.

width: optional vector of relative bar widths.

names: optional character vector of names for the bars.

space: how much space (as a fraction of the average bar width) should be left before each bar. This may be given as a single number or one number per bar. A value of .2 is the default spacing. If **beside** is TRUE, **space** may be specified by 2 numbers where the first is the space between bars in a set, and the second is the space between sets.

inside: if TRUE (default), the internal lines which divide adjacent bars will be drawn.

beside: if TRUE and if **height** is a matrix, the bars for different rows will be plotted beside each other, not as a single divided bar. Default FALSE. If **width** is given in the case beside=TRUE, there must be as many widths as the number of values in **heights**, not the number of columns. The same is true of **names**, if this is supplied.

horiz: if TRUE, the graph will be drawn horizontally, with the first bar at the bottom. Default is FALSE.

legend: a vector of names to be correlated with the bar shading which should be plotted as a legend. The legend is put in the upper right of the plot by default; use the **legend** function if more control over legend positioning is required.

angle: optional vector giving the angle (degrees, counter-clockwise from horizontal) for shading each bar division. (Defaults to 45 if **density** is supplied.)

density: optional vector for bar shading, giving the number of lines per inch for shading each bar division. (Defaults to 3 if **angle** is supplied).

col: optional vector giving the colors in which the bars should be filled or shaded. If **col** is specified and neither **angle** nor **density** are given as arguments, bars will be filled solidly with the colors. If **angle** or **density** are given, **col** refers to the color of the shading lines.

blocks: if TRUE (the default), the **height** matrix is treated according to the "blocks" model in which each value in **height** is the height of a block to be stacked above or below the axis. If **blocks** is FALSE, values in **height** give the coordinates at which the bar dividing lines are to be drawn. In this case, the values in any column of **height** should be monotonically increasing.

Solid filling of bars is dependent on the area-filling capability of the device driver. For devices without explicit area-filling capability, solid filling can be simulated by specifying a very high density shading.

Graphical parameters may also be supplied as arguments to this function (see **par**).

VALUE:

a vector, non-printing, which contains x-coordinates of centers of bars (y-coordinates if bars are horizontal). The returned value is useful if the user wants to add to the plot.

EXAMPLES:

    x ← barplot(height) # do plot, save x coordinates of bar centers
    text(x,height+1,height) # label the tops of the bars

    # The example plot was produced by
    barplot(
        t(telsam.response[1:5,]),
        ylim=c(0,200),
        col=2:5,angle=c(5,40,80,125),density=(4:7)*3,
        legend=telsam.collab,
        names=encode(telsam.rowlab[1:5]),
        xlab="Interviewer",
        ylab="Number of Responses",
        main="Response to Quality of Service Question"
)

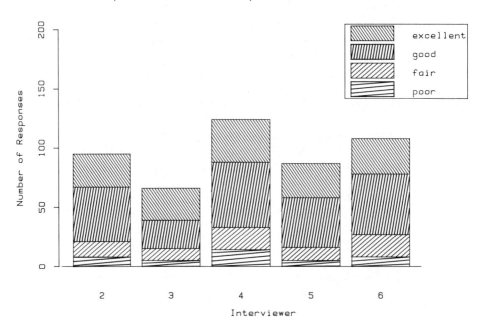

?bartable   Produce Bar Plot of a Category   **?bartable**

USAGE:

?bartable(category)

ARGUMENTS:

category: a category, i.e., created with **cut**, or **code**.

EXAMPLES:

x ← **code(occupation)**   # create category
bartable(x)   # bar plot of numbers in each occupation

?barycentric Compute Barycentric Coordinates for Mixtures **?barycentric**

USAGE:

?barycentric(a, b, c, x, y, size)

ARGUMENTS:

a,b,c: vectors giving amounts of three components of mixture.

x,y: vectors of coordinates at which each mixture should be plotted. x-coordinates range from 0 to 1, y-coordinates from 0 to .866. Plotting is typically done in a triangle with vertices at (0,0), (1.0), and (.5,.866).

size: vector giving relative size of each mixture, proportional to square root of the total amount **a+b+c**, scaled so that the maximum value of **size** is 1.

| **BATCH** | Batch (non-interactive) execution of S | **BATCH** |

USAGE:

!S BATCH in out # UNIX command

ARGUMENTS:

in: the name of a file containing S commands to be executed. The file may contain macro invocations, source and sink commands, etc. Also, deferred graphics may be carried out in batch mode. Interactive graphics devices (**tek14**, **hp72h**, etc.) produce no output when run in non-interactive mode.

out: the name of a file which will receive all of the output from the run. By default, **out** will contain a listing of each expression in **in**, followed by any printed output produced by the expression.

WARNING:

execution begins immediately after **BATCH** is invoked. Processes running in batch are counted toward the total number of processes any one user is allowed. Running several versions of S (either batch or interactive) at the same time can cause you to exceed the process limit, and abort all of the runs.

| **beta** | Beta Probability Distribution | **beta** |

USAGE:

pbeta(q, par1, par2)
qbeta(p, par1, par2)
rbeta(n, par1, par2)

ARGUMENTS:

q: vector of quantiles. Missing values (NAs) are allowed.

p: vector of probabilities. Missing values (NAs) are allowed.

n: sample size.

par1: vector of shape parameters for numerator (>0).

par2: vector of shape parameters for denominator (>0).

VALUE:

probability (quantile) vector corresponding to given quantile (probability) vector, or random sample (**rbeta**).

---

box	Add a Box Around a Plot	**box**

USAGE:
> **box(n)**

ARGUMENTS:
n: number of times the box is drawn (i.e., heaviness of line). Default 1.

Graphical parameters may also be supplied as arguments to this function (see **par**).

EXAMPLES:
> **box(5,col=2)**   # draw a thick color 2 box around plot

---

boxplot	Box Plots	**boxplot**

USAGE:
> **boxplot(arg1, arg2, ...,**
> **range=, width= varwidth=, notch=, names=, plot=)**

ARGUMENTS:
argi: numeric structure or a structure containing a number of numeric components (e.g., the output of **split**). Missing values (NAs) are allowed.

range=:   controls the strategy for the whiskers and the detached points beyond the whiskers. By default, whiskers are drawn to the nearest value not beyond a standard range from the quartiles; points beyond are drawn individually. Giving **range=0** forces whiskers to the full data range. Any positive value of **range** multiplies the standard range by this amount. (The standard range is 1.5*(inter-quartile range).)

width=:   vector of relative box widths. See also argument **varwidth**.

varwidth=:   logical flag. If TRUE, box widths will be proportional to the square-root of the number of observations for the box. Default FALSE.

notch=:   logical flag. If TRUE, notched boxes are drawn, where non-overlapping of notches of boxes indicates a difference at a rough 5% significance level. Default FALSE.

names=:   optional character vector of names for the groups. If omitted, names used in labelling the plot will be taken from the names of the arguments and from component names of structures.

plot=:   logical flag. If TRUE, the box plot will be produced. If false, only the calculated summaries of the arguments are returned. Default TRUE.

Graphical parameters may also be supplied as arguments to this function (see **par**).

VALUE:

if **plot** is FALSE, a structure with the components listed below. Otherwise function **bxp** is invoked with these components, plus optional **width**, **varwidth** and **notch**, to produce the plot. Note that **bxp** returns a vector of box centers.

stats: matrix (5 by number of boxes) giving the upper extreme, upper quartile, median, lower quartile, and lower extreme for each box.

n: the number of observations in each group.

conf: matrix (2 by number of boxes) giving confidence limits for median.

out: optional vector of outlying points.

group: vector giving the box to which each point in **out** belongs

names: names for each box (see argument **names** above)

REFERENCES:

J. M. Chambers, W. S. Cleveland, B. Kleiner and P. A. Tukey, *Graphical Methods for Data Analysis*, Wadsworth, 1983.

McGill, Tukey and Larsen, "Variations of Box Plots", *The American Statistician*, February 1978, Vol 32, No 1, pp 12-16.

EXAMPLES:

boxplot(group1,group2,group3)

boxplot(split(salary,age),varwidth=TRUE,notch=TRUE)

the example plot is produced by:

boxplot(
    split(lottery.payoff,lottery.number%/100),
    main=lottery.label,
    sub="Leading Digit of Winning Numbers",
    ylab="Payoff")

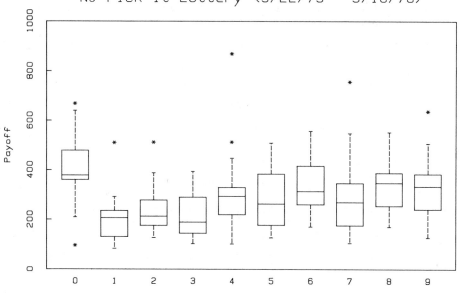

bxp	Boxplots From Processed Data	bxp

USAGE:

    bxp(z, width, varwidth, notch)

ARGUMENTS:

z: structure whose components define the boxplot statistics, normally the result of a call to **boxplot** (which see for the components of **z**), but can be built up in any other way.

width: optional vector of box widths.

varwidth: logical flag, if TRUE, variable width boxes are drawn based on the **n** component of **z**. Default FALSE.

notch: logical flag, if TRUE, use the **conf** component of **z** to produce notched boxes. Default FALSE.

VALUE:

vector, non-printing, which contains x-coordinates of centers of the boxes. This vector is useful if the user wants to add to the plot.

Graphical parameters may also be supplied as arguments to this function (see **par**).

c	Combine Values	c

USAGE:

c(arg1, arg2, ...)

ARGUMENTS:
argi: vector or structure. Missing values (NAs) are allowed.

VALUE:

vector which is the combination of all values from all arguments to the function.

Precisely, if **argi** is a vector structure, its data values are used. Otherwise, if **argi** is a structure, all data from all components of **argi** is used. In this way, **c** performs a function that is the opposite of **split**.

Arguments that are NULL do not contribute anything to the result. See the last example.

EXAMPLES:

c(1:10,1:5,1:10)

c(1,2,3,5,7,11,13)

c(states,"Washington DC")

z ← split(data,group) # split data by group then ..
sorted ← c(sapply(z,"sort")) #sort each component, combine result

x←NULL; for(i in seq(10)) x ← c(x,fun(i))

call	see **help**	call

cancor	Canonical Correlation Analysis	**cancor**

USAGE:

cancor(x, y)

ARGUMENTS:

x,y: two matrices of data. The number of rows (number of observations) must be the same in each.

VALUE:

structure representing the canonical correlation analysis; i.e., a set of pairs of linear combinations of the variables in **x** and in **y** such that the first pair has the largest possible correlation, the second pair has the largest correlation among variables uncorrelated with the first pair, etc.

cor: the correlations between the pairs of variables.

xcoef: the matrix of linear combinations of the columns of **x**. The first column of **xcoef** is the linear combination of columns of **x** corresponding to the first canonical correlation, etc.

ycoef: the matrix of linear combinations of the columns of **y**. The first column of **ycoef** is the linear combination of columns of **y** corresponding to the first canonical correlation, etc.

| **cauchy** | Cauchy Probability Distribution | **cauchy** |

USAGE:

        pcauchy(q, par1, par2)
        qcauchy(p, par1, par2)
        rcauchy(n, par1, par2)

ARGUMENTS:

q: vector of quantiles. Missing values (NAs) are allowed.
p: vector of probabilities. Missing values (NAs) are allowed.
n: sample size.
par1:   vector of location parameters. Default is 0.
par2:   vector of scale parameters (>0). Default is 1.

VALUE:

        probability (quantile) vector corresponding to given quantile (probability) vector, or random sample (**rcauchy**).

EXAMPLES:

        rcauchy(20,0,10) #sample of 20, loc. 0, scale 10

| **cbind** | Form Matrix from Columns or Rows | **cbind** |

USAGE:

        cbind(arg1, arg2, ...)
        rbind(arg1, arg2, ...)

ARGUMENTS:

argi: vector or matrix. Missing values (NAs) are allowed.

VALUE:

        matrix composed by adjoining columns (**cbind**) or rows (**rbind**).

        If several arguments are matrices, they must contain the same number of rows (**cbind**) or columns (**rbind**). Vector arguments are treated as row vectors by **rbind** and column vectors by **cbind**. If all arguments are vectors, the result will contain as many rows or columns as the length of the longest vector. Shorter vectors will be repeated.

If arguments are time-series, no attempt is made to relate their time parameters. See **tsmatrix** for this case.

Arguments that are NULL are ignored. See the last example.

EXAMPLES:

cbind(1,xmatr) %* regr$coef   # add column of ones
        # to matrix multiply with coef vector

rbind(matrix,newrow1,newrow2)   # add 2 new rows

x←NULL
for(i in seq(5)){
    ... # compute z
    x ← cbind(x,z)
    }

---

ceiling	Integer Values	**ceiling**

USAGE:

ceiling(x)
floor(x)
trunc(x)

ARGUMENTS:

x: numeric structure. Missing values (NAs) are allowed.

VALUE:

structure of same mode as **x**. If **x** is an integer vector, the functions have no effect and **x** is returned. **trunc** truncates fractions toward zero, e.g., **trunc(1.5)** is 1., and **trunc(−1.5) is −1**. **floor** is largest integer $<= x$. **ceiling** is smallest integer $>= x$.

---

chapter	Include a User Chapter of Functions	**chapter**

USAGE:

    chapter(file, detach)

ARGUMENTS:

file: character string, the name of the user chapter to be searched for S functions (and for their documentation). The most recently specified chapter is searched first (relevant in the case that two chapters have a function of the same name). Default is the user's own chapter (".").

detach: if TRUE, the named chapter is detached, rather than attached. Default FALSE.

    Details on the creation chapters of new functions are given in the reference.

REFERENCE:

    R. A. Becker and J. M. Chambers, *Extending the S System*, AT&T Bell Laboratories Technical Report, 1984.

EXAMPLES:

    chapter("/usr/abc/testfuns")

---

chisq	Chi-Square Distribution	**chisq**

USAGE:

    pchisq(q, par1)
    qchisq(p, par1)
    rchisq(n, par1)

ARGUMENTS:

q: vector of quantiles. Missing values (NAs) are allowed.
p: vector of probabilities. Missing values (NAs) are allowed.
n: sample size.
par1:   degrees of freedom (>0).

VALUE:

probability (quantile) vector corresponding to given quantile (probability) vector, or random sample (**rchisq**).

chol	Triangular Decomposition of Symmetric Matrix	**chol**

USAGE:
> **chol(x)**

ARGUMENTS:
x: symmetric, positive definite matrix (e.g., correlation matrix or cross-product matrix).

VALUE:
> upper-triangular matrix, **y**, such that **t(y) %*** **y** equals **x**.

chull	Convex Hull of a Planar Set of Points	**chull**

USAGE:
> **chull(x, y, peel, maxpeel, onbdy, tol)**

ARGUMENTS:
x,y: the co-ordinates of a set of points. A structure containing components **x** and **y** may also be given for **x**.

peel: should successive convex hulls be peeled from the remaining points, generating a nested set of hulls? Default FALSE.

maxpeel: maximum number of hulls that should be peeled from the data, default is to peel until all points are assigned to a hull. If **maxpeel** is given, it implies **peel=TRUE**.

onbdy: should points on the boundary of a convex hull (but not vertices of the hull) be included in the hull? Default FALSE, unless **peel** is TRUE in which case **onbdy** is TRUE by default.

tol: relative tolerance for determining inclusion on hull, default .0001.

VALUE:
> vector giving the indices of the points on the hull. If **peel** is TRUE, returns a data structure with components **depth**, **hull** and

          count.

depth: a vector which assigns a depth to each point, i.e., the number of hulls surrounding the point. Outliers will have small values of **depth**, interior points relatively large values.

hull: vector giving indices of the points on the hull. Along with **count**, determines all of the hull peels. The first **count[1]** values of **hull** determine the outermost hull, the next **count[2]** are the second hull, etc.

count: counts of the number of points on successive hulls.

EXAMPLES:

```
hull ← chull(x,y)
plot(x,y)
hatch(x[hull],y[hull],space=0)  # draw hull

p ← chull(x,y,peel=T)   # all hulls
which ← rep(seq(p$count),p$count)   # which peel for each pt
s ← split(p$hull,which)
for(i in seq(ncomp(s))){     # plot all peels
    j ← s$[i]   # indices of points on ith peel
    hatch(x[j],y[j],space=0,lty=i)
    }
```

?cleanup	Selectively Remove Datasets	**?cleanup**

USAGE:

       ?cleanup(pattern)

ARGUMENTS:

pattern: (unquoted) pattern that describes which datasets should be targeted to remove. Default is **$T***.

EFFECT:

      The name of each dataset found is printed, and a reply of **yes** causes the dataset to be deleted; **no** leaves the dataset alone.

EXAMPLES:

      ?cleanup # get rid of all macro temporary datasets

clorder	Re-Order Leaves of a Cluster Tree	**clorder**

USAGE:

> **clorder(tree, x)**

ARGUMENTS:

tree: structure with components named **merge, height,** and **order,** typically a hierarchical clustering produced by function **hclust.**

x: numeric vector with one value for each individual involved in the cluster tree.

VALUE:

> cluster tree structure with the **merge** and **order** components permuted so that at any merge, the cluster with the smaller average **x** value is on the left. This reorders **tree** so that the leaves are approximately in order by the associated **x** values.
>
> In hierarchical cluster displays, a decision is needed at each merge to specify which subtree should go on the left and which on the right. Since, for **n** individuals, there are **n−1** merges, there are $2^{(n-1)}$ possible orderings for the leaves in a cluster tree. The default algorithm in **hclust** is to order the subtrees so that the tighter cluster is on the left (the last merge of the left subtree is at a lower value than the last merge of the right subtree). Individuals are the tightest clusters possible, and merges involving two individuals place them in order by their observation number.

EXAMPLES:

> **h ← hclust(dist(votes.repub))**
> **ave.repub ← ?row(votes.repub,mean)**
> **h2 ← clorder(h,ave.repub)** #leaves ordered by
>   # average republican vote
> **plclust(h2, lab=state.abb)** #cluster plot

cmdscale	Classical Metric Multi-dimensional Scaling	**cmdscale**

USAGE:
> **cmdscale(d, k, eig, add)**

ARGUMENTS:

d: distance matrix structure of the form returned by **dist** or a full, symmetric matrix. Data is assumed to be dissimilarities or relative distances.

k: desired dimensionality of the output space. Default 2.

eig: logical flag which controls return of eigenvalues computed by the algorithm. They can be used as an aid in determining the appropriate dimensionality of the solution. Default FALSE.

add: logical flag which controls computation of additive constant (see component **ac** below); default FALSE.

VALUE:
> a structure potentially with three components named **points**, **eig** and **ac**.

points: a matrix with **k** columns and as many rows as there were objects whose distances were given in **d**. Row i gives the coordinates in **k**-space of the i-th object.

eig: vector of eigenvalues (as many as original data points), returned only if argument **eig** is TRUE.

ac: constant added to all data values in **d** to transform dissimilarities (or relative distances) into absolute distances. The Unidimensional Subspace procedure, (Torgerson, 1958, p.276) is used to determine the additive constant.

REFERENCE:
> Warren S. Torgerson, *Theory and Methods of Scaling*, pp. 247-297, Wiley, 1958.

EXAMPLES:
```
x←cmdscale(dist)  #default 2-space
coord1←x[,1]; coord2←x[,2]
par( pty="s" )  #set up square plot
r←range(x)   #get overall max, min
plot(coord1,coord2,type="n",xlim=r,ylim=r) #set up plot
   # note units per inch same on x and y axes
text(coord1,coord2,seq(coord1))  #plot integers
```

---

code	Create Category from Discrete Data	code

USAGE:

> **code(x, level, labels)**

ARGUMENTS:

x: data vector. Missing values (NAs) are allowed.

level: optional vector of levels that the category should have. Any data value that does not match a value in **level** is coded in the output **Data** vector as NA. Missing values (NAs) are allowed. Default is the set of unique values in **x**, sorted in ascending order, with NAs omitted.

labels: optional vector of values to use as labels for the levels of the category. By default, the vector of levels is used.

VALUE:

> category structure with components **Data** and **Label**.

Data: integer vector indicating which level the category took on for each data value. NAs are returned where a data value in **x** did not match any value in **level**.

Label: character names for the levels of the category. If the argument **labels** (either given or default) was not of mode CHAR, it will have been encoded.

EXAMPLES:

> **code(occupation)** # "doctor", "lawyer", etc.
>
> # make readable labels
> **occ ← code(occupation,level=c("d","l"),label=c("Doctor","Lawyer"))**
>
> # turn category into character vector
> **occ$Label [occ]**
>
> **colors ← code(color,c("red","blue","green"))**
> **tprint(table(colors))**   #print table counting occurrences of colors

---

coerce	Change Mode	**coerce**

USAGE:

> coerce(x, mode)
> x %m mode   #special operator

ARGUMENTS:

x: vector structure. Missing values (NAs) are allowed.
mode: desired mode for data: LGL, INT, REAL or CHAR.

VALUE:

> vector structure whose data elements have the new mode.

SEE ALSO:

> **encode** converts to CHAR in a more flexible way.

EXAMPLES:

> print(matrix(1:20,5,4), rowlab=(1:5)%m CHAR)
>
> mydata ← matrix(read("myfile",mode=CHAR),ncol=5)
> ident ← mydata[,1]   # character identifier in first column
> data ← coerce(mydata[,-1], REAL)   # other columns numeric

---

col	Matrix Column and Row Numbers	**col**

USAGE:

> col(x)
> row(x)

ARGUMENTS:

x: matrix. Missing values (NAs) are allowed.

VALUE:

> integer matrix, containing the row number or column number of
> each element. If z←row(x), $z[i,j]$ is i; if z←col(x), $z[i,j]$ is j.
>
> See also **diag** for diagonal of matrix.

EXAMPLES:

      x[row(x)>col(x)]   # get strict lower triangle of x
      x[row(x)−col(x)==1]   # first sub-diagonal

?col	Apply a Function to the Columns of a Matrix	**?col**

USAGE:

      ?col(x, fun, arg1, ...)

ARGUMENTS:

x: name of matrix. Missing values (NAs) are allowed if **fun** accepts them.

fun: (unquoted) name of function to be applied to the columns of the matrix.

argi: any other arguments that should be given to **fun**.

VALUE:

vector or matrix of results. If each invocation of **fun** produces a vector of length **n** (**n>1**), the result will be an **n** by **ncol(x)** matrix. Otherwise, the result will be a vector of length **ncol(x)**.

SEE ALSO:

Function **apply**.

EXAMPLES:

      ?col(y,median)   # column medians of matrix y
      ?col(y,mean,trim=.25)   #25% trimmed column means

color	Modify AED512 Color Table	color

USAGE:

color(table, interact)

ARGUMENTS:

table: optional matrix with 4 columns, in which each row specifies a color number and its associated hue, lightness, and saturation. Colors not defined by **table** are left unchanged. If **table** is missing and **interact** is TRUE, then no default colors are set up. Otherwise, color 0 (the background) is dark blue and colors 1 through 8 are cyan, green, yellow, blue, magenta, red, white, and black. In addition colors 20 through 35 form a gray scale from black to white and colors 50 through 89 form a spectrum of fully saturated colors beginning and ending with blue.

interact: optional logical flag indicating whether to invoke the interactive mode in which the AED512 color table may be dynamically modified. Default is FALSE.

VALUE:

the input table modified (and perhaps expanded) by the results of any interaction.

COMMENTS:

During interaction there is always a **current color** (specified by its number in the color table) and a **current mix** (a new color potentially "under construction"). These are both displayed in the upper right corner of the screen. Initially the current color is 1 and the current mix has the same formula as 1.

The relabelled numeric key pad on the terminal is used repeatedly to initiate one of eleven functions during interaction. A brief description (usually one word) of the currently selected function always appears at the top left of the screen. The two chief activities during interaction are "picking a number" and "mixing a color", each of which may be accomplished in one of three ways.

To pick a new current color the user may hit "number", in which case the terminal prompts for a number. This should be an integer in the range 0 to 127 followed by a carriage return. Pushing "picture" will cause a cursor to be displayed which can be moved around using the joystick. When "quit" is depressed the color at the current cursor position becomes the current color. Finally, "palette" causes a display of the current color table to appear with a large cursor overlaid. Each square of the table is filled in a different color, numbered consecutively beginning at 0 in the upper left and proceeding across the rows. The rows are ten squares wide so that, for example, the first column contains color numbers 0,10,20,....,120. Again, "quit" causes the color at the current cursor position to become the current color.

In order to mix a new color the user may hit "circle", in which case a color circle is displayed with several colors on the perimeter, and with the interior filled with the current mix. (See the documentation for **ctable** for a description of this circle and of the hue-lightness-saturation model.) Again a cursor is displayed and its position around the circle represents the hue of the current mix while its distance from the center of the circle represents saturation (the perimeter being full saturation). "show color" will cause these two qualities to be determined from the cursor position and the current mix will be appropriately updated. The third quality of the mix--its lightness--is read from the vertical position of the cursor when "lightness" is depressed (the bottom of the screen corresponds to black and the top to white.) Two small white squares may appear in this display and represent the definition of the current color, when it is known. The square inside the circle defines the hue and saturation of the current color while the square at the right of the screen defines the lightness of the current color. As usual, "quit" will cause the cursor (and, in this case, the circle) to disappear.

A second method of mixing a new color is to use "virtual". In this case "show color" and "lightness" work exactly as before with respect to the displayed cursor, but the color circle no longer appears--it is only "virtually" present. In this mode any portions of the screen colored in the current color will temporarily be modified to be colored in the current mix, and will assume their original color when "quit" is hit (provided the current color is "known".) Finally, a color may be mixed by using "same as". The user then selects a color number according to one of the three methods described above and the current mix becomes the same as the color corresponding to that number.

If at any time "save" is depressed, the current color becomes defined by the current mix. Pushing "specify" allows entry of a specification string (see function **ctable**) from the keyboard.

During interaction, the user may determine what is expected by examining the ending of the description at the top left of the screen:

'?'   hit a function key

':'   type in a line followed by carriage return

null   position cursor and hit a function key

'...'   program is talking to AED: do NOTHING!

During cursor positioning there is a one function read-ahead in effect so that striking a function other than "quit" has the same effect as hitting "quit" followed by that function.

Color only "knows" the colors defined in **table** or created during interaction. Only "known" colors are returned by the function or restored after using "virtual".

REFERENCE:
R. A. Becker and A. R. Wilks, "Raster Color Graphics: Principles and an Implementation", AT&T Bell Laboratories Memorandum.

EXAMPLES:
color( ) # set up default table with no interaction
mytable ← color(ctable(oldtable),interact=T)
set up oldtable, interact, and save result

?comment	Add a Comment to a Dataset	**?comment**

USAGE:
?comment(x,comment)

ARGUMENTS:
x: a dataset name.
comment: (unquoted) character comment to be attached to **x**.

Component **comment** is added to dataset **x**. The comment will be printed whenever **x** is printed.

EXAMPLES:

?comment(gnp,Gross national product from government figures)

---

compare	Comparison Operators	**compare**

USAGE:

expr1 op expr2

ARGUMENTS:

expri: numeric structure.

VALUE:

vector structure with FALSE or TRUE in each element according to the truth of the element-wise compare of the operands. For vector operands, the value is as long as the longer of **expr1** and **expr2**. For time-series operands, the time domain of the result is the intersection of the domains of the operands.

See also **xor** for exclusive-or function.

EXAMPLES:

a>b   # true when **a** greater than **b**

x[x>100]   # all **x** with values larger than 100

state=="Wyoming"   # TRUE or FALSE

---

compname	Component Names	**compname**

USAGE:

    compname(x)

ARGUMENTS:

x: hierarchical structure

VALUE:

    character vector giving the names of the components of **x**.

    Time-series have components named **Tsp** and **Data**. Matrices and arrays have components named **Dim** and **Data**.

EXAMPLES:

    labels ← compname(struct)
    compname(y)   #find out structure of y without
                       #looking at all the data

---

contour	Contour Plotting	**contour**

USAGE:

    contour(x, y, z, v, nint=, add=, labex=)

ARGUMENTS:

x: vector containing x coordinates of grid over which **z** is evaluated.

y: vector of grid y coordinates.

z: matrix **len(x)** by **len(y)** giving surface height at grid points, i.e., $z[i,j]$ is evaluated at $x[i]$, $y[j]$. The **rows** of z are indexed by **x**, and the **columns** by y. Missing values (NAs) are allowed.

v: vector of heights of contour lines. By default, approximately **nint** lines are drawn which cover most of the range of z. See the function **pretty**.

nint=:   the approximate number of contour intervals desired, default 5. Not needed if **v** is specified.

add=:   flag which if TRUE causes contour lines to be added to the current plot. Useful for adding contours with different **labex** parameter, line type, etc. Default FALSE.

labex=:   the desired size of the labels on contour lines, default is same as standard (**cex**) character size. If **labex** is 0, no labels are used.

Graphical parameters may also be supplied as arguments to this function (see **par**).

The first argument may be a structure containing components named **x**, **y**, and **z**.

EXAMPLES:

```
rx←range(ozone.xy$x)
ry←range(ozone.xy$y)
usa(xlim=rx,ylim=ry,lty=2,col=2)
i←interp(ozone.xy$x,ozone.xy$y,ozone.median)
contour(i,add=T,labex=0)
text(ozone.xy,ozone.median)
title(main="Median Ozone Concentrations in the North East")
```

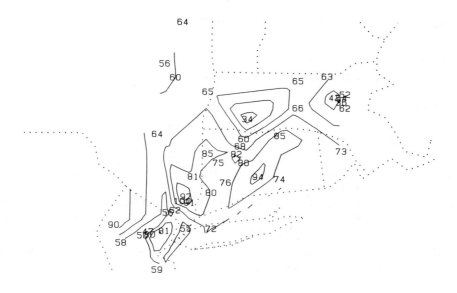

cor	see **var**	cor

cos	Trigonometric Functions	**cos**

USAGE:

> **sin(x)**
> **cos(x)**

ARGUMENTS:

x: numeric structure, in radians. Missing values (NAs) are allowed.

VALUE:

> structure with data transformed by function.

?Cp	All Subsets Regression and Cp Plot	**?Cp**

USAGE:

> **?Cp(x, y, wt, int, names, identify)**

ARGUMENTS:

x: matrix of possible independent variables.

y: vector of dependent variable.

wt: optional vector of weights for each observation.

int: logical flag, should intercept term appear in regression equations. Default TRUE.

names: optional character vector of names for the variables. Default is to label the columns of x (variables) with successive letters from the set "123456789ABCDEFG...Z".

identify: logical flag, should user identify regressions from plot which will then be used to produce full regression statistics? Default TRUE.

EFFECT:

> Produces Cp plot and full regression statistics for regressions optionally selected from the plot.

EXAMPLES:

?Cp(stack.x,stack.loss)   # all-subsets regression and Cp plot

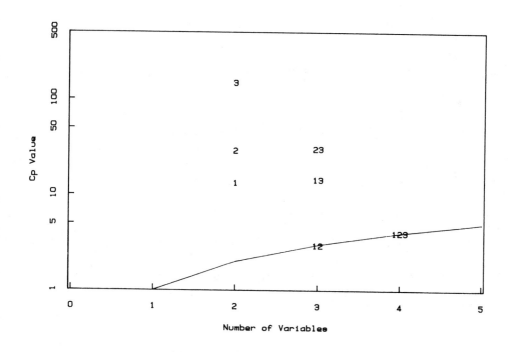

| **crossprod** | Matrix Cross Product Operator | **crossprod** |

USAGE:

mat1 %c mat2
crossprod(mat1, mat2)

ARGUMENTS:

mati: matrix or vector. If **mat2** is omitted, it defaults to **mat1**.

VALUE:

matrix representing the cross product of **mat1** and **mat2**, defined as **t(mat1) %* mat2**, where **%*** is matrix multiplication and **t** is transposition. Thus the **[i,j]**th element of the result is **sum(mat1[,i]*mat2[,j])**.

---

cstr	Create Structure from Components	**cstr**

USAGE:

cstr(arg1, arg2, ...)

ARGUMENTS:

argi:　　　arbitrary value. Missing values (NAs) are allowed.

VALUE:

structure with each argument as a component. The component name is the name given in name=value form. If no name is given in the argument, the dataset name of the argument is used. (If the argument in question has no dataset name, the component generated has an empty component name.)

See **array, matrix,** and **ts** for creating arrays, matrices, or time-series.

EXAMPLES:

cstr(Data=read("gnp"),comment="gross national product")
cstr(a,d=b,1:50)　　#components named **a, d,** and no name

?csweep	Sweep Out Column Effects from Matrix	**?csweep**

USAGE:

?csweep(x, summary, fun)

ARGUMENTS:

x:　　　　　matrix. Missing values (NAs) are allowed.
summary:　values to be swept out of the columns of **x**. Generally **summary** is the result of a **?col** macro or applying a function to the columns of **x**. Missing values (NAs) are allowed.
fun:　　　(unquoted) name of function to be used in sweeping out **summary**. Default is − (minus).

VALUE:

matrix **x** with the summary values swept out.

SEE ALSO:

Function **sweep**.

EXAMPLES:

?csweep(x, ?col(x,median))   # subtract off column medians

---

ctable	Create Color Table for Use on the AED512	**ctable**

USAGE:

ctable(table1, spec1, ..., table2, spec2, ...)

ARGUMENTS:

tablei: optional numeric matrix with 4 columns. Each row gives a color number and its associated hue, lightness, and saturation.

speci: optional character vector, in which each element defines one or more colors. The syntax for these strings is **nrange,hrange[,lrange[,srange]]** where each range is either a single non-negative integer or a colon-separated pair of such integers. **nrange** specifies the range of colors to be defined and **hrange**, **lrange** and **srange** define respectively the hues, lightnesses and saturations to use in the definitions. **nrange** fixes how many colors will be defined and the values implied by **hrange**, **lrange** and **srange** are then determined accordingly. When any of **hrange**, **lrange** or **srange** is a single number, this value is used repeatedly in the generated definitions. Certain symbolic names are recognized in **hrange** in place of numbers (see below).

VALUE:

matrix with 4 columns, containing all the color definitions occurring in the arguments, in the same form as **tablei** above; if more than one definition is given for a particular color number, the last one is used.

COMMENTS:

Valid color numbers range from 0 to 127.

Hues are placed around the perimeter of a circle with blue at 0 degrees, red at 120 degrees and green at 240 degrees. Intermediate values are magenta (60), orange (150), yellow (180) and cyan (300). In each color definition, the hue value is interpreted as an angle in degrees around this circle. In a **speci** argument the particular colors just mentioned may be specified by giving their name in place of their angle around the color circle. Thus "12,red" defines color 12 as red and "20:30,red:yellow" defines colors 20 through 30 as a smooth transition from red to yellow. These specifications are just alternate forms for "12,120" and "20:30,120:180", respectively. If the two values (or names) specified in **hrange** are equal, then an entire circuit around the circle is generated rather than a simple repetition of the value. Thus "2:10,blue:blue" specifies colors 2 through 10 as a complete spectrum beginning and ending with blue.

Lightness is a numerical description of the gray value of the color and ranges from 0 (black) to 100 (white). The default value for lightness is 50.

Saturation is a measure of the intensity of the given hue and ranges from 0 (no intensity--gray) to 100 (full intensity). Its default value is 100.

In addition to the symbolic color names mentioned above, three others are available in the **hrange** position. "white" changes the default lightness to 100, "black" changes it to 0, and "gray" to "0:100".

EXAMPLES:

    mycol ← ctable("10:50,blue:red,50,60")
      # set up a range of pastel colors from blue to red
    color(mycol)   # set up the AED color map
    mycol ← ctable(mycol,"60:70,gray")
      # add a gray scale

| **cumsum** | Cumulative Sums | **cumsum** |

USAGE:

> **cumsum(x)**

ARGUMENTS:

x: numeric structure.

VALUE:

> structure like **x** where the ith value is the sum of the first i values in **x**.

| **cut** | Create Category by Cutting Continuous Data | **cut** |

USAGE:

> **cut(x, breaks, labels)**

ARGUMENTS:

x: data vector. Missing values (NAs) are allowed.

breaks: either a vector of breakpoints, or the number of equal-width intervals into which the data in **x** should be cut. If a vector of breakpoints is given, the category will have **len(breaks)−1** groups, corresponding to data in the intervals between successive values in **breaks** (after **breaks** is sorted). Data less than or equal to the first breakpoint or greater than the last breakpoint is returned as NA.

labels: character vector of labels for the intervals. Default is to encode the breakpoints to make up interval names (if **break** is a vector), or to use the names **Range 1**, etc., if **breaks** is a single integer.

VALUE:

> a category structure with components **Label** and **Data**.

Data: vector of integers telling which group each point in x belonged to.

Label: vector of character names for each group.

EXAMPLES:

> **cut(x,3)**   # cut into 3 groups

cut(x, breaks) # cut based on given breakpoints
cut(x, pretty(x))   # approx 5 "pretty" intervals

---

cutree	Create Groups from Hierarchical Clustering	**cutree**

USAGE:

cutree(tree, k, h)

ARGUMENTS:

tree: hierarchical clustering tree structure, typically the output of **hclust**.

k: optional, the desired number of groups.

h: optional, the height at which to cut **tree** in order to produce the groups. Groups will be defined by the structure of the tree above the cut.

Exactly one of **k** or **h** must be supplied.

VALUE:

structure with two components, giving the assignment vector and the heights of the resulting clusters.

Data: vector with as many elements as there are leaves in the tree. The ith element of the vector gives the group number that individual i is assigned to. Individuals not in the current tree (if **tree** was a subtree from a larger original problem) are assigned group 0.

height: vector with as many values as there are resulting groups, the ith value gives the height of the last merge making up the group. Singleton clusters are given height 0.

EXAMPLES:

cutree(tr←hclust(dist),k=5)   #produce 5 groups

---

cycle	see **time**	**cycle**

defer	Control Deferred Graphics	**defer**

USAGE:

      defer(on, ask)

ARGUMENTS:

on: logical which enables (TRUE) or disables (FALSE) deferred graphics. Default TRUE.

ask: logical flag, if TRUE the user is asked "Save last frame?" before each new frame. Answers beginning with "N" or "n" cause the previous frame to be removed from the graphics save file. Default, FALSE, all frames are saved without user interaction.

      All deferred graphics output goes to the file "sgraph". If "sgraph" already exists, new material will be appended.

      The graphics save file may be replotted on devices other than the device in effect during the **defer** function. This includes batch plotting, if available.

      The graphics save file is in character format, and contains commands with the names "points", "text", "lines", "segs", and "eject" followed by the appropriate data values. The names "par" and "diff" reflect the graphical parameter changes made during plotting.

EXAMPLES:

      defer   # turn on deferred graphics
      defer(FALSE)   # turn off deferred graphics
      defer(ask=TRUE)   # turn on, ask about saving each frame

define	Define a Macro	**define**

USAGE:

define(file, pos, print)

ARGUMENTS:

file: optional character string giving the name of a file which contains definitions for one or more macros. If file is omitted, definitions are read from the terminal.

pos: data directory position on which macros are to be saved. Default 2.

print: logical, should names of defined macros be printed? Default TRUE for macros defined from **file**, FALSE for macros defined from the terminal.

define is used to create new S macro definitions. When **define** reads from terminal, it initially prompts with the string "N>", indicating that it expects the line giving the macro name and arguments. At this point, the user should give a line of the form

MACRO name(arg1/default value/,arg2/default/,...)

The prompt then turns to "D>", indicating a request for the definition of the macro. The definition continues until a line that just contains "END".

Define then prompts "N>" for another macro name. An empty line in response to this prompt terminates the definition process. Definitions from a file are terminated by the end of file.

See also **mprint** and **medit** to print and edit macro definitions.

The keywords **MACRO** and **END** must appear in upper case.

EXAMPLES:

```
> define     #define a macro incr
N> MACRO incr(x,y/1/)
D> (x+y)
D> END
N>
> ?incr(5)*2
```

?define	Define a Macro at Execution Time	?define

USAGE:

?define(name,definition)

ARGUMENTS:

name: name of macro to be defined, i.e., defines macro in dataset
mac.name.

definition: unquoted definition for the macro.

EXAMPLES:

?define(remember,abc,def,ghi)
    # creates dataset mac.remember that can be used
    # to remember values from one invocation of the macro
    # processor to another

---

density	Estimate Probability Density Function	density

USAGE:

density(x, n, window, width, from, to)

ARGUMENTS:

x: vector of observations from distribution whose density is to be estimated.

n: the number of equally spaced points at which to estimate density. Default 50.

window: character string giving the type of window used in computation "cosine", "gaussian", "rectangular", "triangular". Default is "g" (one character is sufficient)

width: width of the window. Default is width of histogram bar constructed by Doane's rule. The standard error of a Gaussian window is **width/4**.

from:
to: the n estimated values of density are equally spaced between **from** and **to**. Default is range of data extended by **width*3/4** for gaussian window or **width/2** for other windows.

VALUE:

plotting structure with two components, **x** and **y**.

x: vector of **n** points at which density is estimated.

y: density estimate at each **x** point.

REFERENCE:

Wegman, E. J. (1972), "Nonparametric Probability Density Estimation", *Technometrics*, Vol 14, pp 533-546.

EXAMPLES:

plot(density(x),type="b")

---

detach	Detach a Data Directory from Search List	**detach**

USAGE:

detach(file, pos)

ARGUMENTS:

file: optional character string giving the name of the data directory to
be detached.

pos: optional position (instead of name) of data directory in search list.
If both arguments are omitted, the last directory on the search list
is detached.

EXAMPLES:

detach("abc")
detach(pos=3) #detach shared data directory

devices	List of Graphical Devices	**devices**

hp2623:   Hewlett-Packard 2623 scope.

hp2627:   Hewlett-Packard 2627 color scope.

hp2647:   Hewlett-Packard 2647 scope.

hp2648:   Hewlett-Packard 2648 scope.

hp7220h,hp7220v: Similar to **hp72h**, etc. for the Hewlett-Packard 7220 series.

hp7221:   Hewlett-Packard 7221 pen plotter, 11x17 paper.

hp7221h: Hewlett-Packard 7221 pen plotter, 8.5x11 paper (horizontal).

hp7221v: Hewlett-Packard 7221 pen plotter, 8.5x11 paper (vertical).

hp7225,hp7225v: Hewlett-Packard 7225 plotter, 8.5x11 paper, horizontal or vertical plots.

hp7470,hp7470v: Hewlett-Packard 7470 plotter, 8.5x11 paper, horizontal or vertical plots.

hpgl: Hewlett-Packard HP-GL plotters.

printer: Any printing terminal.

ram6211:   Ramtek 6211 color scope.

tek10:   Tektronix 4010, 4006 scopes (upper case only).

tek12:   Tektronix 4012 scope.

tek14:   Tektronix 4014 scope.

tek14q: Tektronix 4014 scope with lower resolution, higher speed.

tek4112:   Tektronix 4112 raster-scan scope.

tek46:   Tektronix 4662 pen plotter, 11x17 paper.

tek46h: Tektronix 4662 pen plotter, 8.5x11 paper (horizontal).

tek46v: Tektronix 4662 pen plotter, 8.5x11 paper (vertical).

unixplot: Produces file "unixplot.out" which can be sent to any of the UNIX
                 PLOT(1) device filters.

---

dget	Retrieve a General Structure from File	**dget**

USAGE:
    dget(file)

ARGUMENTS:
file: character string giving file name to read from. Default is the
                standard input.

VALUE:
    data structure corresponding to contents of **file**.

    Normally, the file was created through the function **dput**, possibly so that the data could be transmitted from another user and/or machine. The form of the data is a list notation:

    (name mode length value1 value2 ...)

    where **name** is a character string, and there are **length** values. In the case of a vector (**mode** being REAL, INT LGL or CHAR), the values are data of the corresponding mode. In the case of a hierarchical structure **valuei** is itself a list of the same form. In this case, the **mode** may be followed by the structure name.

    Since the file is readable, one may edit the data on the file, although keeping the file in legal form may not be trivial.

SEE ALSO:
    dput, dump, and **restore**.

EXAMPLES:

      dget("copydata")

---

diag	Diagonal Matrices	**diag**

USAGE:

      diag(x, nrow, ncol)

ARGUMENTS:

x: matrix or vector. Missing values (NAs) are allowed.
nrow: optional number of rows of output matrix.
ncol: optional number of columns of output matrix.

VALUE:

      the vector of diagonal elements of **x**, if **x** is a matrix. Otherwise, a matrix with **x** on its diagonal and zeroes elsewhere. If **x** is a scalar (vector of length 1), the value is an **x** by **x** identity matrix. By default, the matrix is square with zeros off the diagonal, but it can be made rectangular by specifying **nrow** and **ncol**.

EXAMPLES:

      diag(xmat)   # extract diagonal

      diag(diag(xmat))   # matrix of just diagonal of xmat

      diag(5)   # 5 by 5 identity matrix

      x[row(x)==col(x)] ← diag(y) # put diagonal of y
            # into diagonal of x

---

diary	Keep Diary of S Commands	**diary**

USAGE:

> **diary(on, file)**

ARGUMENTS:

on: if TRUE(the default), diary-keeping is turned on.

file: optional file name for diary entries, default is file "diary".

> Each line that the user types while diary-keeping is active is entered into the diary file. Each line that comes from a source file or from a macro is also entered, preceded by the comment character and indented to tell the level of source file nesting. Thus, the diary file can be used as a source file for re-execution of previously entered lines. (Lines that caused errors will also be present in the diary file, so some editing may be necessary.)

> Each time the diary keeping is turned on, a comment line is entered into the diary file giving the current date and time. Also, note that user-typed comments are entered into the diary file.

SEE ALSO:

> **sink** and **stamp**.

EXAMPLES:

> **diary**   # turn on diary file
> **diary(FALSE)**   # turn off diary file

---

diff	Create a Differenced Series	**diff**

USAGE:

> **diff(x, k, n)**

ARGUMENTS:

x: a time-series or vector. Missing values (NAs) are allowed.

k: the lag of the difference to be computed (default is 1).

n: the number of differences to be done (default is 1).

VALUE:

a time-series which is the nth difference of lag **k** for **x**. For first differences where **y** ← **diff(x,k)**, **y[i]** is **x[i]** − **x[i−k]**, and **len(y)** is **len(x)** − **k**. To construct an nth difference, this procedure is iterated **n** times. Any operation on an NA produces an NA.

EXAMPLES:

d2x ← **diff(x,1,2)**   # second difference of lag 1

diff(range(x)) # max(x) − min(x)

The start date of **d2x** will be 2 periods later than **x** due to the differencing.

---

discr	Discriminant Analysis	**discr**

USAGE:

discr(x, k)

ARGUMENTS:

x: matrix of data

k: either the number of groups (if groups are equal in size), or the vector of group sizes; i.e., first **k[1]** rows form group 1, next **k[2]** group2, etc.

Note that the rows of **x** must be ordered by groups.

VALUE:

a structure describing the discriminant analysis, with the following components:

cor: vector of discriminant correlations (cor. between linear combination of variables and comb. of groups)

groups: matrix of linear combinations of groups predicted.

vars: matrix of linear combinations of variables.

Columns of **vars** give discriminant variables; i.e., **x %* vars** produces the matrix of discriminant variables.

SEE ALSO:

the macro **?discr** uses a categorical variable to define groups for function **discr**.

?discr	Discriminant Analysis with Grouping Vector	**?discr**

USAGE:

?discr(x, group)

ARGUMENTS:

x: matrix of data for discriminant analysis.

group: vector or category of length **nrow(x)** telling which group each row of **x** belongs to.

VALUE:

structure like that returned by the **discr** function.

EXAMPLES:

?discr(mydata,group)

dist	Distance Matrix Calculation	**dist**

USAGE:

dist(x, metric)

ARGUMENTS:

x: matrix (typically a data matrix). The distances computed will be among the rows of **x**. Missing values (NAs) are allowed.

metric: character string specifying the distance metric to be used. The currently available options are "euclidean" (the default), "maximum", "manhattan", and "binary". Euclidean distances are root sum-of-squares of differences, maximum is the maximum difference, manhattan is the sum of absolute differences, and binary is the proportion of non-zeroes that two vectors have in common.

VALUE:

the distances among the rows of **x**. Since this structure can be very large, and since the result of **dist** is typically an argument to **hclust**, a special structure is returned, rather than a matrix.

size: the number of objects (that is, rows of **x**).

data: the **size*(size−1)/2** values.

Missing values in a row of **x** are not included in any distances involving that row. Such distances are then inflated to account for the missing values. If all values for a particular distance are excluded by this rule, the distance is NA.

EXAMPLES:

dist(x,"max") # distances among rows by maximum
dist(t(x)) # distances among cols by euclidean

?dist2full	Distance Structure to Full Symmetric Matrix	**?dist2full**

USAGE:

?dist2full(x)

ARGUMENTS:

x: distance structure, as produced by the **dist** function.

VALUE:

full symmetric matrix containing the distances from point **i** to point **j** in the **i,j**th element.

EXAMPLES:

xx ← ?dist2full(dist(datamatrix))

dput	Save a Dataset on a File	**dput**

USAGE:

dput(x, file)

ARGUMENTS:

x: the data structure to be saved. Missing values (NAs) are allowed.

file: character string giving the file name where the data structure is
        to be stored. Default is the user's terminal.

VALUE:

the dataset **x**.

The function **dget** can re-create the dataset from **file**.

SEE ALSO:

dget, **dump**, and **restore**.

EXAMPLES:

dput(abc,"oldabc") # store data structure abc on file oldabc
def ← dget("oldabc") # read in data structure

dump	Dump Datasets to a File	**dump**

USAGE:

dump(list, file, pos)

ARGUMENTS:

list: character vector of the names of the S datasets to dump. Typical-
        ly, this is the result of the **list** function.

file: character name of file where datasets are to be dumped. Default
        "dumpdata".

pos: position on search list from which datasets are to be taken. De-
        fault 0, use first dataset with the correct name encountered on
        search list.

The use of **dump** is usually to dump datasets for transmission to
another machine, or for archive purposes. One useful technique is

to dump all the data under a prefix; e.g., when you want to share the data on a specific project with someone on another machine.

The function **restore** is used to recreate datasets written by **dump**. The functions **dump** and **restore** are not affected by the current prefix.

EXAMPLES:

dump(list("longley.*"),"longleydump") #all longley.*

dump(list(pos=2),"dbdump",pos=2) #the WHOLE save directory
need to use pos= for both dump( ) and list( )

---

edit	Edit Dumped Expressions or Character Vectors	**edit**

USAGE:

edit( )   #for editing the dumped expression
again( )   #to re-evaluate dumped expression

edit(x, pos=)   #for character vector

ARGUMENTS:

x: optional character vector to be edited. If **x** is omitted, **edit** operates on the last expression dumped because of a syntax or an execution error.

pos=:   optional data directory position (on search list) where the edited vector should be stored. Default 1, the working directory.

EFFECT:

If **x** is given, the edited data is assigned, with the same name, on the working directory. Note that this implies that it is impossible to edit an expression (as opposed to a dataset), since this has no name for the assignment: **edit(encode(1:5))** will not work.

After **edit** is invoked, the data from the dumped expression or character vector is written to a hidden file, one element of the vector per line of the file. A text editor (the default is **ed**) is invoked and the file is read.

At this point, editor commands can be used to modify the text. When a suitable version of the text is produced, execute commands to write the edited text back to the file, and to exit from the editor (**w** and **q** in **ed**).

After exiting from the editor, the file is read by **edit**. If the data was a dumped expression, a hidden call to **source** causes the edited expression to be executed. Notice that this edited version stays around, for another call to **edit** to make further changes. If **x** was given, the edited expression is stored on the working directory.

To suppress execution of the edited expression, DO NOT just leave the editor without writing (this leaves the expression as before). Instead, use a command to delete everything and then write the file (the **ed** command **1,$s/.*//**, then **w**); this leaves nothing to execute.

The function **again** is useful to re-execute expressions that were dumped because of execution errors that have been remedied, e.g., when a graphic device has been activated, when a missing dataset has been created, or when a prefix has been specified.

Use the function **medit** to edit macros.

The user can specify which editor should be invoked by setting a Unix shell variable prior to invoking S. For example, the line **EDITOR=vi; export EDITOR** will cause the **edit** to use the **vi** screen editor rather than **ed**.

EXAMPLES:

    edit(state.name) #edit the vector of state names.

eigen	Eigen Analysis of Symmetric Matrix	eigen

USAGE:

eigen(x, n, large)

ARGUMENTS:

x: matrix to be decomposed. Must be square and symmetric.

n: number of eigenvalues and corresponding eigenvectors wanted from **x**. Default is to return all eigenvalues.

large: logical flag; if TRUE (the default), returns the **n** largest eigenvalues; otherwise, returns the **n** smallest eigenvalues.

VALUE:

a structure containing the eigenvalues and eigenvectors.

values: vector of **n** eigenvalues; note that the values are always in ascending order, regardless of **large**.

vectors: matrix with **nrow(x)** rows and **n** columns. Each column is the eigenvector corresponding to the eigenvalue which is the corresponding element of **values**.

SEE ALSO:

svd.

EXAMPLES:

cors ← cor(x,y,trim=.1)
pprcom ← eigen(cors)

else	see **syntax**	else

encode	Encode Text and Numeric Data	**encode**

USAGE:

> **encode(arg1, arg2, ..., sep=)**

ARGUMENTS:

argi: vectors which may be either numeric, logical, or character. If **argi** is of mode character, its elements are concatenated with the next arguments as is. Otherwise, the values are encoded into character strings first. Missing values (NAs) are allowed.

sep=:   the character string to be inserted between successive arguments. Can be "" for no space. Default single space (" ").

VALUE:

> character vector, with length equal to the maximum of the lengths of the arguments. The i-th element of the result is the concatenation of the i-th elements of the arguments (encoded if not originally character strings). If the length of the argument is less than the maximum, elements of that argument are repeated cyclically. In particular, an argument can be a single element, to appear in each element of the result.

EXAMPLES:

> **encode("no.",1:10)**   # gives "no. 1", "no. 2" ...
>
> **encode(state.name,"pop=",pop)**   # "Alabama pop= 12.345"...

end	see **start**	**end**

exp	Math Functions	**exp**

USAGE:

> **exp(x)**
> **log(x)**
> **log10(x)**
> **sqrt(x)**

ARGUMENTS:

x: numeric structure. Missing values (NAs) are allowed.

VALUE:

> structure like **x** with data transformed by the function. **exp** is exponential, **log** natural logarithm, **log10** common log, **sqrt** square root.
>
> Argument values which cannot be handled (e.g., negative values for square root) evaluate to NA and generate a warning message.

extract	Extract Columns or Fields as S Datasets	**extract**

USAGE:

> **!S extract [flag] file [desfile extfile]**   # UNIX command

ARGUMENTS:

file: the name (unquoted) of a file containing input data, either in specific columns or else as fields separated by a field separator character.

flag: selects warning levels: **−f** (fatal) causes **extract** to terminate on the first error encountered. **−w** (warning, the default) prints a warning message whenever a field is not present in a record or does not contain valid information. Numeric fields with errors are set to NA; character fields are set to null strings. **−s** (silent) suppresses warning messages.

desfile: the name of a description file specifying how to extract data from **file**. Defaults to **file.des**.

extfile: the name of the file onto which the external form of the extracted

S data structures will be written. Defaults to **file.ext**.

The macro form, **?extract**, extracts the datasets, writes them to the extract file and uses the **restore** function to bring these datasets into the user's working data directory. The utility version, **S extract**, just forms the extract file, which could then be shipped to another computer system, etc.

The **description file** lists the name of each of the datasets to be extracted and their position in the records of the file. In **column** form of description, each line of the description file is of the form:

name mode start length

with **name** the name of the extracted dataset, **mode** the desired mode of the extracted data (R, I, C for REAL, INTEGER and CHARACTER), **start** the beginning column for the data items on each record, and **length** the number of columns in the item. In the **field** form of input, the first line of the description file should be of the form

−f%

where "%" is the desired field separator character (white-space, tabs or spaces, by default). Remaining lines of the description file are then of the form:

name mode field

with **name** and **mode** as before, and **field** defining which field of the record becomes the data item.

SEE ALSO:

        ?extract.

?extract	Extract Columns of Data from File	?extract

USAGE:

> ?extract(name, print, options)

ARGUMENTS:

name: unquoted name of data file. A file of the name **name.des** must have been pre-defined, in the manner described for utility **extract**.

print: logical, should names of extracted datasets be printed as they are placed on database? Default TRUE.

options: any further options to the **restore** function, e.g., pos=.

EFFECT:

> Columns of the file **name** are read and assigned in a database. The descriptor file determines the names of the datasets created.

EXAMPLES:

> ?extract(mydata,print=T) # extract from file mydata
> # based on descriptor file mydata.des
> # print names of extracted datasets as they are
> # placed on the work file

SEE ALSO:

> **extract** for a description of how to construct the descriptor file.

F	f Probability Distribution	F

USAGE:

> pf(q, par1, par2)
> qf(p, par1, par2)
> rf(n, par1, par2)

ARGUMENTS:

q: vector of quantiles. Missing values (NAs) are allowed.

p: vector of probabilities. Missing values (NAs) are allowed.

n: sample size.

par1: vector of degrees of freedom for numerator.

par2:   vector of degrees of freedom for denominator.

VALUE:

probability (quantile) vector corresponding to the given quantile (probability) vector, or random sample of length **n** (**rf**).

EXAMPLES:

rf(10,5,15) #sample of 10 with 5 over 15 df

---

faces	Plot Symbolic Faces	**faces**

USAGE:

faces(x, which, labels, head, max, nrow, ncol, fill, scale, byrow)

ARGUMENTS:

x: matrix of data values. Missing values (NAs) are allowed.

which: the columns of **x** to be used as the first, second, etc. parameter in the symbolic face. Default **1:min(15,ncol(x))**. See NOTE below for meaning of parameters.

labels: optional character vector of labels for the faces (i.e., for the rows of **x**).

head: optional character vector to use as the heading for the plot.

max: a suggested value for the number of rows and columns to go on each page. By default, all the faces will be fitted onto one page.

nrow:
ncol: optionally, may be given to specify exactly the number of rows and columns for the array of plots on each page.

fill: if TRUE (the default), all unused parameters of the face will be set to their nominal (midpoint) value. If FALSE, all features corresponding to unused parameters will not be plotted.

scale: if TRUE (the default), the columns of **x** will be independently scaled to (0,1). If FALSE no scaling will be done. The data values should then be scaled to the same overall range by some other means; e.g., by scaling the whole of **x** to the range (0,1).

byrow: if TRUE, plots produced in row-wise order; if FALSE (the default) plots are produced in column-wise order.

NOTE: the feature parameters are: 1-area of face; 2-shape of face; 3-length of nose; 4-loc. of mouth; 5-curve of smile; 6-width of mouth;

7,8,9,10,11-loc., separation, angle, shape and width of eyes; 12-loc. of pupil; 13,14,15-loc., angle and width of eyebrow.

REFERENCE:

H. Chernoff, "The use of Faces to Represent Points in k-Dimensional Space Graphically", *Journal of the American Statistical Association*, Vol. 68, pp 361-368, 1973.

EXAMPLES:

faces(chernoff2,head="Chernoff's Second Example")

Chernoff's Second Example

| **fatal** | Informative Message Functions | **fatal** |

USAGE:

> **fatal(msg)**
> **warning(msg)**
> **message(msg)**

ARGUMENTS:

msg: character vector that is printed one element per line. The function **fatal** prepends "Fatal Error:" to the first line of **msg**, and causes abnormal termination of the expression it is involved in. This can be used to terminate the execution of macros, source files, etc. The function **warning** prepends "Warning:" to the first line of **msg**. Function **message** simply prints **msg**.

EXAMPLES:

> **message(instructions)**
> **if(any(NA(x))) fatal("NAs are not allowed in x")**

| **flood** | Flood Areas with Color using Seed Fill | **flood** |

USAGE:

> **flood(x, y, col, bdy, erase)**

ARGUMENTS:

x,y: coordinates of "seed" points within (not on the boundary of) the areas to be flooded. A structure containing components **x** and **y** may also be given.

col: color numbers to be used to flood the areas. If **col** is shorter than **x** and **y**, values will be taken cyclically from **col**.

bdy: optional vector giving the colors of lines which bound the region. If **bdy** is given, the ith region is assumed to be bordered with color **bdy[i]**. If **bdy** is shorter than **x** and **y**, values will be taken cyclically from **bdy**.

erase: should alpha overlay be erased prior to filling? Default is TRUE with the **aed512** device driver, since if the alpha overlay is not erased, alpha characters that overlap the flooded areas may be mistaken for boundaries of the areas.

COMMENTS:

Flooding begins at the specified point and spreads outward until stopped by a boundary. If **bdy** is not given, any color different from the color of the initial point is treated as a boundary. If there is no boundary between the specified point and the edge of the screen, the background area of the screen will be flooded.

It is difficult to fully flood an area that is very narrow, e.g., a thin wedge of a pie chart. This is because numerous "islands" may be created by the discretization of the lines into pixels.

WARNING:

This function is experimental and applies only to the aed512 and ram6211 color scopes; **flood** may be superseded by a more general function at a later date.

EXAMPLES:

usa(fifty=T); flood(state.center,1:50)
note Michigan's upper peninsula not filled
AK and HI drawn as boxes

flood(rdpen( ),10)   # fill pointed-to areas with color 10

hatch(x,y,space=0)   # draw polygon outline
flood(mean(x),mean(y),7)   # fill with color 7

---

floor	see **ceiling**	**floor**

?fmin	Macro to minimize an S expression	**?fmin**

USAGE:

?fmin(pars, f, g, start, radius, partol, gtol, iter, compute=)

ARGUMENTS:

pars: Name of the S dataset to be used for the parameters in the function to be minimized.

f: S expression whose value is the function of **pars** to be minimized.

g: S expression whose value is the derivative of **feval** with respect to **pars**.

start: Vector of starting values for **pars**

radius: Optional, starting value for the radius of trust in the optimization algorithm. Default is 1.0.

partol: Optional, value such that a relative change in **pars** smaller than **partol** implies convergence of the iteration. Default is 1e−3.

gtol: Optional, value such that gradient norm less than **gtol** implies convergence. Default is 1e−3.

iter: Optional, maximum number of iterations allowed. Default is 200.

compute=:   Optional, preliminary expression to be computed to simplify computation of function values.

VALUE:

     A structure describing the function at the computed minimum.

pars: The vector of parameters at the minimum. NOTE: the actual name of this component will be the name supplied as the first argument to **fmin**.

value: The value of the function at the minimum.

gradient: The value of the gradient at the minimum.

variance: The estimate of the variance of the parameter estimates (namely, the inverse of the approximate Hessian).

halt: The reason for halting.

nf: The number of function evaluations in the iteration.

ng: The number of gradient evaluations in the iteration.

EXAMPLES:

     ?fmin(a,a^2,a*2,4) #min of y=x^2 starting at 4

for	see **syntax**	**for**

---

frame   Advance Graphics Device to Next Frame or Figure   **frame**

USAGE:

  frame( )

  Upon receipt of this command, the graphics device will eject to a
  new page, clear the screen, or do whatever is appropriate for the
  device.

  In multiple figure mode, **frame** will advance to the next figure. It
  takes two calls to **frame** to skip a figure: the first goes to the next
  figure, and the second call skips to the next figure.

?full2dist   Full Symmetric Matrix to Distance Structure   **?full2dist**

USAGE:

  ?full2dist(x)

ARGUMENTS:

x: square symmetric matrix.

VALUE:

  structure like that produced by the **dist** function, containing components **Data** and **Size**, and containing just the lower-triangle of
  data from **x**.

EXAMPLES:

  mydist ← ?full2dist(1−cor(x,x))   # turn 1 − correlation of
    # cols of x into distance structure

gamma	Gamma Function (and its Natural Logarithm)	gamma

USAGE:

> **gamma(x)**
> **lgamma(x)**

ARGUMENTS:

x: vector structure. Missing values (NAs) are allowed.

VALUE:

> gamma or natural log of gamma function evaluated for each value in **x**. For positive integral values, **gamma(x)** is **(x−1)!**. For negative integral values, **gamma(x)** is undefined. Function **lgamma** allows only positive arguments. NAs are returned when evaluation would cause numerical problems.
>
> Note that **gamma(x)** increases very rapidly with **x**. Use **lgamma** wherever possible to avoid overflow.

EXAMPLES:

> **gamma(6)**   # same as 5 factorial

Gamma	Gamma Probability Distribution	Gamma

USAGE:

> **pgamma(q, par1)**
> **qgamma(p, par1)**
> **rgamma(n, par1)**

ARGUMENTS:

q: vector of quantiles. Missing values (NAs) are allowed.
p: vector of probabilities. Missing values (NAs) are allowed.
n: sample size.
par1: vector of shape parameters (>0).

VALUE:

> probability (quantile) vector corresponding to the given quantile (probability) vector, or random sample (**rgamma**) from the gamma

distribution.

EXAMPLES:

rgamma(20,10) #sample of 20 with shape parameter 10

---

get	Access Data on a Data Directory	**get**

USAGE:

get(name1, name2, ..., pos=)

ARGUMENTS:

namei: the character strings which give the main name and the component names of the data; i.e., **name1** is the name of the dataset and **name2, name3**, etc. are the names of the component, subcomponent, etc.

pos=:   the position of the data directory from which the data is to be retrieved. If **pos** is zero, all data directories are searched for the dataset. Default 0.

VALUE:

the data found.

The function **get** is the explicit analogue to what happens automatically when the name of a dataset or a component appears in an expression. Occasionally **get** is used explicitly because the names must be constructed or read in, as in the example below.

EXAMPLES:

```
for(i in 1:10){
name ← encode("y",i,sep="")
plot(x,get(name)) }
#scatter plot (x,y1), (x,y2),...(x,y9)
```

gs	Gram-Schmidt Decomposition	gs

USAGE:

      gs(x)

ARGUMENTS:

x: matrix to be decomposed.

VALUE:

      orthogonal decomposition of the matrix $x == r$ $\%^*$ **q**, using the Gram-Schmidt decomposition with iterative re-orthogonalization.

r: upper-triangular matrix, of order **ncol(x)**.

q: matrix of the same dimensions as **x**, whose columns are orthogonal and of unit euclidean norm.

hardcopy	On-line Copy of Current Display	hardcopy

USAGE:

      hardcopy(delay)

ARGUMENTS:

delay: number of seconds to delay while copy is being made. Default is 60 seconds on AED 512, 30 seconds on Tektronix.

      hardcopy currently works with Tektronix 4000 series scopes and Advanced Electronics Design 512 color scopes with an Image Resource Videoprint copier.

SEE ALSO:

      defer.

---

hat	Hat Matrix Regression Diagnostic	**hat**

USAGE:

   hat(x)

ARGUMENTS:

x: matrix of independent variables in the regression model **y=xb+e**.

VALUE:

   vector with one value for each row of **x**. These values are the diagonal elements of the least-squares projection matrix **H**. (Fitted values for a regression of **y** on **x** are determined by **Hy**.) Large values of these diagonal elements correspond to points with high leverage: see the reference for details.

REFERENCE:

   D. A. Belsley, E. Kuh, and R. E. Welsch, *Regression Diagnostics*, Section 2.1, Wiley, 1980.

EXAMPLES:

   h ← hat(longley.x)

---

hatch	Shade in a Polygonal Figure	**hatch**

USAGE:

   hatch(x, y, space, angle, border, fill)

ARGUMENTS:

x,y: coordinates of the vertices of a polygon, listed in order, i.e., the ith point is connected to the i+1st. It is assumed the polygon closes by joining the last point to the first. A structure containing components **x** and **y** can also be given.

space: inches of space between adjacent shading lines. Default .1. If **space** is zero, no hatch lines will be drawn.

angle: angle of shading lines in degrees measured counterclockwise from horizontal. Default 45.

border: should border of polygonal region be drawn? Default TRUE.

fill: should hardware polygon-filling algorithm be invoked to fill the polygon? Default FALSE. If **fill** is TRUE, arguments **space**, **angle**, and **border** are ignored.

Graphical parameters may also be supplied as arguments to this function (see **par**).

EXAMPLES:

    hatch(rdpen( )) # read pen positions, draw and shade polygon
    hatch(c(x1,rev(x2)), c(y1,rev(y2)))
       # shade the area between lines x1,y1 and x2,y2

---

hclust	Hierarchical Clustering	**hclust**

USAGE:

    hclust(dist, method, sim)

ARGUMENTS:

dist: a distance structure or distance matrix. Normally this will be the result of the function **dist**, but it can be any data of the form returned by **dist**, or a full, symmetric matrix.

method: character string giving the clustering method. The three methods currently implemented are "average", "connected" (single linkage) and "compact" (complete linkage). The default is "compact". (The first three characters of the method are sufficient.)

sim=: optional structure replacing **dist**, but giving similarities rather than distances. Exactly one of **sim** or **dist** must be given.

VALUE:

a "tree" representing the clustering, consisting of the following components:

merge: an $(n-1)$ by 2 matrix, if there were **n** objects in the original data. Rows $1,2,...,n-1$ of **merge** describe the merging of clusters at steps $1,2,...,n-1$ of the clustering. If an element in the row is of the form $-j$, then object **j** was merged at this stage. If the element of **merge** is of the form $+i$, then the merge was with the cluster formed at the (earlier) stage **i** of the algorithm.

height: the clustering "height"; that is, the distance between clusters merged at the successive stages.

order: a vector giving a permutation of the original objects suitable for
            plotting, in the sense that a cluster plot using this ordering will
            not have crossings of the branches.

SEE ALSO:

            The functions **plcust** and **labclust** are used for plotting the result
            of a hierarchical clustering. Functions **cutree, clorder,** and **sub-
            tree** can be used to manipulate the tree data structure.

EXAMPLES:

            h ← hclust(dist(x))
            plclust(h)

            hclust(dist(x),"ave")

HELP	Print S Function Documentation	**HELP**

USAGE:

            !S HELP function ...   # UNIX command

ARGUMENTS:
function: names of functions for which documentation should be printed.

            In some cases several functions may be documented in the same
            documentation file. Only one copy of each documentation file is
            printed. Documentation files are printed in alphabetical order.

---

help	On-line Documentation	**help**

USAGE:

            help("name")   # information about functions
            help(macro="name")   # information about macros
            help(dataset="name")   # information about datasets

            call("name")   # just give title and argument names
            call(macro="name")

ARGUMENTS:

name: character string, giving the name of a function, operator, macro, or dataset. If omitted, documentation on "help" is given (this documentation).

If information is found on **name**, the documentation will be printed. This is the same documentation as in the S manual.

An optional argument **width** can be given to specify the maximum width of printed lines. Default is the smaller of 80 and the value of **width** as set in the **options** function.

The function **call** prints only the title and usage for a function or macro.

EXAMPLES:

        help("cstr") # documentation for function cstr
        help("+") #documentation on addition (and other arithmetic)

        help(macro="row")   # ?row documentation
        help(dataset="longley")   # Longley dataset info

        call("stem") # describe arguments for stem

---

hist	Plot a Histogram	**hist**

USAGE:

        hist(x) #simple form
        hist(x, nclass, breaks, scale, plot, angle, density, col, inside)

ARGUMENTS:

x: numeric vector of data for histogram.

nclass: optional recommendation for the number of classes the histogram should have. Default is of the order of log to the base 2 of length of x.

breaks: optional vector of the break points for the histogram classes. The first value of breaks should be smaller than any data point; the last value of breaks should be larger than any data point. If omitted, evenly-spaced break points are determined from **nclass** and the extremes of the data.

scale: if TRUE, y axis will be on a density scale (fraction of total count per fraction of x-range), otherwise y will be counts. Default FALSE. This is useful for comparing histograms, because histograms based on varying numbers of observations or linearly transformed values will have similar y-axes.

plot: if TRUE (default), the histogram will be plotted; if FALSE, the vectors **counts** and **breaks** will be returned.

The **hist** function computes breakpoints, counts the number of points within intervals, and then chains to the function **barplot** to do the actual plotting. Consequently, arguments to the **barplot** function that control shading, etc., can also be given to **hist**. See **barplot** documentation for details of arguments **angle, density, col,** and **inside.**

Graphical parameters may also be supplied as arguments to this function (see **par**).

VALUE:

if **plot** is FALSE, **hist** returns a structure with components **counts** and **breaks.**

counts: count or density in each bar of the histogram.
breaks: break points between histogram classes.

EXAMPLES:

  hist(x)

  hist(x,nclass=15)

  # the example plot is produced by:
  my.sample ← rt(50,5)
  lab ← "50 samples from a t distribution with 5 d. f."
  hist(my.sample,main=lab)

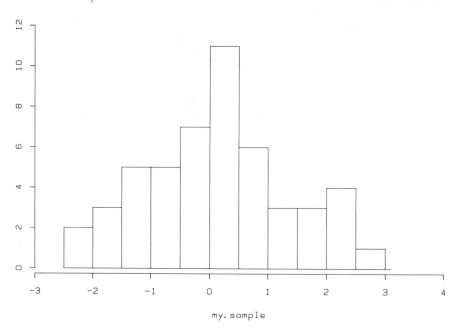

50 samples from a t distribution with 5 d. f.

my.sample

hp2623	see **hp2648**	**hp2623**

hp2627	Hewlett-Packard Color Graphics Terminal	**hp2627**

USAGE:

> **hp2627(ask)**

ARGUMENTS:

ask: logical, should the driver print the message "GO?" to ask permission to clear the screen? Default TRUE. This allows further viewing, making a hard copy, etc., before the screen is erased. When ready for plotting, simply hit return.

When graphic input (**identify, rdpen**) is requested from the terminal, the cursor lines appear. Position the cursor at the desired point and then type any single printing character. Hitting return will terminate graphic input without transmitting the point.

There are eight character sizes (1 to 8) and characters can be rotated by multiples of 90 degrees. Colors 1 through 8 are: white, cyan, magenta, blue, yellow, green, red, and black. These colors can be used for lines, text, and areas. There are 7 line types (1 to 7) giving solid, dotted, short-dashed, long-dashed, dot-space-dash, dot-dot-dot and dot-dashed lines. The hp2627 can fill polygons of up to 148 sides in solid colors (1-8), mixed (dithered) colors (9 - ...) as violet, brown, orange, gold, lime, turquoise, ..., , or patterns (colors −1 thru −7): short dashed hatch, long dashed hatch, hatch, cross hatch, fine cross hatch, checkerboard, fine checkerboard.

A device must be specified before any graphics functions can be used.

---

hp2647   see **hp2648**   **hp2647**

---

hp2648   Hewlett-Packard Graphics Terminals   **hp2648**

USAGE:

> **hp2648(ask)**
> **hp2647(ask)**
> **hp2623(ask)**

ARGUMENTS:

ask: logical, should the driver print the message "GO?" to ask permission to clear the screen? Default TRUE. This allows further viewing, making a hard copy, etc., before the screen is erased. When ready for plotting, simply hit return.

When graphic input (**identify, rdpen**) is requested from the terminal, the cursor lines appear. Position the cursor at the desired point and then type any single printing character. Hitting return will terminate graphic input without transmitting the point.

There are eight character sizes (1 to 8) and characters can be rotated by multiples of 90 degrees. There are 7 line types (1 to 7) giving solid, dotted, short-dashed, long-dashed, dot-space-dash, dot-dot-dot and dot-dashed lines. These devices can fill rectangles (as generated by barplots) in eight patterns: solid, small checkerboard, large checkerboard, one white/3 black, very large checkerboard, spiral, diamond, and lattice.

A device must be specified before any graphics functions can be used.

hp72	Hewlett-Packard 7221a Pen Plotter	hp72

USAGE:

> **hp7221(width, height, ask, auto, with2621, speed)**
> **hp7221h**
> **hp7221v**

ARGUMENTS:

width: width of plotted surface in inches. Default 8 for **hp7221v**, 10 for **hp7221h**, 15 otherwise.

height: height of plotted surface in inches. Default 8 for **hp7221h**, 10 otherwise.

ask: logical, should user be prompted by "GO?" prior to advancing to new frame? Default TRUE.

auto: logical, can device automatically advance the paper (H-P 7221S)? Default FALSE.

with2621: logical, is the plotter configured with H-P 2621 terminal? Default FALSE. If TRUE, special output is provided to make the plotter and terminal work together.

speed: maximum allowed axis pen velocity ranging from 10 to 360 mm/sec. Default 360. Slower speeds are useful for high-quality work or for plotting on special paper or film.

These commands identify Hewlett-Packard 7221 series plotters. The plotter may be used with standard 8.5 by 11 inch paper with the long dimension horizontally (**hp7221h**) or vertically (**hp7221v**), or with 11 by 17 paper (**hp7221**). The arguments listed after **hp7221** can also be given with **hp7221h** or **hp7221v**.

Whenever a new plot is about to be produced, the message "GO?" appears on the terminal. At this time, new paper may be loaded, pens changed, etc. When ready for plotting to proceed, simply hit carriage return.

When graphic input (**identify, rdpen**) is requested, the **enter** button will light. The pen should be positioned by means of the 5 directional buttons. When the desired pen position is reached, depress the **enter** button. To terminate the input, hit carriage return on the terminal (the coordinates of the pen are not transmitted).

Character sizes may be changed to any desired value. Color can be changed to any of the values 1 to 8, indicating the pens in stables 1 to 8. Characters can be rotated to any orientation. Different line styles (1, 2, ...) are available, with patterns becoming more spread-out as the line style is larger.

ALWAYS BE SURE that the pen is back in the holder when you are done with the terminal. Either the S function **q** or hitting in succession the **enter** button followed by the button below the currently empty holder will do this. Users who fail in this respect will be devoured by the dry-pen dragon.

NOTE:

The switches on the back of the device should be set to reflect the correct speed and parity settings for the computer system. Incorrect parity settings may cause failures of digitizing functions, e.g., **rdpen** and **identify**.

A device must be specified before any graphics functions can be used.

hp7220h	see **hpgl**	**hp7220h**

hp7220v	see hpgl	hp7220v

hp7225	see hpgl	hp7225

hp7225v	see hpgl	hp7225v

hp7470	see hpgl	hp7470

hp7470v	see hpgl	hp7470v

hpgl	Hewlett-Packard HP-GL Plotters	hpgl

USAGE:

> **hpgl(width, height, ask, auto, color)**
> **hp7220h**
> **hp7470**
> **hp7225**
>
> **hpglv(width, height, ask, auto, color)**
> **hp7220v**
> **hp7470v**
> **hp7225v**

ARGUMENTS:

width: width of plotted surface in inches. Default 8 for **hp7220v**, **hp7225v**, and **hpglv**; 7.25 for **hp7470v**; 10 otherwise.

height: height of plotted surface in inches. Default 10 for **hp7220v**, **hp7225v**, **hp7470v** and **hpglv**; 7.25 for **hp7470**; 8 otherwise.

ask: logical, should user be prompted by "GO?" prior to advancing to new frame? Default TRUE.

auto: logical, can device automatically advance the paper (H-P 7220S)?

Default FALSE.

color: integer reflecting the degree of color-plotting support provided
by the device. 0=color changes ignored, 1=prompt user when
color changes (to allow manual pen changes, etc), 2=device has
automatic color changing capability (hp7220, hp7470).

These commands identify Hewlett-Packard plotters which accept
the HP-GL instruction set. This includes pen plotter models 7220,
7470, and 7225. The plotters may be used with standard 8.5 by 11
inch paper with the long dimension horizontally (**hp7220h**) or
vertically (**hp7220v**), or with other paper sizes by means of arguments **width** and **height**. The arguments listed after **hpgl** and
hpglv can also be given with the other device functions.

Whenever a new plot is about to be produced, the message "GO?"
appears on the terminal. At this time, new paper may be loaded,
pens changed, etc. When ready for plotting to proceed, simply
hit carriage return.

When graphic input (**identify, rdpen**) is requested, the **enter** button will light. The pen should be positioned by means of the 4
directional buttons. When the desired pen position is reached,
depress the **enter** button. To terminate the input, hit carriage return on the terminal. The coordinates of the terminating point
are not transmitted.

Character sizes may be changed to any desired value. Color can
be changed to any of the values 1 to 4, indicating the pens in
stables 1 to 4. Characters can be rotated to any orientation.
Different line styles (1, 2, ...) are available, with patterns becoming more spread-out as the line style is larger.

The **hpglv, hp7470v**, and **hp7225v** devices actually produce their
"vertical" plots by interchanging the roles of the plotter x and y
axes. This is done since the 7470 and 7225 plotters cannot accommodate standard sized paper in vertical orientation.

ALWAYS BE SURE that the pen is capped when you are done
with the plotter.

NOTE:

The switches on the back of the device should be set to reflect the correct speed and parity settings for the computer system. Incorrect parity settings may cause failures of digitizing functions, e.g., **rdpen** and **identify**.

A device must be specified before any graphics functions can be used.

---

identify	Identify Points on Plot	**identify**

USAGE:

> **identify(x, y, labels, n, plot, atpen, offset)**

ARGUMENTS:

x,y: co-ordinates of points that may be identified. A time-series or structure containing **x** and **y** may also be given in place of **x,y**. Missing values (NAs) are allowed.

labels: optional vector giving labels for each of the points. If supplied, must have the same length as **x** and **y**. Default is **seq(x)**.

n: maximum number of points to identify.

plot: flag, if TRUE (the default), **identify** plots the labels of the points identified. In any case, the subscripts are returned.

atpen: flag, if TRUE (the default) plotted identification is relative to pen position when point is identified; otherwise, plotting is relative to the identified **x,y** value. Useful for controlling position of labels when points are crowded.

offset: identification is plotted as a text string, moved **offset** character widths from the point (default .5). If the pen was left (right) of the nearest point, the label will be offset to the left (right) of the point.

Graphical parameters may also be supplied as arguments to this function (see **par**).

VALUE:

> indices (in **x** and **y**) corresponding to identified points.

See the documentation for the specific graphics device for details on graphical input techniques.

The nearest point to pen position is identified, but must be at most half-inch from pen. In case of ties, earliest point is identified.

EXAMPLES:

identify(x,y,encode(z))   # plot z values when x,y points identified

bad ← identify(x,y,plot=FALSE)
xgood ← x[-bad]   #eliminate identified "bad" points
ygood ← y[-bad]   #from x and y

?idplot	Scatter Plot with Identifiers for Each Point	**?idplot**

USAGE:

?idplot(x, y, xlab=, ylab=, cex=)

ARGUMENTS:

x: x-coordinates of data points.
y: y-coordinates of data points.
xlab=:   unquoted label for x axis, default is the expression used for **x**.
ylab=:   unquoted label for y axis, default is the expression used for **y**.
cex=:   character size to be used to plot the point identification numbers. Default is 1.

EFFECT:

Produces a scatter plot, but with the points on the plot labelled by 1,...,n.

EXAMPLES:

?idplot(sqrt(gnp),log(income))

---

if	see **syntax**	**if**

ifelse	Conditional Data Selection	ifelse

USAGE:

ifelse(cond, true, false)

ARGUMENTS:
cond: logical structure.
true: vector containing values to be returned if **cond** is TRUE.
false: vector containing values to be returned if **cond** is FALSE.

VALUE:

vector like **cond**. The result is made up element-by-element from the values from **true** or **false** depending on **cond**. If **true** or **false** are not as large as **cond**, they will be repeated cyclically. Missing values (NAs) are allowed. NA values in **cond** cause NA to be returned.

EXAMPLES:

ifelse(x>1,1,x) # gives the value 1 if x>1 else gives x
              # equivalent to pmin(1,x)

Evaluation of arguments is done before execution of ifelse. Divide-by-zero will already have occurred in **ifelse(x==0,NA,1/x)** by the time **ifelse** is executed. Avoid the zero-divide by using **1/ifelse(x==0,NA,x)**.

The **if** syntax construct of the S language provides conditional evaluation.

index	Compute Position in Array	index

USAGE:

index(cat1, cat2, ...)

ARGUMENTS:
cati: categories (as produced by **code** or **cut**). Missing values (NAs) are allowed.

VALUE:

structure with components **Data** and **Label**:

Data: vector of integers, giving the position in a contingency table corresponding to each value of the input categories.

Label: structure with one component for each input category; the component corresponding to **cat1** is named "cat1" and is a character vector containing the values from **cat1$Label**, etc.

The **index** function is typically used in conjunction with **tapply** in order to handle "ragged arrays". A ragged array is an array data structure in which the last dimension is of variable length. For example, suppose that there is a measurement of strength done on a number of individuals in different age groups. There may be 10 observations corresponding to the first age group, 12 corresponding to the second group, etc. This is a 2-dimensional ragged array.

Suppose there is a further categorical variable for each individual giving height. A 3-way ragged array could be constructed which has strength observations categorized by combined age/height.

EXAMPLES:

```
ht ← cut(height,c(5,5.5,6,6.5))  # height intervals in feet
ag ← cut(age,seq(0,100,10))   # age groups
sx ← code(sex,c(1,2),c("Male","Female"))
# we have three categories categorizing strength observations
position ← index(ht,ag,sx)
mn.strength ← tapply(strength,position,"mean")
no.obs ← tapply(strength,position,"len")
tprint(strength=strength,nobs=no.obs)   # print a pretty table

fit ← mn.strength[position]
resid ← strength - fit
```

Notice the use of the position vector (produced by **index**) to subscript the mn.strength table in order to give fitted values. This can be done because even the 3-way array mn.strength can be subscripted as if it were a vector.

The functions **index** and **tapply** can be used to perform multidimensional summaries similar to the one-dimensional grouping done by the **split** and **sapply** functions. Notice that **split** produces a data structure corresponding to a 2-dimensional ragged array, whereas **index** does not.

interp	Bivariate Interpolation for Irregular Data	**interp**

USAGE:

> **interp(x, y, z, xo, yo, ncp, extrap)**

ARGUMENTS:

x: x-coordinates of data points.

y: y-coordinates of data points.

z: z-coordinates of data points.

xo: vector of x-coordinates of output grid. Default, 40 points evenly spaced over the range of **x**.

yo: vector of y-coordinates of output grid. Default, 40 points evenly spaced over the range of **y**.

ncp: number of additional points to be used in computing partial derivatives at each data point. If **ncp** is zero, linear interpolation will be used in the triangles bounded by data points. Otherwise, **ncp** must be 2 or greater, but smaller than the number of data points. (Cubic interpolation is done if partial derivatives used. Default 0.

extrap: logical flag, should extrapolation be used outside of the convex hull determined by the data points? Default FALSE. No extrapolation can be performed if **ncp** is zero.

VALUE:

> structure with 3 components:

x: vector of x-coordinates of output grid, the same as input argument **xo**.

y: vector of y-coordinates of output grid, the same as input argument **yo**.

z: matrix of fitted z-values. The value $z[i,j]$ is computed at the x,y point $x[i], y[j]$.

> If **extrap** is FALSE, z-values for points outside the convex hull are returned as NA. The resulting structure is suitable for input to the function **contour**.

REFERENCE:

Hiroshi Akima, "A Method of Bivariate Interpolation and Smooth Surface Fitting for Irregularly Distributed Data Points", *ACM Transactions on Mathematical Software,* Vol 4, No 2, June 1978, pp 148-164.

EXAMPLES:

fit ← interp(x,y,z)   #fit to irregularly spaced data
contour(fit)   #contour plot

?intersect	Intersection of Two Lists	**?intersect**

USAGE:

?intersect(x, y)

ARGUMENTS:
x: vector containing one list.
y: list.

VALUE:
vector of (unique) values which appear in both lists.

---

knapsack	see **napsack**	**knapsack**

?kronecker	Form Kronecker Product of Matrices	**?kronecker**

USAGE:

?kronecker(a, b, fun)

ARGUMENTS:
a,b: matrices. The kronecker product is the matrix with a[1,1]*b in the upper-left block, a[2,1]*b in the block just below this, and so on. (Presumably you have some use for this matrix, since you are reading the documentation.) The kronecker product is similar to the result of **outer**, but permuted and turned into a matrix.

fun: optionally, the operator applied to elements of a and b, rather than the default multiplication. Thus **?kronecker(a,b,+)** forms kronecker sums. Default is *****.

VALUE:

a matrix as described, with dimension **nrow(a)*nrow(b)** by **ncol(a)*ncol(b)**.

---

l1fit	Minimum Absolute Residual (L1) regression	**l1fit**

USAGE:

l1fit(x, y, int)

ARGUMENTS:

x: X matrix for fitting Y=Xb+e with variables in columns, observations across rows. Should not contain column of 1's (see argument **int**). Number of rows of **x** should equal the number of data values in **y**. There should be fewer columns than rows.

y: numeric vector with as many observations as the number of rows of **x**.

int: flag for intercept; if TRUE (default) an intercept term is included in regression model.

VALUE:

structure defining the regression (compare function **regress**).

coef: vector of coefficients with constant term as first value (optional depending on argument **int**)

resid: residuals from the fit, i.e. **resid** is Y−Xb.

REFERENCE:

Barrodale and Roberts, "Solution of an Overdetermined System of Equations in the L1 Norm". *CACM*, June 1974, pp 319-320.

---

labclust	Label a Cluster Plot	**labclust**

USAGE:

> labclust(x, y, labels)

ARGUMENTS:

x,y: coordinates of the leaves of the tree, as produced by **plclust**. Missing values (NAs) are allowed.

labels: optional control of labels on the cluster leaves. By default, leaves are labelled with object number. If **labels** is a character vector, it is used to label the leaves. Otherwise it is interpreted as a logical flag (in particular, the value FALSE suppresses any labelling of leaves, as is appropriate with large trees).

> Graphical parameters may also be supplied as arguments to this function (see **par**).

EXAMPLES:

> xy←plclust(tree,plot=FALSE)   # save coords of tree
> plclust(tree,label=FALSE)   #plot it
> lablcust(xy,label=names)   #now label it

---

lag	Create a Lagged Time-Series	**lag**

USAGE:

> lag(x, k)

ARGUMENTS:

x: a time-series. Missing values (NAs) are allowed.

k: the number of positions the new series is to lag the input series (default=1); negative value will lead the series.

VALUE:

> a time-series of the same length as **x** but lagged by **k** positions. Only the start and end dates are changed; the series still has the same number of observations.

> See also **tsmatrix** for aligning the time domains of several series.

EXAMPLES:

l12gnp←lag(gnp,12) # gnp lagged by 12 months

| **leaps** | All-subset Regressions by Leaps and Bounds | **leaps** |

USAGE:

leaps(x, y, wt, int, method, nbest, names, df)

ARGUMENTS:

x: matrix of independent variables. Each column of **x** is a variable, each row an observation. There should be a maximum of 31 columns and fewer columns than rows.

y: vector of dependent variable with the same number of observations as the number of rows of **x**.

wt: optional vector of weights for the observations.

int: logical flag, should an intercept term be used in the regressions? Default TRUE.

method: character string describing the method used to evaluate a subset. Possible values are "Cp", "r2", and "adjr2" corresponding to Mallows Cp statistic, r-square, and adjusted r-square. Only the first character need be supplied. Default, "Cp".

nbest: integer describing the number of "best" subsets to be found for each subset size. In the case of r2 or Cp methods, the **nbest** subsets (of any size) are guaranteed to be included in the output (but note that more subsets will also be included). Default 10.

names: optional character vector giving names for the independent variables. Default, the names are 1, 2, ... 9, A, B, ...

df: degrees of freedom for **y**; default is **nrow(x)**. Useful if, for example, **x** and **y** have already been adjusted for previous independent variables.

VALUE:

structure with four components.

Cp: the first returned component will be named "Cp", "adjr2", or "r2" depending on the method used for evaluating the subsets. This component gives the values of the desired statistic.

size: the number of independent variables (including the constant term if **int** is TRUE) in each subset.

label: a character vector, each element giving the names of the variables in the subset.

which: logical matrix with as many rows as there are returned subsets.
               Each row is a logical vector that can be used to select the columns
               of **x** in the subset.

REFERENCE:

George M. Furnival and Robert W. Wilson, Jr., "Regressions by
Leaps and Bounds", *Technometrics*, Vol 16, No 4, November 1974,
pp 499-511.

EXAMPLES:

```
r←leaps(x,y)
plot(r$size,r$Cp,type="n")
text(r$size,r$Cp,r$label)      # produces Cp plot
regress( x[,r$which[3,]], y )    #regression corresponding
                               # to third subset
```

---

legend	Put a Legend on a Plot	**legend**

USAGE:

legend(x, y, legend, angle, density, col, lty, marks, pch)

The function **legend** draws a box at specified coordinates and puts
inside examples of lines, points, marks, and/or shading, each
identified with a user-specified text string.

ARGUMENTS:

x,y: coordinates of two opposite corners of the rectangular area of the
               plot which is to contain the legend. A structure containing x and
               y values may be supplied.

legend: vector of text strings to be associated with shading patterns, line
               types, plotting characters or marks.

angle: optional vector giving the angle (degrees, counter-clockwise from
               horizontal) for shading each bar division. Defaults to 45 if **densi-
               ty** is supplied.

density: optional vector for bar shading, giving the number of lines per
               inch for shading each bar division. Defaults to 3 if **angle** is sup-
               plied.

col: optional vector giving the colors in which the bars should be
               filled or shaded. If **col** is specified and neither **angle** nor **density**

are given as arguments, bars will be filled solidly with the colors.

lty: vector of line types.

marks: vector of mark numbers (see documentation for function **points**).

pch: character string of plotting characters. Single characters from **pch** will be used.

Graphical parameters may also be supplied as arguments to this function (see **par**).

SEE ALSO:

barplot which contains a sample legend.

EXAMPLES:

```
# use rdpen to point at lower-left and upper-right corners
# of area to contain the legend -- draw colored boxes
legend(rdpen(2),legend=c("IBM","AT&T","GM"),col=2:4)

# draw legend with different line styles and plotting chars
legend(c(0,5),c(5,10),names,lty=1:5,pch="O+*")
```

---

len	Length of a vector	**len**

USAGE:

len(x)

ARGUMENTS:

x: vector structure. Missing values (NAs) are allowed.

VALUE:

the number of data values in **x**.

---

lgamma	see **gamma**	**lgamma**

lines	Add Lines or Points to Current Plot	**lines**

USAGE:

> **lines(x, y)**
> **points(x, y, mark=)**

ARGUMENTS:

x,y: the co-ordinates for lines or points. A time-series or structure containing **x** and **y** may also be given for **x**. Missing values (NAs) are allowed.

mark=:   mark number for plotting special symbols at the points. Basic marks are: square (0); octagon (1); triangle(2); cross(3); X (4); diamond(5) and inverted triangle(6). To get superimposed versions of the above use the following arithmetic(!): 7==0+4; 8==3+4; 9==3+5; 10==1+3; 11==2+6; 12==0+3; 13==1+4; 14==0+2.

> Graphical parameters may also be supplied as arguments to this function (see **par**).

> Data values with an NA in either **x** or **y** are not plotted by **points**. Also, **lines** does not draw to or from any such point, thus giving a break in the line segments.

> If a log scale was specified for either the x- or y-axis, **points** and **lines** will plot the corresponding values on a log scale.

SEE ALSO:

> **segments, arrows,** and **symbols**.

EXAMPLES:

```
par(usr=c(-1,15,0,1)) # produce a plot of all the marks
for(i in 0:14){
    points(i,.5,mark=i)
    text(i,.35,i)}
text(7,.75,'Samples of "mark=" Parameter')
```

Samples of "mark=" Parameter

□ ○ △ + × ◇ ▽ ⊠ ✳ ✦ ⊕ ⊠ ⊞ ⊠ ◪

0 1 2 3 4 5 6 7 8 9 10 11 12 13 14

---

list	see **ls**	list

?listall	List Contents of all Current Databases	**?listall**

USAGE:

    ?listall(pattern)

ARGUMENTS:

pattern: optional character string giving a pattern of dataset names to be listed.

EFFECT:

    Prints each database name, followed by a listing of its contents. If a prefix is in effect, only datasets within the prefix are listed.

EXAMPLES:
>   ?listall
>   ?listall("mac.*")   #list macros on all databases

?listfun	List Names of all S Functions	?listfun

USAGE:
>   ?listfun()

VALUE:

character vector containing names of all S functions accessible on any of the current chapters. Functions which are part of the S executive (normally arithmetic, subscripting, etc.) are not listed.

EXAMPLES:
>   ?listfun   # get list of all functions

---

lnorm	Log-normal Probability Distribution	lnorm

USAGE:
>   plnorm(q, par1, par2)
>   qlnorm(p, par1, par2)
>   rlnorm(n, par1, par2)

ARGUMENTS:

q: vector of quantiles. Missing values (NAs) are allowed.
p: vector of probabilities. Missing values (NAs) are allowed.
n: sample size.
par1,par2: vectors of means and standard deviations of the distribution of the log of the random variable. Thus, exp(par1) is a scale parameter and par2 a shape parameter for the lognormal distribution. Default par1=0,par2=1.

VALUE:

probability (quantile) vector corresponding to the given quantile (probability) vector, or random sample (**rlnorm**).

log	see **exp**	**log**

log10	see **exp**	**log10**

logic	Logical Operations; And, Or, Not	**logic**

USAGE:

> *expr1* **&** *expr2*
> *expr1* **|** *expr2*
> **!** *expr1*

ARGUMENTS:
expri: logical structures. Missing values (NAs) are allowed.

VALUE:

> logical result of and-ing, or-ing or negation. In case of the first two the result is as long as the longer of the operands. When time-series operands are used, the time domain of the value is the intersection of the time domains of the operands.

SEE ALSO:

> **xor**, exclusive or.

EXAMPLES:

> **x[a>13 & b<2]**   #elements of x corresp. to a>13 and b<2

| **logistic** | Logistic Probability Distribution | **logistic** |

USAGE:

> plogis(q, par1, par2)
> qlogis(p, par1, par2)
> rlogis(n, par1, par2)

ARGUMENTS:

q: vector of quantiles. Missing values (NAs) are allowed.
p: vector of probabilities. Missing values (NAs) are allowed.
n: sample size.
par1: vector of location parameters. Default is 0.
par2: vector of scale parameters. Default is 1.

VALUE:

> probability (quantile) vector corresponding to the given quantile (probability) vector, or random sample (**rlogis**).

| **loglin** | Contingency Table Analysis | **loglin** |

USAGE:

> **loglin(table, margin, start, eps, iter, print)**

ARGUMENTS:

table: contingency table (array) to be fit by log-linear model. Table values must be non-negative.

margins: vector describing the marginal totals to be fit. A margin is described by the factors not summed over, and margins are separated by zeroes. Thus c(1,2,0,3,4) would indicate fitting the 1,2 margin (summing over variables 3 and 4) and the 3,4 margin in a four-way table.

start: starting estimate for fitted table. If **start** is omitted, a start is used that will assure convergence. If structural zeroes appear in **table**, **start** should contain zeroes in corresponding entries, ones in other places. This assures that the fit will contain those zeroes.

eps: maximum permissible deviation between observed and fitted marginal totals. Default 0.1.

iter: maximum number of iterations, default 20.

print: flag; if TRUE (the default), the final deviation and number of iterations will be printed.

VALUE:

structure like **table**, but containing fitted values.

REFERENCE:

S. J. Haberman, "Log-linear Fit for Contingency Tables - Algorithm AS51", *Applied Statistics*, Vol 21, No 2, pp 218-225, 1972.

---

logo	Draw the S Logo	**logo**

USAGE:

logo(x, y, size)

ARGUMENTS:

x,y: optional co-ordinates for the center of the logo. By default, plotted in lower right of figure.

size: optional size of logo relative to the height of a character. Default 2.5.

Graphical parameters may also be supplied as arguments to this function (see **par**).

EXAMPLES:

logo   # logo in lower right hand corner of figure

---

lowess	Scatter Plot Smoothing	**lowess**

USAGE:

lowess(x, y, f, iter, delta)

ARGUMENTS:

x,y: vectors of data for scatter plot.

f: fraction of data used for smoothing at each **x** point. The larger the **f** value, the smoother the fit. Default 2/3.

iter: number of iterations used in computing robust estimates. Default 3.

delta: interval size (in units corresponding to **x**). If **lowess** estimates at two **x** values within **delta** of one another, it fits any points between them by linear interpolation. Default 1% of the range of **x**. If **delta=0** all but identical **x** values are estimated independently.

VALUE:

plot structure containing components named **x** and **y** which are the x,y points of the smoothed scatter plot. Note that **x** is a sorted version of the input **x** vector, with duplicate points removed.

This function may be slow for large numbers of points; execution time is proportional to (**iter*f*n^2**). Increasing **delta** should speed things up, as will decreasing **f**.

REFERENCE:

W. S. Cleveland, "Robust Locally Weighted Regression and Smoothing Scatterplots", *JASA, Vol 74, No 368, pp 829-836.*, December 1979.

EXAMPLES:

plot(x,y)
lines(lowess(x,y))   #scatter plot with smooth

fit ← lowess(x,y)
resid ← y-approx(fit,x)$y #residual from smooth

---

ls	List of Datasets in Data Directory	**ls**

USAGE:

ls(pattern, pos)
list(pattern, pos)

ARGUMENTS:

pattern: an optional character string describing the dataset names of interest. For example, **list("abc")** or **ls("abc")** would only return the name of the dataset **abc** (if it existed). The character "*" at the end of the pattern matches any number of characters following. **list("abc*")** would return the names of any datasets whose names began with **abc**. The default pattern is "*", which matches all dataset names.

pos: the position on the data directory search list of the data directory to be searched. Position 1 (default) is the working directory. Normally, positions 2 and 3 are the save and shared data directories.

VALUE:

a character vector which is the alphabetical list of the datasets (matching pattern) on the specified data directory.

EXAMPLES:

list   #list all working dataset names
list("lottery*",pos=3)   #shared db names beginning lottery

If a prefix is in effect, all matching takes place within datasets whose name begins with the prefix. This can be changed by using an initial "$" in the pattern, thus matching "fully qualified" dataset names. The pattern "$*" matches all dataset names, regardless of current prefix, while "*" matches all names within the prefix.

BUGS:

Unlike general regular expressions, for example in the UNIX shell, the pattern matching capability of **ls** is limited: the "*" character may only be used at the end of the pattern.

---

macros	The S Macro Facility	**macros**

Macros can be used to reduce the amount of typing to perform repetitive tasks within S. A macro is simply a set of commands which will be executed as if they were typed from the terminal. The power of macros lies in the ability to give them arguments, so that similar tasks may be carried out by a single macro.

A macro is often used as a repository for a set of commands that is long or complicated, hence hard to type interactively. Because it involves a certain amount of overhead, the macro processor is ordinarily not used when input is typed to S. The special character "?" before a name signals a macro and causes the current line to be sent through the macro processor before being executed.

Macros are defined by means of the **define** function. The **medit** function can be used to edit macros and the **mprint** function to print them.

---

mail	Suggestions, Questions and Reports	**mail**

USAGE:

> **mail**
> **mail(file)**

ARGUMENTS:

file: optional character string giving the name of a file containing a letter. Without an argument, **mail** will prompt (with the characters "M>") for a multi-line letter to be sent to the local S support staff. The letter is terminated by a line consisting only of a period.

The **mail** functions sends UNIX mail to a system dependent userid which is supposed to be read by a local S expert.

?mat2vecs	Change Columns of Matrix to Individual Vectors	**?mat2vecs**

USAGE:

> **?mat2vecs(x, names)**

ARGUMENTS:

x: matrix of data. Missing values (NAs) are allowed.

names: character vector of names for the columns of **x**. Should not contain embedded blanks, special characters, etc.

VALUE:

This function creates on the work file datasets named by **names**, one for each column of **x**.

EXAMPLES:

?mat2vecs(data,c("age","height","weight")) # creates 3 vectors
named age, height, and weight

---

match	Match Items in Vector	**match**

USAGE:

match(x, list)

ARGUMENTS:

x: vector of items that are to be looked for in **list**.
list: the possible values in **x**.

VALUE:

vector like **x** giving the index in **list** of the item that matches
each element of **x**. If an element of **x** is not in **list**, NA is returned.

EXAMPLES:

match(data,primes)
state.abb[match(names,state.name)] #change names to abbrevs

?matedit	Text Editing of Matrices	**?matedit**

USAGE:

?matedit(x)

ARGUMENTS:

x: matrix.

The **?matedit** macro writes the matrix **x** to a file and then invokes
a text editor on that file. After modifying the file with the editor,
write the modified file and then quit the editor. The modified

file will be read back in as a matrix and will be assigned the name **x** on the working database.

The number of columns of the matrix cannot be changed during editing.

The text editor is **ed** by default, but can be specified by means of the shell variable EDITOR. To use the **vi** screen editor, execute the following shell commands

USAGE:

EDITOR=vi
export EDITOR
prior to invoking S.

EXAMPLES:

?matedit(mymatrix)   # edit dataset named mymatrix
1,$s/999/NA/
w
q
mymatrix changed

---

matlines	see **matplot**	**matlines**

---

matrix-multiply	Matrix Multiplication Operator	**matrix-multiply**

USAGE:

mat1 %* mat2

ARGUMENTS:
mati: matrix or vector.

VALUE:

matrix product of **mat1** and **mat2**.

Vector results are returned as vectors, not as (n by 1) or (1 by n) matrices.

The last extent of **mat1** must be the same size as the first extent of **mat2**. Vectors are not oriented, therefore a vector of length n can multiply an n by n matrix on the left or right.

---

matplot	Plot Columns of Matrices	**matplot**

USAGE:

matplot(x, y, type=, lty=, pch=, col=)

ARGUMENTS:

x,y: vectors or matrices of data for plotting. The first column of **x** is plotted against the first column of **y**, the second column of **x** against the second column of **y**, etc. If one matrix has fewer columns, plotting will cycle back through the columns again. (In particular, either **x** or **y** may be a vector, against which all columns of the other argument will be plotted.) Missing values (NAs) are allowed.

type=:   an optional character string, telling which type of plot (points, lines, both, none or high-density) should be done for each plot. The first character of **type** defines the first plot, the second character the second, etc. Elements of **type** are cycled through; e.g., "pl" alternately plots points and lines. Default is "p" (points) for **matplot** and **matpoints** and "l" for **matlines**.

lty=:   optional vector of line types. The first element is the hardware line type for the first line, etc. Line types will be used cyclically until all plots are drawn. Default is 1,2,3,4,5.

pch=:   optional character vector for plotting-characters (only the first element is used). The first character is the plotting-character for the first plot, the second for the second, etc. Default is the digits (1 through 9, 0) then the letters.

col=:   optional vector of colors. Default is 1,2,3,4. Colors are also cycled if necessary.

Graphical parameters may also be supplied as arguments to this function (see **par**).

matplot generates a new plot; **matpoints** and **matlines** add to the current plot.

EXAMPLES:

matplot(x,y,type="pl") #points for 1st col, lines for 2nd
matpoints(x,y,pch="*") # points with "*" for all plots
the example plot is produced by:
matplot(iris[,1,],iris[,2,],xlab="Petal Length",
ylab="Petal Width",
sub="1=Setosa, 2=Virginica, 3=Versicolor",
main="Fisher's Iris Data")

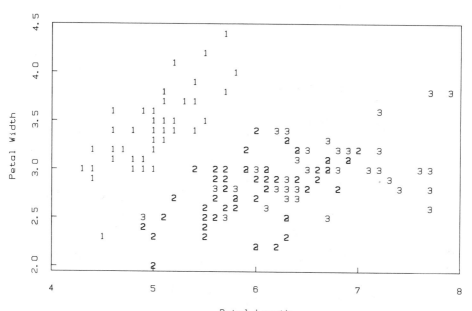

Fisher's Iris Data

| matpoints | see **matplot** | matpoints |

matrix	Create Matrix	**matrix**

USAGE:

> **matrix(data, nrow, ncol, byrow)**

ARGUMENTS:

data: vector containing the data values for the matrix in normal array order: the first subscript varies most rapidly. Missing values (NAs) are allowed.

nrow: first subscript, number of rows.

ncol: second subscript, number of columns

byrow: flag, if TRUE the data values are assumed to be the first row, then the second row, etc. If FALSE (the default) the values are assumed to be the first column, then the second column, etc. (this is how the data is stored internally). Should be TRUE if the data values were read from a file, arranged by rows.

> If one of **nrow** or **ncol** is omitted, the dimensions of the matrix are determined by the other argument and the length of **data**.

VALUE:

> an **nrow** by **ncol** matrix with the same mode as **data**. The matrix is filled column-wise with the data vector given. If the vector does not fill the matrix, it is repeated until the matrix is filled (giving a warning if it does not take an integral number of repetitions of **data**).

EXAMPLES:

> **m ← matrix(0,4,5)** #a 4 by 5 matrix of zeros
>
> **mm ← matrix(read("mfile"), ncol=5, byrow=TRUE)**
> #read all rows from the file

---

max	Extremes	**max**

USAGE:

> **max(arg1, arg2, ...)**
> **min(arg1, arg2, ...)**

ARGUMENTS:

argi: numeric structure. Missing values (NAs) are allowed.

VALUE:

> the single maximum or minimum value found in any of the args.

SEE ALSO:

> **pmax** and **pmin** for generating a vector of parallel extremes, and **range** for computing both **max** and **min**.

?mdsplot	Plot the Results of Multidimensional Scaling	**?mdsplot**

USAGE:

> **?mdsplot(mds)**

ARGUMENTS:

mds: matrix representing data points, such as that produced by the function **cmdscale**.

EFFECT:

> Produces a plot where the x and y axes have the same coordinates per inch, thus distances are represented correctly. Each point is identified by its sequence number.

EXAMPLES:

> **?mdsplot(m ← cmdscale(distmat))**

---

mean	Mean Value	**mean**

USAGE:

        mean(x, trim)

ARGUMENTS:

x: numeric structure (missing values, NAs, are ignored).

trim: fraction of values to be trimmed off each end of the ordered data. Default value 0.

VALUE:

        (Trimmed) mean of x.

EXAMPLES:

        mean(scores,trim=.25)   # trims outer 50% of data

---

median	Median	**median**

USAGE:

        median(x)

ARGUMENTS:

x: numeric structure. Missing values (NAs) are allowed.

VALUE:

        median of data (excluding the NAs).

---

medit   Edit a Macro   **medit**

USAGE:

medit(defin, pos)

ARGUMENTS:

defin: dataset containing the macro to be edited. Macros are datasets with names beginning "mac.". If **defin** is omitted, the most recent previously edited text is edited (this allows recovering if the edited macro is rejected by the macro-definition process).

pos: optional, data directory position on which edited macro is to be saved; default 2.

medit invokes a text editor (default is **ed**) on a copy of the macro text. Use the editor to make whatever changes are desired in the macro. Then write the macro, and exit from the editor (**w** and **q** in **ed**). The macro will automatically be re-processed through the **define** function, and stored on the specified data directory.

The user can specify which editor should be invoked by setting a UNIX shell variable prior to invoking S. For example, the shell commands **EDITOR=vi; export EDITOR** will cause **medit** to use the **vi** screen editor rather than **ed**.

EXAMPLES:

medit(mac.abc) # edits macro **abc** and replaces it

?medit   Edit a Macro   **?medit**

USAGE:

?medit(name)

ARGUMENTS:

name: unquoted name of macro.

EFFECT:

Prepends **$mac.** to **name** and invokes the **medit** function. This makes sure that macro names are correct even when a prefix is in effect.

EXAMPLES:

> **?medit(probsamp)** # edit the probsamp macro

menu	Menu Interaction Function	**menu**

USAGE:

> **menu(items, actions, how, prompt, force, return)**

ARGUMENTS:

items: character vector giving the list of choices on the menu.

actions: optional character vector of the same length as **items** giving expressions to be executed if the corresponding item is chosen.

how: integer describing the method of menu use: 1, the default, is to display a printed menu with associated numbers. The user types the appropriate number to make a selection. 2 through 4 specifies graphical presentation of the menu followed by user choice using a graphical input device. Choice 2 draws the menu along the rightmost side of the plot, 3 allows selection from a previously displayed menu, and 4 displays the menu and redisplays it with graphical parameter **col=0** after the user interaction. On some devices this will overwrite the menu with the background color, thus erasing it.

prompt: Character string to be printed prompting the user to make a choice. Default "Which one?".

force: logical, should user be forced to make a choice of one of the menu items. Default TRUE.

return: logical, after completing the actions associated with **menu**, should execution return to the context it was in prior to executing **menu**. Default TRUE. Using **return=FALSE** is recommended if **menu** is the last action of a macro or source file, since it prevents a (growing) stack of source files, which may eventually use up all available file descriptors.

Graphical parameters may also be supplied as arguments to this function (see **par**).

VALUE:

If **action** is not supplied by the user, **menu** returns the integer corresponding to the item the user selected. If **force** is FALSE and the user makes an invalid selection, 0 is returned.

EXAMPLES:

```
items ← c("List Datasets in Work Directory",
  "List Datasets in Save Directory",
  "quit")
actions ← c("ls( )","ls(pos=2)","q")
menu(items,actions)

menu(ls( ),encode("rm",ls( )),
  prompt="Which dataset should be removed?")
```

message	see **fatal**	message

min	see **max**	min

?missing	Find Observations Containing NAs	**?missing**

USAGE:

?missing(x, y, ...)

ARGUMENTS:

x: matrix, typically the x matrix for a regression. Missing values (NAs) are allowed.

y: vector with **nrow(x)** values. Missing values (NAs) are allowed.

VALUE:

logical vector with **nrow(x)** values, where TRUE indicates a missing value in the corresponding row of x or value of y.

The macro actually accepts any number of matrices or vectors as arguments, as long as the number of rows of all matrices are equal and are also equal to the lengths of all vectors.

EXAMPLES:

> i ← ?missing(x,y)
> regress(x[!i], y[!i])   # vector x
> regress(x[!i,], y[!i])   # matrix x

?mixplot	Barycentric Plot of Mixture	**?mixplot**

USAGE:

> **?mixplot(x, y, z, xlab, ylab, zlab, largest)**

ARGUMENTS:

x,y,z: vectors giving the amounts of the three components of each mixture.

xlab: label for x-component, default is the name of **x**.

ylab: label for y-component, default is name of **y**.

zlab: label for z-component, default is name of **z**.

largest: radius in inches of the largest circle to be plotted. Circle area is proportional to the sum of the component amounts, x+y+z. Default is 1.

EXAMPLES:

> # given vectors named **police**, **fire** and **education**
> # which give dollars spent on these services by different cities
> **?mixplot(police,fire,education)**
> **title("Area of circle proportional to total budget")**

?mlist	List Names of Macros	**?mlist**

USAGE:

> **?mlist(pos)**

ARGUMENTS:

pos: position on database search list of database containing macros. Default is 2.

VALUE:

character vector of macro names.

EXAMPLES:

> ?mlist   # names of all macros on user's database
> ?mlist(pos=3)   # names of macros on system database

---

mode	Mode (Data Type) of the Values in a Vector	**mode**

USAGE:

> mode(x)

ARGUMENTS:

x: vector structure. Missing values (NAs) are allowed.

VALUE:

the mode of x. Returns 1 for logical data, 2 for integer, 3 for real, 5 for character.

NOTE:

this function has nothing to do with the statistical concept of the mode of a distribution.

---

monthplot	Seasonal Subseries Plot	**monthplot**

USAGE:

> monthplot(y, labels, prediction)

ARGUMENTS:

y: a time-series, typically the seasonal component of a seasonal adjustment procedure such as **sabl**. The series y is broken into **nper(y)** subseries, e.g., the January subseries, the February subseries, ...

labels: optional vector of labels for each subseries. Default values are: (1) month names if **nper(y)** is 12; (2) quarter number, if **nper(y)** is 4; (3) digits from 1 to **nper(y)**, for all other cases.

prediction: number of years at the end of the series which are predictions. These are drawn as dotted lines (graphical parameter **lty=2**). Default 1.

Graphical parameters may also be supplied as arguments to this function (see **par**).

EFFECT:

For each subseries, the plot shows variation about the subseries midmean (25% trimmed mean).

REFERENCE:

William S. Cleveland and Irma J. Terpenning, "Graphical Methods for Seasonal Adjustment", *Journal of the American Statistical Association*, Vol. 77, No. 377, pp. 52-62, 1982.

EXAMPLES:

fit ← sabl(ship)
monthplot(fit$seasonal,
    main="Seasonal Component of Manufacturing Shipments")

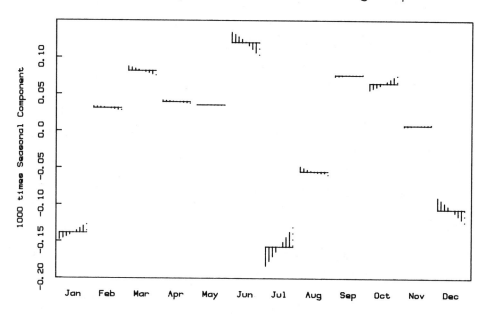

Seasonal Component of Manufacturing Shipments

---

mprint	Print Macros	**mprint**

USAGE:

> **mprint(mac1, mac2, ..., list=, file=, pos=)**

ARGUMENTS:

mac1: macro to be printed. Note that macros all have names beginning "mac.".

list=: character vector giving the names of macros to be printed.

file=: character string name of file to which the macro printing is directed. Default is the user's terminal.

pos=: position on data directory search list where datasets given in **list** will be found. Default 0, search for each dataset on all attached data directories.

The arguments **mac1, ...** and **list=** are mutually exclusive.

The printed macros have macro substitution characters **$1**, etc. in the body of the macro instead of the named arguments.

EXAMPLES:

mprint(mac.abc) # print macro **abc** on terminal
mprint(list=list("mac.*",pos=2),file="abcd",pos=2)
# print all macros in save directory onto file **abcd**

| **mprompt** | Produce Shell of Documentation for Macros | **mprompt** |

USAGE:

mprompt(macro, file)

ARGUMENTS:
macro: dataset containing the macro to be documented, i.e., mac.abc.
file: optional name of output documentation file, default is **abc.d**, where **abc** is the name of the macro.

This function is ordinarily used via the system macro "?prompt".

EXAMPLES:

mprompt(mac.def)   #document macro def, produces file def.d

| **mstr** | Modify Structure | **mstr** |

USAGE:

mstr(str, arg2, arg3, ...)

ARGUMENTS:
str: arbitrary structure to be modified.
argi: arbitrary name=value argument.

VALUE:

the structure **str** modified by the named arguments. If a name matches a component name of **str**, that entry in **str** is changed to the value of the argument. An argument with a null value will delete the entry. If a name does not match any entries in **str**, the entry will be added to **str** in the result. Missing values (NAs) are allowed.

| **mstree** | Minimal Spanning Tree and Multivariate Planing | **mstree** |

USAGE:

    mstree(x, plane)

ARGUMENTS:

x: matrix of data where rows correspond to observations, columns to variables. Should be scaled so that values on all variables are roughly comparable (the algorithm computes Euclidean distances from one observation to another).

plane: logical, should multivariate planing and lining information be returned? If TRUE (the default), all components listed below are returned; if FALSE, only the minimum spanning tree **mst** is returned.

VALUE:

     structure with components **x**, **y**, **mst**, and **order**, describing the planing, minimal spanning tree, and two versions of lining. If **plane** is FALSE, only the **mst** vector is returned.

x,y: coordinates of the observations computed by the Friedman-Rafsky algorithm.

mst: vector of length **nrow(x)−1** describing the edges in the minimal spanning tree. The ith value in this vector is an observation number, indicating that this observation and the ith observation should be linked in the minimal spanning tree.

order: matrix, **nrow(x)** by 2, giving two types of ordering: The first column presents the standard ordering from one extreme of the MST to the other. The second column presents the radial ordering, based on distance from the center of the MST.

REFERENCE:

     J. H. Friedman and L. C. Rafsky, "Graphics for the Multivariate Two-Sample Problem", *Stanford Linear Accelerator Corp.*, SLAC PUB-2193, 1978.

EXAMPLES:

```
plot(x,y)   # plot original data
   # compute minimal spanning tree
mst ← mstree(cbind(x,y),plot=F)
   # show tree on plot
segments(x[seq(mst)],y[seq(mst)],x[mst],y[mst])

i ← rbind( iris[,,1],iris[,,2],iris[,,3])
tree ← mstree(i)   # multivariate planing
plot(tree,type="n")   # plot data in plane
text(tree,seq(nrow(i)))   # identify points
```

---

mtext	Text in the Margins of a Plot	**mtext**

USAGE:

mtext(text, side, line, outer, at)

ARGUMENTS:

text: character string to be plotted.

side: side (1,2,3,4 for bottom, left, top, or right).

line: line (measured out from the plot in units of standard-sized character heights). By default (unless parameter **mar** is changed by the **par** function), standard-sized characters can be placed at values of **line** less than or equal to 4 on the bottom, 3 on left and top, and 1 on the right.

outer: logical flag, should plotting be done in outer margin? Default FALSE.

at: optional vector of positions at which **text** is to be plotted. If **side** is 1 or 3, **at** will represent x-coordinates. If **side** is 2 or 4, **at** will represent y-coordinates. **at** can be used for constructing specialized axis labels, etc.

mulbar	Multiple Bar Plot	mulbar

USAGE:

mulbar(width, height, rowlab, collab, gap)

ARGUMENTS:

width: matrix of bar widths, all values must be positive.

height: matrix of bar heights, values may be negative. **width** and **height** must have the same number of rows and columns.

rowlab: optional character vector for row labels.

collab: optional character vector for column labels.

gap: fraction of plot used for gap between rows and columns, default .1

Graphical parameters may also be supplied as arguments to this function (see **par**).

EXAMPLES:

```
counts ← telsam.response[1:5,]
fit ← loglin(counts,c(1,0,2))  # fit independence model
resid ← counts - fit
par(mar=c(7,4.1,4.1,2))
mulbar(
    sqrt(fit),
    resid/sqrt(fit),
    collab=telsam.collab,
    rowlab=encode(telsam.rowlab[1:5]),
    ylab="Interviewer",
    main="Chi-Plot for Fit to Interviewer Data"
    )
mtext(side=1,line=3,
  "Height proportional to Signed Contribution to Chi Statistic")
mtext(side=1,line=4,"Width proportional to Root-Fitted Value")
mtext(side=1,line=5,"Area proportional to Fitted Value")
```

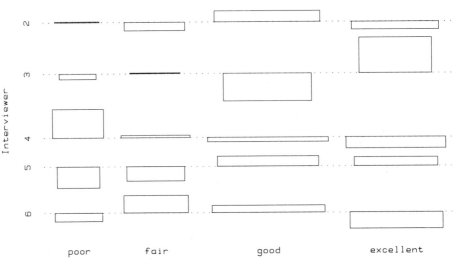

Chi-Plot for Fit to Interviewer Data

Height proportional to Signed Contribution to Chi Statistic
Width proportional to Root-Fitted Value
Area proportional to Fitted Value

na	Pick Out Missing Values	**na**

USAGE:

> **na(x)**
> **NA(x)**

ARGUMENTS:

x: vector structure.

VALUE:

> logical structure like **x**, with TRUE wherever a missing value
> (NA) appeared in **x**, and FALSE elsewhere.

NOTE:

> Whenever an NA appears in either operand of any logical opera-
> tion, the corresponding value is NA. In particular, the expression
> **x==NA** will always produce a vector of all NAs; the **na** function
> must be used to test for missing values.

NA	see **na**	NA

napsack	Solve Knapsack Problems	**napsack**

USAGE:

napsack(x, target, best)

ARGUMENTS:

x: vector of generators for knapsack problem. The function attempts
to find subsets of **x** whose sums are equal (or close to) **target**.

target: scalar target value.

best: the desired number of solutions (or approximate solutions) to the
problem. Default 10.

VALUE:

matrix of logical values, of size **len(x)** by **best**. Each column tells
which elements of **x** are contained in one of the **best** subsets.
Thus, a column of the result can be used to subscript **x** to obtain a
subset.

The knapsack problem is NP-complete, and the algorithm is exponential in time and space complexity. If **n** is the length of **x**,
then the algorithm requires time $O(\ 2^{(n/2)}\ )$ and space $O(\ 2^{(n/4)}\ )$. Problems with **n** < 30 can be readily solved by this
function. Remember that both time and space requirements in-
crease very rapidly with problem size!

The solutions produced may not include all subsets that can generate solutions with a certain error. It is guaranteed to produce
an exact solution if one is possible, but may not find all of a
number of exact solutions.

REFERENCE:

Richard Schroeppel and Adi Shamir, "A $T*S^2 = O(2^n)$
Time/Space Tradeoff for Certain NP-Complete Problems", *Twen-
tieth Symposium in Foundations of Computer Science,* October 1979.

EXAMPLES:

given areas of counties of Nevada, find subsets of the

counties with approximately 1/2 the total state area
subsets ← napsack(nevada,sum(nevada)/2)
subsets %c nevada   # areas of the subsets

---

ncol	Extents of a Matrix	**ncol**

USAGE:

ncol(x); nrow(x)

ARGUMENTS:
x: matrix. Missing values (NAs) are allowed.

VALUE:

the number of rows or columns in **x**.

---

ncomp	Number of Components	**ncomp**

USAGE:

ncomp(x)

ARGUMENTS:
x: hierarchical structure.

VALUE:

the number of components in **x**.

Time-series and arrays have 2 components.

SEE ALSO:

Function **compname** returns a vector of the component names.

EXAMPLES:

for(i in seq(ncomp(x)))
    tsplot(x$[i])   # time-series plot of each component of x

norm	Normal Probability Distribution	norm

USAGE:

pnorm(q, par1, par2)
qnorm(p, par1, par2)
rnorm(n, par1, par2)

ARGUMENTS:

q: vector of quantiles. Missing values (NAs) are allowed.
p: vector of probabilities. Missing values (NAs) are allowed.
n: sample size.
par1: vector of means. Default is 0.
par2: vector of standard deviations. Default is 1.

VALUE:

probability (quantile) vector corresponding to the given quantile (probability) vector, or random sample (**rnorm**).

EXAMPLES:

rnorm(20,0,10) #sample of 20, mean 0, standard dev. 10

nper	see **start**	nper

nrow	see **ncol**	nrow

option	see **options**	option

options	Set or Print Options	**options**

USAGE:

options(echo=, width=, length=, space=, max=,
    macsize=, dump=, sourcexit=)

ARGUMENTS:

echo=:   controls the amount of information echoed by S. If **echo=0** (the default) there is no printing of commands or macro expansions. If **echo=1**, there is printing of all macro expansions and input from source files. If **echo=2** (the default for batch mode) all commands are printed before being executed. If **echo=3**, commands, input from source files, and macro expansions are printed.

width=:   specifies the width (in print positions) of the user's terminal. Default 80, minimum 20, maximum 133.

length=:   the length of an output page, currently used only for the length of printer plots. Default 56, must be between 20 and 100.

space=:   the number of characters separating printed numbers. Default 2.

max=:   the maximum number of elements of any vector that will be printed. If the length of a vector to be printed exceeds this number, only the first **max** will be printed, followed by a message telling what the actual length of the vector was. Default is 1000000.

macsize=: the size (in characters) of the macro workspace. Should be large enough to accommodate the largest macro used, and larger if macros are nested. Default 10000.

dump=:   how long should an expression be (in characters) before an error or interrupt generates an automatic dump (see **edit**). Default 20.

sourcexit=: if TRUE, exit from a source file if an error occurs (the default); if FALSE, continue execution of the source file regardless of errors.

If **options** is called without arguments, it prints the current options values.

EXAMPLES:

options( ) # print current option values
    echo = 0 macsize = 10000 dump = 20
    width = 80 length = 56 space = 2
    max = 1000000 sourcexit = T

| **order** | Ordering to Create Sorted Data | **order** |

USAGE:

    order(x1, x2, ...)

ARGUMENTS:

xi: vector. All arguments must have the same length.

VALUE:

integer vector with same number of elements as data elements in **xi**. Contains the indices of the data elements in ascending order, i.e., the first integer is the subscript of the smallest data element, etc. For character vectors, the sorting order is lexicographic.

Sorting is primarily based on vector **x1**. Values of **x2** are used to break ties on **x1**, etc. All sorting is done in ascending order.

EXAMPLES:

```
# sort y and z to follow the order of x
list ← order(x)
cbind(x[list],y[list],z[list])
```

This function is often used in conjunction with subscripting for sorting several parallel arrays.

| **outer** | Generalized Outer Products | **outer** |

USAGE:

    x %o y #operator form
    outer(x, y, fun, arg1, arg2, ...) #general form

ARGUMENTS:

x,y: first and second arguments to the function **fun**. Missing values (NAs) are allowed if **fun** accepts them.

fun: in the general form, the name of some S function. This function will be called repeatedly, for all the values of the first subscript of **x** and the last subscript of **y**. For example, if **x** and **y** are vectors, and **fun** returns a single value when its two arguments are single values, the result is a matrix of **len(x)** by **len(y)**. In the operator case, **fun** defaults to multiply (*).

argi: other arguments to **fun**, if needed.

VALUE:

array, whose dimension vector is the concatenation of the dimension vectors of **x** and **y**.

EXAMPLES:

z ← x %o y   # z[i,j] equals x[i] * y[j]

z ← outer(x,y,"/") # z[i,j] equals x[i]/y[j]

---

pairs	Plot Pair-wise Scatters of Multivariate Data	**pairs**

USAGE:

pairs(x, labels, type, head, full, max, text)

ARGUMENTS:

x: matrix of data to be plotted. Missing values (NAs) are allowed. A scatter plot will be produced for each pair of columns of **x**.

labels: optional character vector for labelling the x and y axes of the plots. The strings **labels[1]**, **labels[2]**, etc. are the labels for the 1st, 2nd, etc. columns of **x**. By default, labels are "1", "2", etc. If supplied, the label vector must have length equal to **ncol(x)**.

type: optional character string to define the type of the scatter plot. Possible values are "p", "l", "b" for points, lines or both. Default is "p". See also **text** below.

head: optional character string for a running head. This is plotted as a title at the top of each page. By default, the name of the data is used. If there is more than one page, the page number is included in the title.

full: should the full **ncol(x)−1** by **ncol(x)−1** array of plots be produced? Default FALSE, only the lower triangle is produced, saving space and plot time, but making interpretation harder.

max: optional, suggested limit for the number of rows or columns of plots on a single page. Default forces all plots on 1 page. The algorithm tries to choose an array of plots which efficiently uses the available space on the display.

text: optional vector of text to be plotted at each of the points. If the
               length of the vector is less than **nrow(x)**, elements will be reused
               cyclically. If missing, the plotting is controlled by **type**.

               Graphical parameters may also be supplied as arguments to this
               function (see **par**).

EXAMPLES:

               pairs(longley.x,labels=longley.collab,head="Longley Data")

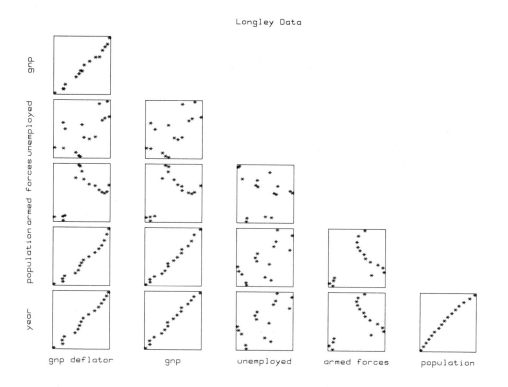

Longley Data

?pairs	All Pair-wise Scatter Plots	**?pairs**

USAGE:

?pairs(x, labels, head)

ARGUMENTS:

x: matrix. Missing values (NAs) are allowed.

labels: character vector of names for the columns of **x**.

head: unquoted character string to be used to title the page of scatter plots.

EFFECT:

This macro is similar to the **pairs** function; however the macro produces all pair-wise scatter plots with labelled axes. All plots will appear on a single page, therefore there should not be too many columns to the matrix **x**.

EXAMPLES:

?pairs(longley.x,label=longley.collab)

par	Graphical Parameters	**par**

USAGE:

par(arg1, arg2, ...)

ARGUMENTS:

argi: arguments in the **name**=value form, where **name** is one of the options below.

These are the graphical parameters which control the appearance of graphical output. There are three classes of parameters: those which are available only through the high-level functions; those which can appear in **par** or any graphical function; and those which are available only through the **par** command.

If given through **par**, the options are set permanently; otherwise, they are reset at the end of the graphical operation.

Some of the most commonly used parameters are: **cex**= character expansion, **log**= logarithmic axes, **lty**= line type, **main**= main title, **pch**= plotting character, **type**= type of plot, **xlab**= x axis label, **ylab**= y axis label, **xlim**= range for x axis, and **ylim**= range for y axis.

In the following lists of parameters, the notation "c" denotes a character, **"string"** denotes a character string, **i, j, m,** and **n** are integers, **L** is a logical value, and **x** is numeric.

HIGH-LEVEL PARAMETERS:

The following parameters may only be used in high-level graphical functions (those that set up coordinate systems, e.g., **plot, qqplot**),

axes=L   if FALSE, suppresses all axis plotting (x, y axes and box). Useful to make a high-level plotting routine generate only the plot portion of the figure.

log="c"   controls logarithmic axes. Values "xy", "x", or "y" produce log-log or log-x or log-y axes.

main="string"
main title for top of plot. Plotted in characters of size 1.5*cex.

sub="string"
sub-title, to be placed below the X-axis label in standard sized characters.

type="c"   type of plot desired. Values are "p" for points, "l" for lines, "b" for both points and lines (lines miss the points), "o" for overlaid points and lines, "n" for no plotting, and "h" for high-density vertical line plot.

xlab="string"
label for plotting below the x-axis.

xlim=c(x1,x2)
approximate minimum and maximum values to be put on x-axis. These values are automatically rounded to make them "pretty" for axis labelling.

ylab="string" see **xlab**.

ylim=c(x1,x2) see **xlim**.

GENERAL PARAMETERS:

The following parameters can be used in any graphical functions, including **par**.

adj=x   string justification parameter. 0 = left justify, 1 = right justify, .5 = center.

bty="c"   character representing the type of box. Characters **o**, **l** (ell), **7**, **c** will produce boxes which resemble the corresponding upper-case letters. The value **n** will suppress boxes.

cex=x   character expansion relative to device's standard size. For example, when **cex** = 2, characters are twice as big as normal for the device.

col=x   color, device dependent. Default 1. Generally, small integers are used to specify pen numbers on pen plotters, color map indices on scope devices, etc.

crt=x   character rotation in degrees measured counterclockwise from horizontal. When **srt** is set, **crt** is automatically set to the same value.

csi=x   character height (interline space) in inches.

err=x   error mode: −1 = do not print any error messages, 0 = print messages.

lab=c(x,y,llen)
desired number of tick intervals on the X and Y axes and the length of labels on both axes. Default **c(5,5,5)**.

las=x   style of axis labels. 0 = always parallel to axis (the default), 1 = always horizontal, 2 = always perpendicular to axis.

lty=x   line type, device dependent. Normally type 1 is solid, 2 and up are dotted or dashed.

mgp=c(x1,x2,x3)
> margin parameters which give the margin coordinate for the axis title, axis labels, and axis line. Default is **c(3,1,0)**.

mkh=x   height in inches of mark symbols drawn by the **mark=** argument to function **points**.

pch="c"   the character to be used for plotting points. If **pch** is a period, a centered plotting dot is used.

smo=x   smoothness of circles and other curves. **smo** is the number of rasters that the straight-line approximation to the curve is allowed to differ from the exact position of the curve. Large values produce more crude approximations to curves, but allows the curves to be drawn with fewer line segments and hence speeds up output. The minimum number of line segments that will be used for a circle is 8, regardless of **smo**.

srt=x   string rotation in degrees measured counterclockwise from horizontal. When specified, sets **crt** to same value.

tck=x   the length of tick marks as a fraction of the smaller of the width or height of the plotting region. If **tck** is negative, ticks are drawn outside of the plot region. If **tck** = 1, grid lines are drawn.

xaxp=c(ul,uh,n)
> axis parameters giving coordinates of lower tick mark **ul**, upper tick mark **uh**, and number of intervals **n** within the range from **ul** to **uh**.

xaxs="c"   style of axis interval calculation. The styles "s" and "e" set up standard and extended axes, where numeric axis labels are more extreme than any data values. In addition, extended axes may be extended another character width so that no data points lie very near the axis limit. Style "i" creates an axis labelled internal to the data values. This style wastes no space, yet still gives pretty labels. Style "r" extends the data range by 7% on each end, and then labels the axis internally. This ensures that all plots take up a fixed percent of the plot region, yet keeps points away from the axes. Style "d" is a direct axis, and axis parameters will not be changed by further high-level plotting routines. This is used to "lock-in" an axis from one plot to the next. Default is "e".

xaxt="c"   axis type. Type "n" (null) can be used to cause an axis to be set up by a high-level routine, but then not plotted.

xpd=L   logical value controlling clipping. Values: FALSE means no points or lines may be drawn outside of the plot region. TRUE means points, lines, and text may be plotted outside of the plot region as long as they are inside the figure region.

yaxp=c(ul,uh,n)   see **xaxp**.

yaxs="c"   see **xaxs**.

yaxt="c"   see **xaxt**.

LAYOUT PARAMETERS:

The following parameters may only be used in function **par**, because they change the overall layout of plots or figures.

fig=c(x1,x2,y1,y2)
coordinates of figure region expressed as fraction of device surface.

fin=c(w,h)
width and height of figure in inches.

fty="c"   figure type. Default "m", the maximum size that will fit on the device.

mai=c(x1,x2,x3,x4)
margin size specified in inches. Values given for bottom, left, top, and right margins in that order.

mar=c(xbot,xlef,xtop,xrig)
maximum value for margin coordinates on each side of plot. Margin coordinates range from 0 at the edge of the box outward in units of **mex** sized characters. If the margin is respecified by **mai** or **mar**, the plot region is re-created to provide the appropriate sized margins within the figure. Default value is **c(5.1,4.1,4.1,2.1)**.

mex=x   the coordinate unit for addressing locations in the margin is expressed in terms of **mex**. **mex** is a character size relative to default character size (like **cex**) and margin coordinates are measured in terms of characters of this size.

mfg=c(i,j,m,n)

multiple figure parameters which give the row and column number of the current multiple figure and the number of rows and columns in the current array of multiple figures.

mfrow=c(m,n)

subsequent figures will be drawn row-by-row in an **m** by **n** matrix on the page.

mfcol=c(m,n)

subsequent figures will be drawn column-by-column in an **m** by **n** matrix on the page.

new=L
if TRUE, the current plot is assumed to have no previous plotting on it. Any points, lines, or text will set **new** FALSE.

oma=c(x1,x2,x3,x4)

specification of the outer margin coordinates in terms of text of size **mex**. **oma** provides the maximum value for outer margin coordinates on each of the four sides of the multiple figure region. **oma** causes recreation of the current figure within the confines of the newly specified outer margins.

omd=c(x1,x2,y1,y2)

the region within the outer margins (which is to be used by multiple figure arrays) is specified by **omd** as a fraction of the entire device.

omi=c(x1,x2,x3,x4)

size of outer margins in inches.

pin=c(w,h)

width and height of plot, measured in inches.

plt=c(x1,x2,y1,y2)

the coordinates of the plot region measured as a fraction of the figure region.

pty="c"
the type of plotting region currently in effect. Values: "s" generates a square plotting region; "m" (the default) generates a maxi-

mal size plotting region, which, with the margins, completely fills the figure region.

usr=c(x1,x2,y1,y2)
> user coordinate limits (min and max) on X and Y axes. Default, when device is initialized, is **c(0,1,0,1)**.

---

pardump	Dump of Graphical Parameters	**pardump**

USAGE:
> **pardump(arg1, arg2, ...)**

ARGUMENTS:
argi: names of graphical parameters to be dumped (see **par** for a list of the names).

EFFECT:
> Prints a dump of the graphical parameters currently in effect. If arguments are given to **pardump**, only the parameters with the corresponding names will be printed. If no arguments are given, all parameters are printed. Parameters are listed alphabetically. With no arguments given, **pardump** will produce about a page of output.

---

pbeta	see **beta**	**pbeta**

---

pcauchy	see **cauchy**	**pcauchy**

---

pchisq	see **chisq**	**pchisq**

persp	3-Dimensional Perspective Plots	**persp**

USAGE:

persp(z, eye, ar)

ARGUMENTS:

z: matrix of heights given over a regularly spaced grid of **x** and **y** values, i.e., **z[i,j]** is the height at **x[i],y[j]**. Although **x** and **y** are not input to **persp**, the algorithm plots as if **x** and **y** are in increasing order and equally spaced from −1 to 1.

eye: vector giving the **x,y,z** coordinates for the viewpoint. Default is **c(−6,−8,5)**. Since the implied **x,y** grid ranges from −1 to 1, the **x,y** coordinates of **eye** should not both be in the range from −1 to 1.

ar: aspect   ratio   of   the   actual   **x,y**   grid,   i.e., **(xmax−xmin)/(ymax−ymin)**. Default 1.

persp sets up the plot under the assumption that a unit in the **x**, **y**, and **z** directions represents the same physical size. Thus the values in **z** should ordinarily be scaled to the range (0,1).

The algorithm attempts hidden-line elimination, but may be fooled on segments with both endpoints visible but the middle obscured.

EXAMPLES:

```
persp(z)   #perspective plot of heights z
        #from default viewpoint
#the example plot is produced by:
i ← interp(ozone.xy$x,ozone.xy$y,ozone.median)
i$z ← ifelse(NA(i$z),0,i$z)
persp(i$z/200)
title(main="Median Ozone Concentrations in the North East")
```

Median Ozone Concentrations in the North East

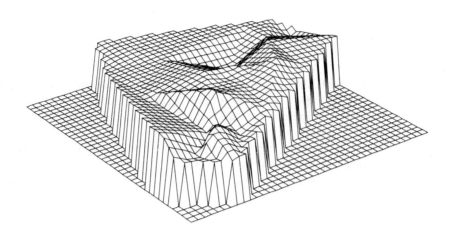

---

pf	see **f**	**pf**

---

pgamma	see **gammadist**	**pgamma**

---

pic	Produce Graphics on Phototypesetter via pic	**pic**

USAGE:

    pic

This function produces a file of commands which are compatible with the pic troff preprocessor. The file, named "pic.out", can be given alone as input to the **pic** program, or included in documents by the construction

.PS <pic.out in the midst of text.

The pic.out file may contain more than one picture, each bracket-
ed by ".PS/.PE" lines. Since the file is ASCII text, it can be read
and edited by the user.

When graphic input (**identify, rdpen**) is requested, the prompt
x,y: appears on the users terminal, and the user should type the
(x,y) coordinate pair of the point to be input. To terminate the
input, hit carriage return. Graphic input is not allowed in batch
mode.

A device must be specified before any graphics functions can be
used.

EXAMPLES:

pic
plot(hstart)
q   # exit from S

pic −Taps pic.out | troff −Taps >troff.out
    # now send file to aps phototypesetter

---

pie	Pie Charts	**pie**

USAGE:

   pie(x, names, size, inner, outer, explode, angle, density, col)

ARGUMENTS:

x: vector of relative pie slice sizes. The ith slice will take the frac-
            tion **abs(x[i])/sum(abs(x))** of the pie. The slices start with a hor-
            izontal line to the right and go counter-clockwise.

names: optional character vector of slice labels. Labels are positioned
            along the center-line of the slice, and are shifted as far in toward
            the center as possible without overlapping into adjacent slices
            (see arguments **inner** and **outer**).

size: optional fraction of the short dimension of the plot taken up by
            the circle. Default .75.

inner: optional fraction giving the innermost position that labels can occupy. The default value of .3 means that labels can go no further toward the center than .3 of the radius.

outer: optional fraction giving the outer limit for starting the labels, default 1.1.

explode: logical vector specifying slices of the pie which should be exploded (moved out from the center).

angle: optional vector giving the angle (degrees, counter-clockwise from horizontal) for shading each slice. (Defaults to 45 if **density** is supplied.)

density: optional vector for pie shading, giving the number of lines per inch for shading each slice. Defaults to 5 if **angle** is supplied. A density of 0 implies solid filling, and is the default if **col** is specified but angle is not. Negative values of density produce no shading.

col: optional vector giving the colors in which the pie slices should be filled or shaded. If **col** is specified and neither **angle** nor **density** are given as arguments, slices will be filled solidly with the colors.

Solid filling of pie slices is dependent on the area-filling capability of the device driver. For devices without explicit area-filling capability, solid filling can be simulated by specifying a very high density shading.

Graphical parameters may also be supplied as arguments to this function (see **par**).

EXAMPLES:

pie(revenues,revenue.class)

pie(expenses,c("Interest","Materials","Payoffs"),inner=.5, outer=.5,size=1) # force labels to start halfway out

pie(revenues, explode= revenues>.1*sum(revenues))
** # explode any piece larger than 10% of the sum**

 # the example plot is produced by:
datatel←?col(telsam.response,sum)
pie(datatel,telsam.collab)
title(main="Response to Quality of Service Questions concerning Telephone Service")

Response to Quality of Service Questions

concerning Telephone Service

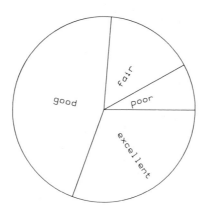

| **plclust** | Plot Trees From Hierarchical Clustering | **plclust** |

USAGE:

      plclust(tree, hang, unit, level, hmin, square, labels, plot)

ARGUMENTS:

tree: a hierarchical clustering tree, of the form returned by function **hclust**.

hang: the fraction of the height of the plot that any individual node will hang below the cluster that it joins. A value of −1 will cause all individuals to start at y-value 0. Default 0.1.

unit: logical flag. If TRUE, the heights of the merges will be ignored and instead merge i will occur at height i. Useful for spreading out the tree to see the sequence of merges. Default is FALSE.

level: logical flag. If TRUE, plotted tree will be "leveled", where merges in different subtrees are arbitrarily assigned the same height in order to compress the vertical scale. Particularly useful with **unit=TRUE**. Default is FALSE.

hmin: optional minimum height at which merges will take place. Can be used to get rid of irrelevant detail at low levels. Default 0.

square: logical flag. If TRUE (default), the tree is plotted with "U" shaped branches, if FALSE, it has "V" shaped branches.

labels: optional character vector of labels for the leaves of the tree. If omitted, leaves will be labelled by number. To omit labels entirely, use **labels=FALSE**.

plot: logical flag. If TRUE (default), plotting takes place. If FALSE, no plotting is done (useful for returned value).

Graphical parameters may also be supplied as arguments to this function (see **par**).

VALUE:

if **plot** is FALSE, a structure containing the coordinates of the leaves of the tree and the interior nodes of the tree.

x,y: x and y coordinates of the leaves of the tree, i.e., **x[i],y[i]** gives the coordinates of the leaf corresponding to the ith individual.

xn,yn: x and y coordinates of the interior nodes of the tree, i.e., **xn[i],yn[i]** gives the coordinates of the node representing the ith merge.

EXAMPLES:

```
plclust(hclust(distances))

plclust(tree,label=FALSE)   # plot without labels
xy←plclust(tree,plot=FALSE)      # no plot, save structure
# allow user to point at leaf and have it identified
identify(xy)

# the example plot is produced by:
$T←?row($author.count,sum)
$T←?rsweep(author.count,$T,/)
par(mar=c(18,4,4,1))
plclust(hclust(dist($T)),label=author.rowlab)
title("Clustering of Books Based on Letter Frequency")
```

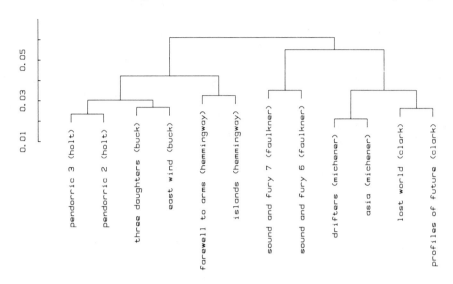

Clustering of Books Based on Letter Frequency

---

plnorm   see **lnorm**   **plnorm**

---

plogis   see **logis**   **plogis**

---

plot   Scatter Plots   **plot**

USAGE:

        plot(x, y)

ARGUMENTS:

x,y: coordinates of points on the scatter plot. A time-series or structure containing components named **x** and **y** may also be given for
x. Missing values (NAs) are allowed.

Graphical parameters may also be supplied as arguments to this function (see **par**).

Graphical parameters that are particularly useful with **plot**:

type= where values of "p", "l", "b", "o", "n", and "h" produce points, lines, both, both (overlaid), nothing, and high-density lines.

log= where values of "x", "y", or "xy" specify which axes are to be on a logarithmic scale.

main= main title

sub= sub-title

xlab= x axis label (defaults to the name of the x dataset).

ylab= y axis label (defaults to the name of the y dataset).

EXAMPLES:

```
plot(x,y)     #simple scatter plot
plot(x,y,type="l")      #connected lines
plot(x,y,log="xy")       #log-log plot
plot(x,y,type="n");text(x,y,seq(x))   # do not plot, then
      # use text to label each point from 1 to n
plot(gnp,type="h")   #high-density plot of time-series
plot(density(x),type="l")  #plot of xy structure
# the example plot is produced by:
plot(corn.rain,corn.yield)
```

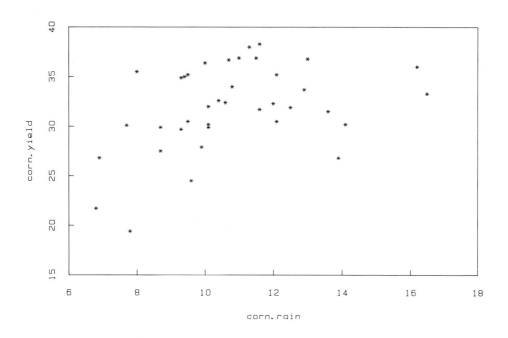

| **plotfit** | Two-way Plot of Fit | **plotfit** |

USAGE:

plotfit(fit, w, c, rowlab, collab, grid)

ARGUMENTS:

fit: structure with components **row**, **col**, **grand**, and **resid**, reflecting a two-way fit to a matrix. See, for example, the output of function **twoway**. Missing values (NAs) are allowed in components of **fit**.

w: interaction term, i.e., the coefficient of the row*col interaction. Default 0. The residuals in **fit** should NOT reflect this term, i.e., **data[i,j]** equals **grand + row[i] + col[j] + resid[i,j]**.

c: residuals larger in magnitude than **c** will be displayed. If $c<0$, no residuals will be displayed; if $c=0$, all residuals will be displayed. Default -1.

rowlab: character vector giving labels for the rows of the matrix. Defaults to "Row i". To omit labels, use **rowlab=""**.

collab: character vector giving labels for the columns of the matrix. Defaults to "Col i". To omit labels, use **collab=""**.

grid: should grid of fitted values be drawn? Default TRUE.

Graphical parameters may also be supplied as arguments to this function (see **par**).

EXAMPLES:

plotfit(twoway(datamat))

\# the example plot is produced by:
vy←votes.year[27:31] #get last five election years
vr←twoway(votes.repub[1:10,27:31]) #first 10 states, last 5 years
plotfit(vr,c=8,rowlab=state.name,collab=encode(vy))
title(main="Twoway Fit to Republican Votes",
        sub="10 States for 1964 - 1972")

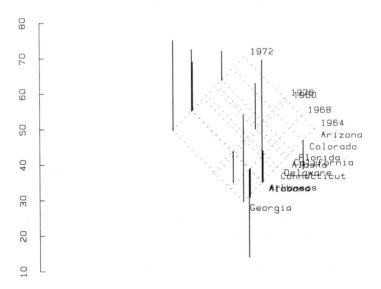

Twoway Fit to Republican Votes

10 States for 1964 - 1972

---

pmax	Parallel Maximum or Minimum	**pmax**

USAGE:

> pmax(arg1, arg2, ...)
> pmin(arg1, arg2, ...)

ARGUMENTS:

argi: numeric structure.

VALUE:

> vector whose first element is the maximum (**pmax**) or minimum (**pmin**) of the first elements of the arguments, and similarly for the second element, etc. The length of the vector is the length of the longest argument. Shorter vectors are reused cyclically. Missing values (NA) are allowed; if an element of any of the arguments is NA, the corresponding element of the result is also NA.

EXAMPLES:

> z ← pmax(x,y,5) # vector as long as larger of x and y
>   # where z[i] is max of x[i], y[i], and 5

> Note the difference between **pmax**, **pmin** and **max**, **min**. The latter give the single element which is the max or min of all the arguments. See also **range**.

---

pmin	see **pmax**	**pmin**

---

pnorm	see **norm**	**pnorm**

---

points	see **lines**	**points**

---

ppoints	Plotting Points for Q-Q Plots	**ppoints**

USAGE:

> **ppoints(n)**

ARGUMENTS:

n: sample size for which plotting points desired (if **n** has only 1 value) or data against which plot is to be made.

VALUE:

> the vector of probabilities, **p**, such that **qdist(p)** plotted against **sort(y)** gives a probability (Q-Q) plot of **y** against the distribution of which **qdist** is the quantile function. Computes $p[i]=(i-.5)/n$ if $n>10$ or $(i-.375)/(n+.25)$ otherwise.

EXAMPLES:

> **plot(qlnorm(ppoints(y)),sort(y))**   #log normal q-q plot

---

prcomp	Principal Component Analysis	**prcomp**

USAGE:

> **prcomp(x, retx)**

ARGUMENTS:

x: data matrix to be decomposed. Principal component analysis defines a rotation of the variables (columns) of **x**. The first derived direction is chosen to maximize the standard deviation of the derived variable, the second to maximize the standard deviation among directions uncorrelated with the first, etc.

retx: logical, if TRUE (the default) the rotated version of the data matrix is returned. Setting to FALSE just saves space in the returned data structure.

VALUE:

> structure describing the principal component analysis:

sdev: standard deviations of the derived variables.

rotation: orthogonal matrix describing the rotation. The first column is the linear combination of columns of **x** defining the first principal component, etc. May have fewer columns than **x**.

x: if **retx** was TRUE, the rotated version of **x**; i.e., the first column is the **nrow(x)** values for the first derived variable, etc. May have fewer columns than **x**.

The analysis will work even if **nrow(x)<ncol(x)**, but in this case only **nrow(x)** variables will be derived, and the returned **x** will have only **nrow(x)** columns. In general, if any of the derived variables has zero standard deviation, that variable is dropped from the returned result.

---

precedence	Order of Expression Evaluation	**precedence**

Precedence determines the order of evaluation of a non-parenthesized expression. Operations of a higher precedence are evaluated before any operations of lower precedence. Equal precedence operators are evaluated left-to-right (except assignment and exponentiation which are right-to-left).

$	component select	HIGH
%x	special operator	
−	unary minus	
:	sequence operator	
^ **	exponentiation	
* /	mult/div	
+ −	add/sub	
< > <= >= == !=	logical	
!	not	
& |	and/or	
← _ →	assignment	LOW

prefix	Set Prefix for Dataset Names	prefix

USAGE:

> **prefix(name)**

ARGUMENTS:

name: character string, consisting of characters which are legal for dataset names, i.e., letters, digits and ".". The function then has the side effect that this string is automatically prepended to all dataset names used in subsequent expressions. The prefix may be overridden by giving a fully-qualified dataset name and starting the name with "$". If the argument is omitted, the prefix is made empty (also the initial state).

VALUE:

> the previous prefix, so that one can store and restore the prefix (useful in macros). The result of **prefix** is not automatically printed.

> It is recommended that **prefix** be used to distinguish datasets generated during a particular analysis. Thus, for example, if we want to analyze some data concerning wages, the prefix may be set to "wages." with the result that commonly used names, such as "x", "y", "label", etc. do not conflict with datasets of the same name, used in analysis of other data.

> Slashes may also be used in prefixes (and in dataset names given in character form to the **get** or **assign** functions), to specify subdirectories under the directory containing the data. The construction of subdirectories corresponding to prefixes is sometimes a good idea, as it avoids the UNIX restriction of dataset names to 14 characters. However, the subdirectory must be created (by the system command mkdir) before datasets can be stored there.

BUGS:

> The **ls** function does not work for subdirectories, and it is difficult to refer to datasets inside subdirectories without using **prefix**.

EXAMPLES:

> **oldp←prefix("wages.")** #set prefix, store old one
> **print(x,y,$z)** #print wages.x wages.y and z
> **print(prefix( ))** #delete prefix and print its previous value

```
!mkdir swork/abc
prefix("abc/")
x ← 1:10      # creates dataset abc/x
prefix( )   # turn off prefix
abc/x       # this means abc divided by x
get("abc/x")  # this reads the dataset
```

---

pretty	Return Vector of Prettied Values	**pretty**

USAGE:

pretty(x, nint)

ARGUMENTS:

x: vector of data; prettied values will cover range of **x**. Missing values (NAs) are allowed.

nint: optional, approximate number of intervals desired. Default 5.

VALUE:

vector of (ordered) values, defining approximately **nint** intervals covering the range of **x**. The individual values will differ by 1, 2 or 5 times a power of 10.

EXAMPLES:

pretty(mydata,10)

---

print	Print Data	**print**

USAGE:

print(arg1, arg2, ...,
    rowlab=, collab=, max=, quote=, head=)

ARGUMENTS:

argi: any structure. Missing values (NAs) are allowed.

rowlab=: character vector of labels for the rows of a matrix.

collab=: character vector of labels for the columns of a matrix.

max=: limit on number of values printed in a single vector or component, default is value from **max** parameter of **options**. Use a small value, say **max=10**, to print just the structure information about large data sets. See also **compname** for this purpose.

quote=:   logical, controls whether quotes appear around character strings. Default TRUE.

head=:   logical, determines if time-series and arrays are preceded by a heading that describes their **Tsp** and **Dim** components. Default, TRUE.

VALUE:

structure composed of all **args**.

The function is invoked automatically if a command line evaluates to a result which is not assigned a name.

If **argi** is given in **name=value** form, **name** is printed to label the value, and is used for the component name in the returned structure.

See **options** for control of line width and spacing.

---

printer	Line Printer Graphics	**printer**

USAGE:

printer
printer(width, length)

ARGUMENTS:

width: number of characters for width of printer plots. Default is value of **width** as set in **options** function (initially 80). Maximum width is 133.

length: desired number of lines for length of plot. Default is value of **length** as set in **options** function (initially 56). Maximum length is 100.

The **printer** device is applicable to any typewriter-like terminal. The quality is not equal to that of true graphics devices, but may be sufficient for exploratory data analysis.

The size of the printer plot is determined by the terminal width and length in effect at the time the function **printer** is invoked.

Each plot is stored in an internal buffer. This enables commands to be given to add to the existing plot. When the next plot is started, the previous plot is printed. The function **show** can be used to print the current plot. After using **show**, the plot can be further augmented. **printer** and **show** can also be used to preview plots that are to be produced by the deferred graphics facility **defer**. Use **show** to view the picture, then decide if it should be saved.

Character size and orientation cannot change. Different line types are available, but the resolution of the printer is generally too coarse to make much distinction.

Any graphic input (**identify**, **rdpen**) done on the printer will prompt for x and y coordinates. Type in the desired coordinates, or hit carriage return to terminate the graphic input.

A device must be specified before any graphics functions can be used.

EXAMPLES:

options(width=120) # for terminal with long line
printer; plot(x,y)
title("a title to be added to the plot")
show # we want to see the plot now

| **?probsamp** | Sample of Data with Specified Probability | **?probsamp** |

USAGE:

?probsamp(x, alpha)

ARGUMENTS:

x: vector of data.

alpha: probability (between 0 and 1) of picking any individual data value in x to be in the generated sample. If **alpha** is not given, its user is prompted for the value.

VALUE:

a sample of the data vector **x**.

SEE ALSO:

Function **sample**.

EXAMPLES:

?probsamp(mydata,.5)   # return sample of mydata
each value in mydata has 50% chance of appearing in sample

---

prod	Products and Sums	**prod**

USAGE:

prod(x, naout)
sum(x, naout)

ARGUMENTS:

x: numeric structure. Missing values (NAs) are allowed.
naout: logical, if TRUE, all NAs are removed from **x** prior to carrying out the operation.

VALUE:

the sum (product) of all elements of **x**. If NAs are present and not removed, the result is NA.

?prompt	Create Documentation for Macros or Datasets	**?prompt**

USAGE:

?prompt(name, pos=)

ARGUMENTS:

name: the name of the macro or dataset(s) to be documented. If there is no macro named **name**, dataset documentation for **name** is produced. If there are several datasets whose names begin with **name**, they will all be included in the documentation file.
pos=:   database number on which the macro or dataset(s) reside. Default

is 2.

EFFECT:

Creates file **name.d** with shell of the documentation. The file should be edited and then installed by the **S NEWDOC** utility.

EXAMPLES:

?prompt(abc)   #documentation for macro or dataset abc

create documentation file for longley data on system database
?prompt(longley,pos=3)

pt	see **tdist**	**pt**

punif	see **unif**	**punif**

q	Terminating Execution	**q**

USAGE:

q

Causes termination of execution and returns to the operating system. If deferred graphics is on at the time, the last frame will be saved (see **defer**). If a graphics device is active, a device-dependent wrapup routine will be executed, for example, to put away plotter pens.

qbeta	see **beta**	**qbeta**

qcauchy	see **cauchy**	qcauchy

qchisq	see **chisq**	qchisq

qf	see **f**	qf

qgamma	see **gammadist**	qgamma

qlnorm	see **lnorm**	qlnorm

qlogis	see **logis**	qlogis

qnorm	see **norm**	qnorm

qqnorm	see **qqplot**	qqnorm

qqnorm	Quantile-Quantile Plots	**qqnorm**

USAGE:

        qqplot(x, y, plot)
        qqnorm(xy, datax, plot)

ARGUMENTS:

x,y: vectors (not necesarily of the same length). Each is taken as a sample, for the x and y axis values of an empirical probability

	plot.
xy:	vector of data for a normal probability plot.
datax:	logical flag, if TRUE, data goes on the **x** axis, if FALSE (default) data goes on the **y** axis.
plot:	logical flag. If **plot** is FALSE, these routines all return a structure with components **x** and **y** which give the coordinates of the points that would have been plotted. Default is TRUE.

Graphical parameters may also be supplied as arguments to these functions (see **par**). These functions can also take arguments **type** and **log** to control plot type and logarithmic axes.

VALUE:

if **plot** is FALSE, a structure with components **x** and **y** are returned, giving coordinates of the points that would have been plotted.

EXAMPLES:

```
zz←qqplot(x,y,plot=F)      #save x and y coords of empirical qq
plot(zz)      #plot it
abline(rreg(zz$x,zz$y))  #fit robust line and draw it

# the example plot is produced by:
my.sample←rt(50,5)
lab←"50 samples from a t distribution with 5 d. f."
qqnorm(my.sample,main=lab,sub="QQ Plot with Normal")
```

50 samples from a t distribution with 5 d. f.

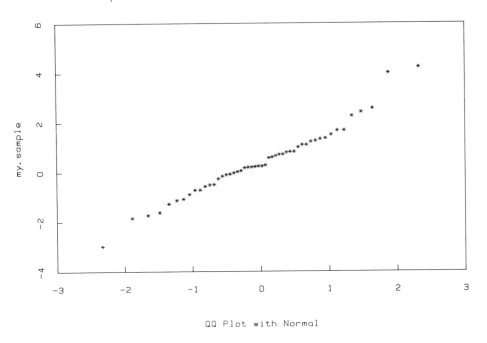

QQ Plot with Normal

?qqplot	Probability Plot for Arbitrary Distribution	**?qqplot**

USAGE:

?qqplot(x, dist,par1,par2)

ARGUMENTS:

x: Data to be plotted.

dist: Name of distribution (e.g., t, beta, chisq) Default is ?PROMPT(What distribution?).

pari: Optionally, parameters to the distribution. (Note that location and scale parameters are not needed; only shape or degrees of freedom).

EFFECT:

Produces a probability (Q-Q) plot from an arbitrary distribution, so long as there is a corresponding quantile function.

EXAMPLES:

?qqplot(rnorm(50), t, 3) #random normals vs t on 3 d.f.

qt	see **tdist**	qt

quickvu	Make Slides with Simple Lists	**quickvu**

USAGE:

    quickvu(head, listtype)

ARGUMENTS:

head: Optionally, the vu commands to start off the slide. By default, will color the title portion in color 1, embolden once and increase the size 20% over the remaining text.

listtype: The command defining the list type (".DL", ".NL", etc.) By default, bullet lists are produced.

VALUE:

Character vector implementing a slide suitably for an argument to the **vu** function. The slide consists by default of a multi-line title, left-justified and emboldened once, followed by a list of items. The **quickvu** function prompts for terminal input giving the title (terminated by an empty line). This is followed by input for the list items (one line per item by default) with an empty line signalling the end of the slide. Options are possible (see below).

NOTE:

The title, list marker and list text are generated in colors 1, 2 and 3 respectively. The limit of one line per item can be overridden by beginning the input line with "&". This character is thrown away and the rest of the line taken as a continuation of the previous line. (This allows also the input of **vu** control lines, but note that color, etc. are set at the beginning of each item; therefore, to change the color of an item make a .C line the first line of the item.

EXAMPLES:

    > **testvu ← quickvu( )**

Title line: **THE FIRST LINE OF THE TITLE**
Title line: **AND THE SECOND**
Title line:
List item: **I have a little list**
List item: **And theyld none of them be missed**
List item: **&˜˜(which can be continued)**
List item: **And a third item.**
List item:

---

qunif	see **unif**	**qunif**

---

ram6211	Ramtek 6211 Color Graphics Terminal	**ram6211**

USAGE:

      ram6211(ask)
      !S RAM6211   # UNIX command

ARGUMENTS:
ask: logical, should device driver ask before clearing graphic display for a new figure? Default TRUE.

COMMENTS:

      The function **ram6211** is the device driver for the Ramtek 6211 color graphics terminal. The **S RAM6211** utility should be used immediately after logging on; it provides initial values for the color table.

      Color 0 (the background) is dark blue and colors 1 through 7 are cyan, green, yellow, black, magenta, red, and white.

      A feature of this terminal is that alphanumeric information is displayed in an overlay separate from graphical information (text in titles, axis labels, etc. is considered graphical). The buttons labelled "view alpha", "view graph", and "view both" cause either the alphanumeric information, graphics, or both to be displayed. When these buttons are pushed in conjunction with the "control" key, the corresponding areas are erased.

If **ask** is TRUE, the message "GO?" appears whenever a new plot is about to be produced. This allows further viewing, making a hard copy, etc., before the screen is erased. When ready for plotting to proceed, simply hit carriage return.

When graphic input (**identify, rdpen**) is requested from this terminal, a "+"-shaped cursor appears on the screen. The cursor can be positioned by the lightpen or by the cursor motion keys on the keyboard (cursor motion keys can be made to move more quickly by holding down the "control" key). When the cursor is at the desired point, press the lightpen against the screen, or hit "control" "enter" to enter the coordinates of the point. A carriage return will terminate graphic input without transmitting the point.

The terminal can change character size and line type. Character sizes are 1 and 2. Line types 1 through 6 are solid, short dash, long dash, dash dot, long dash dot, and dotted. Characters cannot be rotated on this device.

A device must be specified before any graphics functions can be used.

RAM6211	see **ram6211**	**RAM6211**

---

range	Range of Data (minimum, maximum)	**range**

USAGE:

   range(arg1, arg2, ...)

ARGUMENTS:

argi: numeric structure. Missing values (NAs) are allowed.

VALUE:

   vector of two elements, the first the minimum of all the elements of all arguments; the second the maximum. This function is useful as an argument to plotting when it is desired to specify the limits for the x- or y-axes. (See the example.)

EXAMPLES:
      plot(x,y,ylim=range(y,0,1)) # force y-axis to include (0,1)

---

rank	Ranks of Data	**rank**

USAGE:
      rank(x)

ARGUMENTS:
x: numeric structure.

VALUE:
      the ranks; i.e., the i-th value is the rank of **x[i]**. In case of ties, the
      average rank is returned.

?ranktest	Two-sample Rank Test	**?ranktest**

USAGE:
      ?ranktest(x, y)

ARGUMENTS:
x: vector of data from first sample.
y: vector of data from second sample.

VALUE:
      both datasets are ranked together, and the sum of the ranks of the
      values in the first set is returned.

EXAMPLES:
      ?ranktest(heights1,heights2)

rbeta	see **beta**	rbeta

rbind	see **cbind**	rbind

rbiwt	Robust Simple Regression by Biweight	**rbiwt**

USAGE:

> **rbiwt(x, y)**
> **rbiwt(x, y, start, k, tol, iter)**

ARGUMENTS:

x: vector of observations on independent variable.

y: vector of observations on dependent variable.

start: vector giving starting values of intercept and slope. Default, use least-squares **start**.

k: biweight scale parameter, default 6.

tol: convergence criterion. Default .001

iter: maximum number of iterations. Default 20.

VALUE:

> structure containing components **coef**, **resid**, and **wt**.

coef: vector giving intercept and slope.

resid: vector like **y** giving residuals from fit.

wt: vector giving weights used in final weighted least-squares step.

> Does not fit data to more than 1 independent variable (see **rreg**).

REFERENCE:

> Coleman, D., Holland, P., Kaden, N., Klema, V., and Peters, S. C., "A System of Subroutines for Iteratively Re-Weighted Least-Squares Computations", *ACM Trans. Math. Soft.*, Vol. 6, 327-336, 1980.

SEE ALSO:

> Function **rreg** generalizes **rbiwt** to multiple regression.

EXAMPLES:

plot(x,y)
abline(rbiwt(x,y))   #add line to plot

---

rcauchy	see **cauchy**	**rcauchy**

---

rchisq	see **chisq**	**rchisq**

---

rdpen	Get Coordinates from Plot	**rdpen**

USAGE:

rdpen(n, type)

ARGUMENTS:

n: the maximum number of points to identify. Default is 500.

type: character describing interactive drawing option. If **type** is "n", nothing is drawn (the default). Values of "p", "l", "b", and "o" plot points at the digitized coordinates, lines connecting them, or both points and lines.

VALUE:

structure containing vector components **x** and **y** which give coordinates for each point. The length of these vectors is at most **n**, but can be shorter if the user terminates graphic input after fewer than **n** points are given.

See the individual device documentation for the protocol on the device for identifying points and terminating graphic input.

EXAMPLES:

user points at outlier which is then labelled
text(rdpen(1),"outlier")

lines(rdpen( ))   # input a number of points, connect with line
repen(type="l")   # alternative

---

read	Read Data from a File	**read**

USAGE:

    read(file, length, mode, skip, print)

ARGUMENTS:

file: character string, naming the file to read. If the file name is omitted, reading is from the standard input (terminal).

length: the maximum number of values to read, default is to read until end-of-file.

mode: the mode of the data, default REAL.

skip: number of lines of the file to skip before starting to read. Default 0.

print: logical flag, should number of items read be printed? Default TRUE.

VALUE:

    vector comprised of the data read from the file. If fewer than **length** values are read before end-of-file, the length of the result is the number of values read.

    Also prints the number of values read, and warns if end-of-file was not reached.

    Character strings read (**mode=CHAR**) can contain internal blanks only if enclosed by quotes.

EXAMPLES:

    tst ← read("testdata")   # normal usage

---

reg	Regression	**reg**

USAGE:

    reg(x, y, wt, int, print, names, ynames, q)

ARGUMENTS:

x: x matrix for fitting Y=Xb+e with variables in columns, observations across rows. Should not contain column of 1's, (see argument **int**). Number of rows of **x** should equal the number of rows of **y**. There should be fewer columns than rows.

y: y vector (or matrix with one column for each regression).

wt: vector of weights for weighted regression. Should have length equal to the number of rows of **y**. If the different observations have non-equal variances, **wt** should be inversely proportional to the variance.

int: flag, if TRUE (the default) a constant (intercept) term is included in each regression.

print: flag, if TRUE (the default), the regression results are given to **regprt** for printing.

names: optional character vector giving the names of the variables in the matrix **x**. By default, names are "x1", "x2", etc.

ynames: optional character vector giving the names of the columns of the matrix **y**. By default, names are "y1", "y2", etc.

q: logical flag, if TRUE, the **q** matrix from an orthogonal decomposition of **x** is returned. Default FALSE.

VALUE:

structure with the following components:

coef: matrix of coefficients with one column for each regression and (optional) constant terms in first row.

resid: structure like **y** containing residuals.

r,corth: components of orthogonal decomposition of **x** matrix.

names: names of x variables (for later use by **regprt**).

ynames: names of y variables (for later use by **regprt**).

int: records whether intercept was used in this regression.

q: matrix from orthogonal decomposition if **q** argument was TRUE.

sqrtw: vector of the square-roots of the input vector **wt** (only if **wt** was given).

EXAMPLES:

       reg(cbind(a,b,c),y)   #regress y on a, b, and c with intercept

---

| **regress** | Regression Printing | **regress** |

USAGE:

regress(x, y, wt, int, print, names, ynames, q)
regprt(regstr, names, ynames)

The functions **regress** and **regprt** are related to the **reg** function, since they print a formatted result of the regression. **regress(...)** is equivalent to **regprt(reg(...))**. **regprt** takes the structure produced by **reg** and prints a summary of the regression statistics. **regress** and **regprt** can take vectors of names to augment the printing of the regression summary.

See **reg** for details about arguments to **regress**.

EXAMPLES:

regress(stack.x,stack.loss,names=stack.collab)

```
                             Coef       Std Err    t Value
Intercept                 -39.91967    11.89599   -3.355723
air flow                    0.7156403   0.1348582   5.306613
cooling water inlet temp    1.295286    0.3680243   3.519567
acid concentration         -0.1521226   0.1562940  -0.973310

Residual Standard Error = 3.243364  Multiple R-Square = 0.913577
N = 21        F Value = 59.9022 on 3, 17 df

Covariance matrix of coefficients:
                             Intercept      air flow
Intercept                   141.5147
air flow                      0.287587    0.01818673
cooling water inlet temp     -0.651794   -0.03651067
acid concentration           -1.676321   -0.00714352

                             cooling water inlet temp
cooling water inlet temp           0.1354418
acid concentration                 0.00001047646

                             acid concentration
acid concentration                 0.02442783

Correlation matrix of coefficients:
                             Intercept      air flow
air flow                     0.1792632
cooling water inlet temp    -0.1488789   -0.7356413
acid concentration          -0.9015999   -0.3389164

                             cooling water inlet temp
acid concentration                 0.000182136
```

An explanation of some of the printed statistics: "Coef", "Std Err", and "t Value" are the coefficients of the regression, the standard errors of the coefficients, and the t-value (on 17 degrees of freedom) for the hypothesis that the coefficient is zero. "Residual Standard Error" is the square root of the error variance, the sum of squared residuals divided by the degrees of freedom for the residuals. "Multiple R-Square" is the fraction of the variance explained by the non-intercept terms of the model. (If no intercept was fit, this is the fraction of the total variance of **y** explained by the model.) "N" is the total number of observations. "F Value on n, d df" is the F-test for significance of the whole regression, where **n** is the number of degrees of freedom for the numerator (the regression), and **d** is the number of degrees of freedom for the denominator (the residuals). "Covariance and Correlation matrix of coefficients" give the variances and covariances of the coefficients, and correlations between the coefficients.

REFERENCE:

Norman Draper and Harry Smith, *Applied Regression Analysis*, Second Edition, Wiley, 1981.

---

regsum	Regression Summaries	**regsum**

USAGE:

      regsum(z)

ARGUMENTS:

z: regression structure, with components **coef, resid, r, corth, int** and (possibly) **sqrtw**. See documentation for **regress**.

VALUE:

    structure with components summarizing the regression statistics.

names: vector of names for the **x** variables in the regression: either the names provided as arguments to **reg** or "x1", "x2", etc. If the regression has an intercept, "Intercept" is the first name.

ynames: vector of names for the **y** variables in the regression: either the names provided as arguments to **reg** or "y1", "y2", etc.

coef: coefficient matrix.

stderr: matrix of standard errors associated with the coefficients of each regression.

t: matrix of t-statistics associated with the coefficients of each regression.

cor: correlation matrix of the coefficients.

cov: covariance matrix of the coefficients. In the case of more than one y variable, this matrix is standardized so that the actual covariance matrix corresponding to each y variable is the standardized covariance matrix times the square of the corresponding **rms** value.

rsq: vector of multiple R-square statistics (the fraction of the variance of **y** explained by the model).

rms: vector of standard errors of the residuals.

fval: vector of the overall F-statistics for the regressions.

df: the degrees of freedom for numerator and denominator of **fval**.

EXAMPLES:

    z ← reg(x,y) #do the regression
    zsumy ← regsum(z) #get the summary

---

rep	Replicate Data Values	**rep**

USAGE:

rep(x, times, length)

ARGUMENTS:

x: numeric structure. Missing values (NAs) are allowed.

times: how many times to replicate **x**. There are two ways to use **times**. If it is a single value, the whole of **x** is replicated that many times. If it is a vector of the same length as **x**, the result is a vector with **times[1]** replications of **x[1]**, **times[2]** of **x[2]**, etc.

length: the desired length of the result. This argument may be given instead of **times**, in which case **x** is replicated as much as needed to produce a result with **length** data values.

VALUE:

a vector of the same mode as **x** with the data values in **x** replicated according to the argument **times** or **length**.

EXAMPLES:

rep(0,100)   # 100 zeroes
rep(1:10,10)   # 10 repetitions of 1:10
rep(1:10,1:10) # 1, 2, 2, 3, 3, 3, ...

---

replace	see **append**	**replace**

---

replot	Plot Deferred Graphics File	**replot**

USAGE:

replot(file)

ARGUMENTS:
file: optional name of deferred graphics file, default "sgraph".

The contents of **file** are plotted on the current graphics device from first frame to last. There is no provision for plotting single frames.

REPORT	Report writing facility	**REPORT**

USAGE:

!S REPORT [−e] input output   # UNIX cmd - batch
!S REPORT [−e] input >output   # UNIX command
!S REPORT [−e] <input >output   # UNIX command

ARGUMENTS:
input: the name of a file containing the text of the report, along with **troff** macros and embedded S commands. The S commands are executed and their printed results replace them in the body of the input file. The S commands are surrounded in the text by braces ({ and }). Output lines from the S command have leading blanks and all quote characters (") removed. If the S expressions are surrounded by doubled braces ({{ and }}), the printed S output is left alone.

output: the name of a file which will receive all of the output from the run.

If the flag −e is given to REPORT, then single or double braces beginning S expressions must be preceded by a backslash character (\{ or \{{). This allows the input file to contain braces which are not interpreted by REPORT; in particular, the UNIX eqn preprocessor uses braces for grouping.

Warning: execution of REPORT in batch mode begins immediately after the command is typed. Processes running in batch are counted toward the total number of processes any one user is allowed. Running several versions of S (either batch or interactive) at the same time can cause you to exceed the process limit, and abort all of the runs.

EXAMPLES:

Suppose the input file named **myinput** looks like this:

```
.PP
This is a test report generated by S expressions.
For example, the mean winnings of New Jersey Lottery players
was ${ round(mean(lottery.payoff),2)}.
Another thing, there were { len(lottery.payoff)}
players during the time period.
.PP
How about a stem-and-leaf of the winnings?
.DS
{{stem(lottery.payoff)}}
.DE
Well, this is it. { "Guess I'll quit now." }
.PP
What the heck, I'd better include an in-line table to make
people happy.  For example, the first hunk of the iris data
looks like:
{tbl(iris[1:5,,1],file="iris.tbl",head="A Hunk of Iris Data")
!tbl iris.tbl >iris.out}
.so iris.out
So here it is, embedded right in the middle of the text I'm
now composing.  Pretty neat, huh?
```

The report-writing S run can be invoked in batch with

S REPORT myinput myoutput

Once the batch run is complete, the output file can be processed
with the Unix command line:
nroff −ms myoutput

to produce the following document: (The command **S REPORT
myinput|nroff -ms** would produce the same result.)

```
         This is a test report generated by S expressions.   For
example, the mean winnings of New Jersey Lottery players was
$290.36. Another thing, there were 254 players   during   the
time period.

How about a stem-and-leaf of the winnings?

N = 254    Median = 270.25
Quartiles = 194, 365

Decimal point is 2 places to the right of the colon

   0 : 8
   1 : 000011122233333333333344444
   1 : 55555566666777777888888889999999999
   2 : 000000011111111111222222233333333444444444
   2 : 55555666666666677778889999999999999999
   3 : 00000000111111222233333333444
   3 : 555555555666667777777888888899999999
   4 : 0122234
   4 : 55555678888889
   5 : 111111134
   5 : 555667
   6 : 44
   6 : 7

High: 756.0 869.5

Well, this is it. Guess I'll quit now.

     What the heck, I'd better include an in-line  table  to
make  people happy.  For example, the first hunk of the iris
data looks like:
```

```
             ----------------------------
             |_A_Hunk_of_Iris_Data__|
             |5.1 | 3.5 | 1.4 | 0.2 |
             |4.9 | 3.0 | 1.4 | 0.2 |
             |4.7 | 3.2 | 1.3 | 0.2 |
             |4.6 | 3.1 | 1.5 | 0.2 |
             |5.0 | 3.6 | 1.4 | 0.2 |
             |____|_____|_____|_____|
```

```
So here it is, embedded right in the middle of the text  I'm
now composing.  Pretty neat, huh?
```

A few comments are in order concerning the example. The table is produced by two separate steps. First, the function **tbl** writes an ascii file. Then, the Unix **tbl** command reads that file and produces a file suitable for **troff**. The reason for the latter step is that the **tbl** command will not follow **.so** lines in a document.

S expressions enclosed in doubled braces should be processed in **troff** with a constant-width font to preserve spacing.

Plotted output can be included in a phototypeset report by means of the **pic** device driver and Unix **pic** and **troff** commands. Typical embedded S commands might look like this:
```
{ pic; plot(my.x, my.y); printer
!mv pic.out fig.1 }
.PS <fig.1
```

This sequence invokes the **pic** device driver, produces a plot, and then invokes the **printer** device driver to cause **pic** to terminate. Then, the output file is renamed via the Unix **mv** command and called **fig.1**. Sequences such as this can appear many times in the document as long as the **pic.out** files are all given different names.

When the **S report** phase is through, the output file is passed through the Unix **pic** command prior to **troff**:
```
pic myoutput |troff −ms
```

---

restore	Restore Datasets from Dump File	**restore**

USAGE:

      restore(file, pos, print)

ARGUMENTS:

file: character string giving the name of a file onto which the **dump** function dumped datasets.

pos: the position in the data directory search list of the data directory onto which the restore should put the datasets. Default 1 (the work directory).

print: logical flag; when TRUE, dataset names are printed as they are put in the data directory. Default FALSE.

Typically, the pair of functions **dump** and **restore** are used to move data directories from one computer system to another. An entire data directory may be dumped, the resulting file transported to a file on another system, and the data directory restored.

NOTE:

The functions **dump** and **restore** ignore the current prefix.

EXAMPLES:

    restore("dumpfile",2) #restore contents of dumpfile onto **save** db

---

rev	Reverse the Order of Elements in a Vector	**rev**

USAGE:

    rev(x)

ARGUMENTS:

x: vector. Missing values (NAs) are allowed.

VALUE:

vector like **x** but with the order of data values reversed (the last value in **x** is the first value in the result).

EXAMPLES:

    rev(sort(t))   # sort t in descending order

---

rf	see **f**	**rf**

---

rgamma	see **gammadist**	**rgamma**

rlnorm	see **lnorm**	**rlnorm**

rlogis	see **logis**	**rlogis**

rm	Remove Datasets from a Data Directory	**rm**

USAGE:
> rm(arg1, arg2, ..., list=, print=, pos=, value=)

ARGUMENTS:

argi: datasets to be removed. Note that the arguments must actually correspond to datasets in a data directory; i.e., the argument is just a data-set name. Compare argument **list**. Missing values (NAs) are allowed.

list=: character vector of the names of datasets to be removed. The advantage over explicitly naming the datasets as above is that the datasets need not be accessed. Typically, the **list=** argument is used to remove many datasets at once, such as all the datasets under a prefix (see the example below).

print=: should names of removed datasets be printed? Default FALSE.

pos=: position in the data directory search list of the data directory from which the datasets should be removed. Default 1 (the work directory).

value=: the value that the function should return. Useful in macros that generate temporary datasets that should be removed. This allows the **rm** function to be the last function in the macro, and to return a value.

VALUE:
> the **value=** argument, if given.

EXAMPLES:
> rm(x,y,z) #remove datasets x,y,z
> rm(list=list("wages.*")) #remove anything in work directory
>   # whose name starts with "wages."
> rm(list=list("wages.*",pos=2),pos=2) #same for save directory
> rm(wages.x,pos=2) #remove wages.x from save directory

rm($Tx,value=$Tx) #remove $Tx and return its value

rnorm	see **norm**	**rnorm**

round	Rounding	**round**

USAGE:

    round(x, dec)

ARGUMENTS:

x: numeric structure. Missing values (NAs) are allowed.

dec: number of decimal places desired. Default 0. **dec** can be negative, for rounding large numbers to 10, 100, etc.

VALUE:

    structure like **x** with data rounded to the specified number of places.

EXAMPLES:

    round(mydata,dec=2)   #round to 2 decimals

    round(mydata,−1)   #round to nearest 10

row	see **col**	**row**

| **?row** | Apply a Function to the Rows of a Matrix | **?row** |

USAGE:

?row(x, fun, arg1, ...)

ARGUMENTS:

x: name of matrix. Missing values (NAs) are allowed if **fun** accepts them.

fun: (unquoted) name of function to be applied to the rows of the matrix.

argi: any other arguments that should be given to **fun**.

VALUE:

vector or matrix of results. If each invocation of **fun** produces a vector of length **n** (n>1), the result will be an **nrow(x)** by **n** matrix. Otherwise, the result will be a vector of length **nrow(x)**.

SEE ALSO:

Function **apply**.

EXAMPLES:

?row(y,median)   # row medians of matrix y
?row(y,mean,trim=.25)   #25% trimmed row means

| **?rperm** | Random Permutation | **?rperm** |

USAGE:

?rperm(x)

ARGUMENTS:

x: vector of data. Missing values (NAs) are allowed.

VALUE:

random permutation of the values in **x**.

SEE ALSO:

Function **sample**.

EXAMPLES:

>?rperm(mydata)

rreg	Robust Regression	**rreg**

USAGE:

>rreg(x, y)   # simple form
>rreg(x, y, w, int, init, method, wx, iter, k, acc, stop, conv)

ARGUMENTS:

x: matrix of independent variables for regression. Should not include a column of 1's for the intercept.

y: vector of dependent variable, to be regressed on **x**.

w: initial weights for robustness. **w** may be the weights computed from residuals in previous iterations of **rreg**. The argument **wx** should be used for weights that are to remain constant from iteration to iteration.

int: should intercept term be included in the regression? Default TRUE.

init: optional vector of initial coefficient values (normally the result of some other regression, e.g., **reg(x,y)$coef** or **l1fit(x,y)$coef**). When omitted the initial value is computed as follows: if **wx** and/or **w** is supplied, it is the weighted least squares estimate, otherwise it is the ordinary least squares estimate.

method: choice of method (see below). Default is the converged Huber estimate followed by two iterations of Bisquare.

wx: optional weighting vector (for intrinsic weights, not the weights used in the iterative fit).

iter: maximum number of iterations; default 20.

k: constant in the weighting function. This constant is chosen to give the estimate a reasonable efficiency if the errors do come from a normal distribution. (See below for exact values.)

acc: convergence tolerance; default 10*sqrt(machine precision).

stop: method of testing conversion. Values 1 (default),2,3,4 use relative change in residuals, coefficients and weights and an orthogonality test of residuals to x.

conv: should component **conv** be returned as a result? Default FALSE.

VALUE:

structure with the following components:

coef: vector of coefficients in final fit.
resid: vector of final residuals.
w: vector of final weights in the iteration, excluding the influence of
             wx.
int: flag telling whether intercept was used.
method: name of robust weighting rule used.
k: value of k used for method.
conv: vector of the value of the convergence criterion at each iteration.

METHOD:

The routine uses iteratively reweighted least squares to approximate the robust fit, with residuals from the current fit passed through a weighting function to give weights for the next iteration. There are 8 possible weighting functions, all specified by character strings given as method: "andrews", "bisquare", "cauchy", "fair", "huber", "logistic", "talworth", and "welsch". The corresponding default values of k are 1.339, 4.685, 2.385, 1.4, 1.345, 1.205, 2.795, and 2.985. Method "huber" gives more least-squares-like fits usually; the proper choice of method, however, is still a research problem.

REFERENCE:

Coleman, D., Holland, P., Kaden, N., Klema, V., and Peters, S. C., "A system of subroutines for iteratively re-weighted least-squares computations", *ACM Trans. Math. Soft.*, Vol. 6, 327-336, 1980.

| rstab | see stab | rstab |

?rsweep	Sweep Out Row Effects from Matrix	**?rsweep**

USAGE:

    ?rsweep(x, summary, fun)

ARGUMENTS:

x: matrix. Missing values (NAs) are allowed.

summary: values to be swept out of the rows of **x**. Generally **summary** is the result of a **?row** macro or applying a function to the rows of **x**. Missing values (NAs) are allowed.

fun: (unquoted) name of function to be used in sweeping out **summary**. Default is − (minus).

VALUE:

    matrix **x** with the summary values swept out.

SEE ALSO:

    Function **sweep**.

EXAMPLES:

    ?rsweep(x, ?row(x,median)) # subtract off row medians

---

rt	see **tdist**	**rt**

---

runif	see **unif**	**runif**

---

sabl	Seasonal Decomposition	**sabl**

USAGE:

    sabl(x, power, calendar, trend, seasonal, revisions)

ARGUMENTS:

x: the time-series to be decomposed.

power: vector of powers for transforming **x**. Sabl will pick the value from **power** that minimizes a measure of the interaction between trend and seasonal components. For a value **p** in **power**, **p>0** corresponds to the transformation $x\hat{}p$; **p==0** to **log(x)**; and **p<0** to $-x\hat{}p$. If **x** has any zero or negative values, no transformation is made and **power** defaults to 1. Otherwise, the default is **c(−1,−.5,−.25,0,.25,.5,1)**

calendar: if FALSE, (default) no calendar component is computed. Calendar computation can be done only for monthly data.

trend: number of points in the trend smoothing window, an odd integer greater than 2. Default 11.

seasonal: number of points in the seasonal smoothing window, an odd integer greater than 2. Default 15.

revisions: if FALSE, (default) no revisions are calculated. The series must be at least 7 cycles long for revisions to be calculated. A maximum of 5 cycles of revisions are calculated.

VALUE:

structure with the following components:

trend: time-series giving the long term change in level.

seasonal: time-series giving the part of **x** that repeats or nearly repeats every **nper(x)** time units. This series contains predicted seasonal values for one additional cycle.

irregular: time-series giving the noisy variation not explained by **trend** or **seasonal** (or **calendar** if computed).

transformed: the series **x** after power transformation and month length correction, from which the components are extracted.

adjusted: time-series with the seasonal component and calendar component (if computed) removed, on the original (untransformed) scale.

calendar: time-series of variation due to day-of-the-week effect. Returned if argument **calendar** is TRUE. This series contains predicted calendar values for one additional cycle.

power: power that was actually used in transforming the time-series **x**.

tstat: vector of t statistics used to pick the power actually used to transform **x**. Only returned if length of the argument **power** is >1.

revisions: time-series of revisions, if computed.

weights: time-series of final robustness weights used in the decomposition.

The components returned by **sabl** are related as follows:

transformed equals **trend** + **seasonal** + **irregular**
(if no calendar component was computed)

or

transformed equals **trend** + **seasonal** + **calendar** + **irregular**
(if calendar component was computed)

REFERENCES:

William S. Cleveland, Susan J. Devlin, and Irma J. Terpenning (1981), "The SABL Statistical and Graphical Methods", and "The Details of the SABL Transformation, Decomposition, and Calendar Methods", Bell Laboratories Memoranda.

William S. Cleveland and Susan J. Devlin, "Calendar Effects in Monthly Time Series: Modeling and Adjustment", *Journal of the American Statistical Association*, Vol 77, No. 379, pp. 520-528, September 1982.

SEE ALSO:

?sablplot, and **monthplot**.

EXAMPLES:

h ← sabl(hstart)   #decomposition of housing starts series
tsplot(hstart, h$adjusted, type="pl")

?sablplot	Sabl Decomposition - Data and Components Plot	?sablplot

USAGE:

?sablplot(x, title)

ARGUMENTS:

x: data structure containing the result of a call to the function **sabl**.
title: main title for the page of plots.

EFFECT:

This macro produces one page of plots, showing in separate plots the transformed (month-length corrected) series, and the trend, seasonal, calendar (if there is one), and irregular components. To the right of each plot is a bar which portrays the relative scaling of that plot.

REFERENCE:

>William S. Cleveland and Irma J. Terpenning, "Graphical Methods for Seasonal Adjustment", *Journal of the American Statistical Association*, Vol. 77, No. 377, pp. 52-62.

EXAMPLES:

>h ← sabl(hstart)
>?sablplot(h,"Housing Starts")

---

sample	Generate Random Samples or Permutations of Data	**sample**

USAGE:

>sample(x, size, replace)

ARGUMENTS:

x:
: numeric or character vector of data (the population) to be sampled, or a scalar giving the size of the population. Missing values (NAs) are allowed.

size:
: sample size. Default is the same as the population size, and thus (with **replace=FALSE**) will generate a random permutation.

replace:
: logical flag, if TRUE, sampling will be done with replacement. Default FALSE.

VALUE:

>if **x** is a population vector, the result is a sample from **x**; otherwise, the result is a set of integers between 1 and **x** giving the indices of the selected observations.

EXAMPLES:

>**sample(state.name,10)**   # pick 10 states at random
>**sample(1000000,75)** # pick 75 numbers between 1 and one million
>**sample(50)**   # random permutation of numbers 1:50

| **?sample** | Random Sample | **?sample** |

USAGE:

    ?sample(x, n)

ARGUMENTS:

x: vector of data. Missing values (NAs) are allowed.

n: number of values from **x** to be included in the sample. Default is to prompt the user for the desired sample size.

VALUE:

    sample of **n** values randomly chosen from **x**.

SEE ALSO:

    Function **sample**.

EXAMPLES:

    ?sample(mydata,10) # chose sample of 10 values from mydata

---

| **sapply** | Apply a Function to Components of a Structure | **sapply** |

USAGE:

    sapply(x, fun, max=, arg1, arg2, ...)

ARGUMENTS:

x: hierarchical structure. An arbitrary S function will be applied to each component of **x**, and the result will become the corresponding component of the result of **sapply**. Missing values (NAs) are allowed if **fun** accepts them.

fun: character string giving the name of the function.

max=:   optional argument (only in the **name=expression** form) giving the maximum size for the result of any single application of the function. For example, if **fun** returned a result the same size as its argument, the maximum size of the components of **x** is the relevant number. Default 1000.

argi: other arguments to **fun**, if any.

VALUE:

structure whose first component is the result of **fun** applied to the first component of **x**, etc. If all the results are the same length, **sapply** returns a matrix with one column for each component. If all the results are scalars, a vector is returned.

EXAMPLES:

sapply(x,"mean") #vector of means of components of x
sapply(x,"sort") # sort the components

SEE ALSO:

Function **apply** can be used to perform similar operations on the sections of a matrix or array, and **tapply** operates on data classified by categorical variables.

---

save	Save Data in Save Data Directory	save

USAGE:

save(arg1, arg2, ..., pos=)

ARGUMENTS:

argi: structure to be saved in the data directory. If **argi** is given in **name=value** form, the value is saved under the given name. Otherwise, **argi** should be a dataset name.

pos=:   the position in the data directory search list of the data directory into which the data should be saved. Default 2 (normally, the user's save directory).

VALUE:

a structure composed of arg1, arg2, ...

Note that **save** does not automatically print its result.

EXAMPLES:

print(save(a,b,y=1:10))   #saves a, b and y
            # and prints their values

scale	Scale Columns of a Matrix	**scale**

USAGE:

> **scale(x, center, scale)**

ARGUMENTS:

x: matrix to be scaled. Missing values (NAs) are allowed.

center: control over the value subtracted from each column. If TRUE (the default), the mean of (the non-missing data in) each column is subtracted from each column. If given as a vector of length **ncol(x)**, this vector is used; i.e., **center[j]** is subtracted from column **j**. If FALSE, no centering is done.

scale: control over the value divided into each column to scale it. If TRUE (the default) each column (after centering) is divided by the square root of sum-of-squares over **n−1**, where **n** is the number of of non-missing values. If given as a vector of length **ncol(x)**, column **j** is divided by **scale[j]**. If FALSE, no scaling is done.

VALUE:

> matrix like **x** with optional centering and scaling.

> The default values of **center** and **scale** produce columns with mean 0 and standard deviation 1.

EXAMPLES:

> **scale(x)** #scale to correlation (0 mean, 1 std dev)
> **scale(x, center=apply(x,2,"median"),**
>   **scale=FALSE)** #remove column medians, do not scale
>   # see also **sweep**

scandata	Scan fixed format data file	scandata

USAGE:

!S scandata <**file** >**summary**   # UNIX command

ARGUMENTS:

file: the name (unquoted) of a file containing fixed-format input data, a sequence of records with fields in specific columns. A maximum of 500 characters per record is permitted.

summary: the name of the output file which will summarize the data. For each column in **file** a summary will be produced describing the number (and percentage) of occurrences of specific characters in that column (a 1-dimensional contingency table if the column contains categorical data).

Special characters (newlines, control characters, etc.) are represented by the following notation in **summary:** ^A is control-A, ... ^J is control-J (newline), etc.

The summary also includes information on the number of records in the file and the maximum record length.

SEE ALSO:

extract.

EXAMPLES:

If the file "mydata" contains:

```
123
abc
xyz
aaa
321
```

Then the command
S scandata <**junk**
produces:

```
Char    Count      %     Column 1

 '1'       1     20.00
 '3'       1     20.00
 'a'       2     40.00
 'x'       1     20.00
```

Char	Count	%	Column 2
'2'	2	40.00	
'a'	1	20.00	
'b'	1	20.00	
'y'	1	20.00	

Char	Count	%	Column 3
'1'	1	20.00	
'3'	1	20.00	
'a'	1	20.00	
'c'	1	20.00	
'z'	1	20.00	

Char	Count	%	Column 4
'^J'	5	100.00	

```
Longest Record (including final newline) = 4
Total Records  = 5
```

---

search	Print Current Search List	**search**

USAGE:

> **search(which)**

ARGUMENTS:

which: integer defining which search list to return: 1=chapters (for functions); 2=data directories. Default 2.

VALUE:

> character vector giving the list of data directory or chapter names currently being searched. See **chapter, attach** and **detach** for modifying the current search lists.

EXAMPLES:

> **search( )** #list of active data directories

| **segments** | Plot Disconnected Line Segments or Arrows | **segments** |

USAGE:

> **segments(x1, y1, x2, y2)**
> **arrows(x1, y1, x2, y2, size, open, rel)**

ARGUMENTS:

x1,y1,x2,y2:   co-ordinates of the end-points of the segments or arrows. Lines will be drawn from **(x1[i],y1[i])** to **(x2[i],y2[i])**. Missing values (NAs) are allowed.

size: width of the arrowhead as a fraction of the length of the arrow if **rel** is TRUE. If **rel** is FALSE, **size** is arrowhead width in inches. Default 0.2.

open: logical, if TRUE the arrowhead is "v" shaped, if FALSE (default) it is diamond shaped.

rel: logical, should arrowhead be sized relative to the length of the arrow? Default FALSE.

Graphical parameters may also be supplied as arguments to this function (see **par**).

Note the distinction from **lines** which draws a curve (connected line segments). Any segments with any NAs as end-points will not be drawn.

EXAMPLES:

```
# draw arrows from ith to i+1st points
s ← seq(len(x)-1)  # sequence one shorter than x
arrows( x[s], y[s], x[s+1], y[s+1] )
```

| **select** | Component Selection | **select** |

USAGE:

> **str$component**
> **str$[icomp]**
> **select(str, scomponent)**

ARGUMENTS:

str: an expression evaluating to a structure.
component: the name of a component of **str**. This can contain the minimum number of characters adequate to uniquely distinguish the component from other components in **str**.
icomp: an integer giving the position of a single component within **str**.
scomponent: character string giving the name of a component of **str**.

VALUE:

the selected component.

The $ operator has very high precedence, and will be done before any other operators. Thus, if **str** is an expression returning a structure, it should ordinarily be parenthesized.

EXAMPLES:

regress(x,y)$coef

array2$Dim[1]   #first element of Dim component of array2

for(i in seq(ncomp(z))) hist(z$[i])
   #histogram of each of the components of z

---

seq	Sequences	**seq**

USAGE:

from:to   # as operator
seq(from, to, by, length)   # as function

ARGUMENTS:
from: starting value of sequence.
to: ending value of sequence.
by: spacing between successive values in the sequence.
length: number of values in the sequence.

VALUE:

a numeric vector with values (**from, from+by, from+2*by, ... to**). If arguments are omitted, an appropriate sequence is generated; for example, with only one argument a sequence from **1** to the value of the single argument is constructed. **from** may be larger or smaller than **to**. If it is specified, **by** must have appropriate sign to generate a finite sequence. If only **from** is specified, and it is a vector with more than one value, a sequence from **1** to **len(from)** is generated.

When used as an operator, **:** has a very high precedence. Thus to create a sequence from **1** to **n−1**, parentheses are needed **1:(n−1)**.

EXAMPLES:

seq(5) #1,2,3,4,5
1:5 #same thing

5:1 #5,4,3,2,1

seq(0, 1, .01) #0,.01,.02,.03,...,1.

seq(x) # 1, 2, ..., len(x)

seq(−3.14,3.14,len=100) # 100 values from −pi to pi

---

show	Show Current Printer Plot	**show**

USAGE:

show( )

show provides a facility for viewing the current state of a printer plot. When **show** is invoked, the printer plot is produced. It can then be further augmented (with titles, lines, text), saved for deferred graphics, etc. If further graphics commands are used to augment a previously shown plot, the plot will be displayed again when any graphics functions advance to the next frame. If nothing has been added to a previously shown plot, the advance to the next frame will be silent.

EXAMPLES:

printer
plot(x,y)   # scatter plot
show   # preview it
title("an interesting plot")   #add to it

| **sin** | see **cos** | **sin** |

| **sink** | Send S Output to the Diary or to a File | **sink** |

USAGE:

    sink(file)

ARGUMENTS:

file: optional, logical value, or character string giving the name of a file.

If **file** is TRUE, it indicates that S output that is ordinarily typed on the terminal is is to be appended to the diary file. If **file** is a character string, the output diverted to a file named **file**. If **file** is FALSE or omitted, any diversion is ended.

When output is being diverted to the diary or a sink file, nothing appears on the terminal except prompt characters, printer plots, and error messages.

The difference between **sink(TRUE)** and **sink("diary")** is that the first expression causes output to be appended to the diary whereas the second expression overwrites the diary.

EXAMPLES:

    diary #turn on the diary
    sink(T)   # divert output to the diary
    regress(x,y) # save regression output on diary
    sink( ) # revert output to the terminal

| **smatrix** | Print a Symbolic Matrix for Multivariate Data | **smatrix** |

USAGE:

> **smatrix(x, set, plength, nstrike, spread, scale,**
>   **rowlab, collab, head, na)**

ARGUMENTS:

x: matrix of data. One symbol (possibly overstruck) will be printed for each element of **x**. One line of printing is produced for each row. Missing values (NAs) are allowed.

set: optional character string. The number of characters per symbol is given by **nstrike** (below). If **nstrike==1**, the first character of **set** is the first symbol, etc. If **nstrike>1**, the first **nstrike** characters of **set** are overstruck to form the first symbol, the second **nstrike** characters to form the second symbol, etc. By default, if **nstrike==1**, the symbols are ".+*". If **nstrike==2**, the default symbols are ".", "+", "*" and overstruck "$" and "X". The equivalent value of **set** would be ". + * $X".

plength: number of lines to print per page (exclusive of running head). At the end of each page, a page eject is executed. The head string (see below) is printed at the top of each page. Default page length is 50.

nstrike: number of lines superimposed to make the symbols. See the examples under **set** above. Default 1.

spread: number of blanks (maximum) to put between the columns. Default 3.

scale: logical, if TRUE, the columns of **x** will be scaled so that the minimum in each column is 0 and the maximum is 1. If FALSE, the assumption is that **x** has been scaled to the range 0 to 1 by some other procedure. Default TRUE.

rowlab: optional character vector for labelling the rows. By default, labels are "1", "2", etc. and only every fifth label is plotted. If supplied, the label vector must have length equal to **nrow(x)**.

collab: optional character vector for labelling the columns. By default, labels are "1", "2", etc. and only every fifth label is plotted. If supplied, the label vector must have length equal to **ncol(x)**.

head: optional character string for a running head. This is plotted as a title at the top of each page. By default, the name of **x** is used. If there is more than one page, the page number is included in the title.

na: character to be printed corresponding to missing values (NAs) in **x**. Default "N".

The symbols are chosen by dividing the range 0 to 1 into **ns** equal intervals, where **ns** is the number of characters in **set**, divided by **nstrike**.

EXAMPLES:

south ← state.region==2
smatrix(votes.repub[south,],rowlab=state.name[south],
collab=encode(votes.year),head="Southern State Votes")

```
Southern State Votes

Alabama         N N N + + + + + + . + + . + . + + *  . . . . + . + + * . * +
Arkansas        N N N + . + + + + + + * + + + * + + . + + + + + + . + + .
Delaware        * * + + . * * * * * * * * * * * * * * * * * * * * . * . *
Florida         N N N N + * * * * N + . . . . . + + + + + + + * * * . + + * .
Georgia         N N N . . . + * * * * * * * * * * * * * * * * * * . * . *
Kentucky        . . . . + + * * * * * * * * * . + . + * . + . *
Louisiana       N N N . + * + * . + . . . . . . + . . . + . + . *
Maryland        . . + . . * * * * * * * * * * + * * * * * * * + . . * . *
Mississippi     N N N N + . . + . . . . . . : . : . . . . . . + . . * . *
North Carolina  N N N + + * * * * * * * * . * * * + + + + + + * . * * *
Oklahoma        N N N N N N N N N N N N * * + * * * + * * * * * . * + * + +
South Carolina  N N N * * * + . . + . . . . . . . . . . . . * . * + * + +
Tennessee       N N N * . + * * * * * * * * + * * * + * * * * * . * + *
Texas           N N N N . . . . . + + + + + . . + . * . . + . + * * . * + *
Virginia        . . N N . + * * * * * * * + * * * * * * * * * * . * + *
West Virginia   N N * * . + * * * * * * * + * * * * * * * * * * + . * . +

                1 1 1 1 1 1 1 1 1 1 1 1 1 1 1 1 1 1 1 1 1 1 1 1 1 1 1 1
                8 8 8 8 8 8 8 8 8 8 8 9 9 9 9 9 9 9 9 9 9 9 9 9 9 9 9 9
                5 6 6 6 7 7 8 8 8 9 9 0 0 0 1 1 2 2 2 3 4 4 4 5 5 6 6 6 7 7
                6 0 4 8 2 6 0 4 8 2 6 0 4 8 2 6 0 4 8 2 6 0 4 8 2 6 0 4 8 2 6
```

---

smooth	Non-linear Smoothing Using Running Medians	**smooth**

USAGE:

smooth(x, t)

ARGUMENTS:

x: vector (at least 5 points) to be smoothed.

t: logical flag, should twicing be done? Default TRUE. (Twicing is the process of smoothing, computing the residuals from the smooth, smoothing these and adding the two smoothed series together.)

VALUE:

smoothed vector using 4(3RSR)2H twice.

REFERENCE:

J. W. Tukey, *Exploratory Data Analysis,* Chapters 7 and 16, Addison Wesley.

---

solve	Solve Linear Equations and Invert Matrices	**solve**

USAGE:

solve(a, b)

ARGUMENTS:

a: matrix of coefficients. Must be square and non-singular.

b: optional matrix of coefficients. If **b** is missing, the inverse of matrix **a** is returned.

VALUE:

the solution **x** to the system of equations **a** %* **x** = **b**.

EXAMPLES:

ainv ← solve(a)   #invert a

sort	Sort in Ascending Numeric or Alphabetic Order	**sort**

USAGE:

    sort(data)

ARGUMENTS:

data: vector.

VALUE:

    vector with its data sorted in ascending order. Character data is sorted in lexicographic order.

    The result is a vector, even if **data** was a structure (matrix, time-series).

    The ordering of character data depends on the character code. For ASCII code, digits precede upper-case letters, which precede lower-case letters. The position of other characters is unintuitive.

source	Execute S Expressions from a File	**source**

USAGE:

    source(file, return)

ARGUMENTS:

file: character string giving the name of a file. Commands are read from the file, instead of from the terminal. This continues until an end-of-file is encountered, until any error (syntax or execution) is generated, or until **source** with a null argument is encountered. The argument **sourcexit** to function **options** can be used to allow execution to continue after errors in a source file.

return: logical, after completing the actions contained in **file**, should execution return to the context it was in prior to executing **source**. Default TRUE. Using **return=FALSE** is recommended if the **source** invocation is the last action of a macro or source file, since it prevents a (growing) stack of source files, which may eventually use up all available file descriptors.

Source files can themselves invoke source files, macros, etc.

Generally, nothing should follow the source command on the same line or an anomalous effect may occur; for example,
source("abc"); 1+2
will evaluate **1+2** before reading anything from file "abc". Similarly, **source** should not be used inside loops.

---

split	Split Data by Groups	**split**

USAGE:

    split(data, group)

ARGUMENTS:

data: vector structure containing data values to be grouped. Missing values (NAs) are allowed.

group: vector or category giving the group for each data value. For example, if the third value of **group** is 12, the third value in **data** will be placed in a group with all other data values whose group is 12.

VALUE:

    structure in which each component contains all data values associated with a group. Within each group, data values are ordered as they originally appeared in **data**. The name of the component is the corresponding value in **group**, or the corresponding category name.

USAGE:

    The main use for **split** is to create a data structure to give to **boxplot**. A combination of **code** and **tapply** is usually preferred to using **split** followed by **sapply**.

EXAMPLES:

    boxplot(split(income,month))
    split(people,age%/10)   # note integer divide
    split(gnp,cycle(gnp)) #component for each month
    split(student,grade)

sqrt	see **exp**	sqrt

stab	Stable Family of Distributions	stab

USAGE:

rstab(n, par1, par2)

ARGUMENTS:

n: number of random values desired.

par1:   parameter for the member of the stable family desired. These are specified in the form given in the reference. The parameter **par1** is usually called the index of the family: 2 corresponds to the normal, 1 to the Cauchy. Generally, smaller values mean longer tails. Only values between 0 and 2 are legal.

par2:   modified skewness parameter (see the reference). Negative and positive values correspond to skewness left and right.

VALUE:

vector of **n** values from the chosen family.

Stable distributions are of considerable mathematical interest. Statistically, they are used mostly when an example of a very long-tailed distribution is required. For small values of **par1**, the distribution degenerates. See the reference and other works cited there.

Note that there are no probability or quantile functions for this distribution. The efficient computation of such values is an open problem.

REFERENCE:

J. M. Chambers, C. L. Mallows, and B. W. Stuck, "A Method for Simulating Stable Random Variables", *JASA*, Vol 71, pp 340-344, 1976.

EXAMPLES:

hist(rstab(200,1.5,1.5)) #fairly long tails, skewed right

| **stamp** | Time Stamp Output, Graph, and Diary | **stamp** |

USAGE:

stamp(name, print, plot)

ARGUMENTS:

name: optional string to be printed, plotted, and put on diary file. Default is the current date and time in the form "Tue Jun 1 14:00:16 EDT 1982".

print: should **name** be printed on the users terminal? Default TRUE.

plot: should **name** be plotted in the lower right corner of the current plot? Default TRUE if a graphical device has been specified.

Graphical parameters may also be supplied as arguments to this function (see **par**).

Since the current date and time is plotted on the current graphical device as well as appearing on the terminal and in the diary file, the **stamp** function makes it easy to identify the functions used to create a plot or display.

EXAMPLES:

diary
plot(x,y)
this is an interesting plot
stamp

| **stars** | Star Plots of Multivariate Data | **stars** |

USAGE:

stars(x, full, scale, radius, type, labels, head,
**   max, byrow, nrow, ncol)**

ARGUMENTS:

x: matrix of data. One star symbol will be produced for each row of the matrix. Missing values (NAs) are allowed.

full: logical, if TRUE, the symbols will occupy a full circle. Otherwise, they occupy the (upper) semi-circle only. Default TRUE.

scale: logical, if TRUE, the columns of the data matrix are scaled so that the maximum value in each column is 1 and the minimum 0. If FALSE, the presumption is that the data has been scaled by some other algorithm to the range $0<=x[i,j]<=1$. Default TRUE.

radius: logical, if TRUE (default), the radii corresponding to each variable in the data will be drawn (out to the point corresponding to $x[i,j]==1$).

type: optional character string, giving the type of star to draw. Reasonable values are "l", "p", "b" for lines, points and both. Default is "l".

labels: optional character vector for labelling the plots. By default, labels are "1", "2", etc. If supplied, the label vector must have length equal to **nrow(x)**.

head: optional character string for a running head. This is plotted as a title at the top of each page. By default, the name of the data is used. If there is more than one page, the page number is included in the title.

max: optional, suggested limit for the number of rows or columns of plots on a single page. Default forces all symbols to be on one page. The algorithm tries to choose an array of plots which efficiently uses the available space on the display.

byrow: logical flag, should the symbols be plotted row-by-row across the page, or column-by-column? Default FALSE.

nrow:
ncol: optionally may be given to specify exactly the number of rows and columns for the array of plots on each page.

EXAMPLES:

the example plot is produced by:
stars(votes.repub[state.region==1,]/100,radius=T,scale=F,
labels=encode(state.name[state.region==1]),
head="Republican Votes (Northeast) 1856 - 1976")

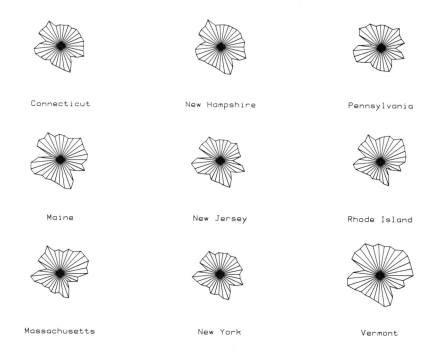

Republican Votes (Northeast) 1856 - 1976

Connecticut	New Hampshire	Pennsylvania
Maine	New Jersey	Rhode Island
Massachusetts	New York	Vermont

starsymb	Plot a Single Star Symbol	**starsymb**

USAGE:

starsymb(x, full, scale, radius, type, collab, sample)

ARGUMENTS:

x: data matrix, as passed to stars. Missing values (NAs) are allowed.

full: logical, TRUE (the default) if full 360 degree symbols wanted.

scale: logical, TRUE (the default), if the columns of the matrix should be scaled independently.

radius: logical, if TRUE (default), radii corresponding to each variable will be drawn.

type: the type of the plotted symbol (see **stars**).

collab: vector of character string labels for the variables (columns of **x**).

sample: which of the star symbols for the data **x** (row of **x**) should be shown to illustrate the symbol? Default 1.

EXAMPLES:

```
#  plot the republican votes from New Jersey
starsymb(votes.repub/100, collab=encode(votes.year),
    sample=30,scale=F)
title(main=encode("Republican Votes for",state.name[30]))
```

Republican Votes for New Jersey

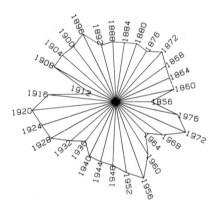

start	Time-Series Parameters	start

USAGE:

        start(x)
        end(x)
        nper(x)

ARGUMENTS:

x: time-series. Missing values (NAs) are allowed.

VALUE:

        vector giving the appropriate time parameter of **x**. In the case of
start and **end**, the result has 2 values, the year and month.

The values of these functions are appropriate for any time-series function taking arguments start, end, nper.

EXAMPLES:

        # assume ts is monthly from Jan. 1970 to Jul. 1975
        start(ts)   #is the vector c(1970,1)
        end(ts)   # is the vector c(1975,7)
        nper(ts) #is 12
        xyz←ts(data,start=start(gnp),nper=nper(gnp))
            # create time-series beginning at same time as gnp

---

stem	Stem and Leaf Display	**stem**

USAGE:

stem(x, nl, scale, twodig, fence, head, depth)

ARGUMENTS:

x: numeric vector to be displayed. Missing values (NAs) are allowed.

nl: number of different leaf values on a stem. Allowed values are 2, 5, 10. Default is to determine an appropriate value automatically.

scale: position at which break occurs between stem and leaf, counting to the right from the decimal point; e.g., -1 would break between the tens and the units digit. By default, a suitable position is chosen from the range of the data.

twodig: number of leaf digits, 1 or 2 (default 1).

fence: the multiple of the inter-quartile range used to determine outliers. By default, any point further than 2 inter-quartile ranges from the nearest quartile is considered an outlier, and is printed separately from the body of the stem and leaf display. If the inter-quartile range is zero, the algorithm performs outlier detection by means of quartiles of the remainder of the data after exclusion of values equal to the median and quartiles.

head: logical, default TRUE, which controls heading giving median, quartiles, and counts of data values and NAs.

depth: logical, default FALSE, controls printing of optional initial columns of **stem** output: depth and count. The count is the number of data values on a line. The depth is the cumulative sum of the counts to the nearer extreme.

EXAMPLES:

stem(lottery.payoff)

```
N = 254   Median = 270.25
Quartiles = 194, 365

Decimal point is 2 places to the right of the colon

   0 : 8
   1 : 0000111222333333333333344444
   1 : 55555566666677777788888889999999999
   2 : 000000011111111111122222223333333344444444
   2 : 5555566666666667777788899999999999999999
   3 : 0000000011111122223333333333444
   3 : 5555555556666667777777788888889999999
   4 : 0122234
   4 : 55555678888889
   5 : 111111134
   5 : 555667
   6 : 44
   6 : 7

High: 756.0 869.5
```

---

subset	Subsets of Data	**subset**

USAGE:

> **data [expr]**   # vector subscripts
> **data [expr1, expr2, ...]**   # multi-dimensional array subscripts

ARGUMENTS:

data: a matrix, array or vector structure. Missing values (NAs) are allowed.

expr: either a logical vector of same length as **data** or an (integer) vector of subscripts, in which case the subscripts must be all positive or all negative. Positive values select those elements; negative values select all except those elements. Missing values (NAs) are allowed if **expr** is logical.

For multi-dimensional data, **expri** should be a logical vector of the same length as the extent of the ith subscript, or an integer vector of subscripts for the ith dimension.

VALUE:

> a vector of same mode as **data** containing either the elements for which **expr** is TRUE (if **expr** is logical), or the elements given by positive integer **expr,** or all elements other than those given by negative integer **expr.**

EXAMPLES:

> **x[x!=999.999]** # x values not equal to 999.999
>
> **x[order(y)]**   # sort x by increasing values of y
>
> **x[−c(1,3)]** # all but the first and third

> If more than one **expri** is given, then **data** must have exactly the same number of dimensions as there are subscripts. The first expression selects on the first dimension, the second on the second, etc.

> All vector structures, including arrays, can appear with one subscript, in which case the result is a vector.

EXAMPLES:

> **A ← array(1:30, c(5,3,2))**   # array with dimension 5 x 3 x 2
>
> **A[1,1,1]**   # a scalar, the first data value of A
> **A[1]**   # the same
>
> **A[,1:2,]**   # a (5,2,2) array
>
> **A[A>3]**   # the vector 4:30

---

subtree	Extract Part of a Cluster Tree	**subtree**

USAGE:

> **subtree(tree, leaves)**

ARGUMENTS:

tree: a cluster tree structure, normally the result of a call to **hclust.**
leaves: vector of objects (i.e., row numbers of the original data matrix)

which should be included in the extracted subtree.

VALUE:

the smallest subtree of **tree** which includes all the **leaves**. The subtree includes the components **merge**, **height** and **order** described under **hclust**. It can be used in the various summaries of cluster results, such as **plclust**.

EXAMPLES:

z ← **hclust(dismat)**
subtree(z,c(1,10)) #subtree including 1st and 10th objects

---

sum	see **prod**	**sum**

---

svd	Singular Value Decomposition	**svd**

USAGE:

svd(y, nu, nv)

ARGUMENTS:

y: matrix of arbitrary size.

nu: optional number of columns wanted in matrix **u**, may be zero. Default is **min(nrow(y),ncol(y))**.

nv: optional number of columns wanted in matrix **v**, may be zero. Default is **min(nrow(y),ncol(y))**.

VALUE:

structure containing the components of the singular value decomposition, **y=u %* d %* t(v)**

u: if **nu>0**, an **nrow(y)** by **nu** matrix of unit orthogonal columns. Missing if **nu=0**.

d: the vector of singular values (diagonal elements of matrix **d**).

v: if **nv>0**, **ncol(y)** by **nv** matrix of unit orthogonal columns. Missing if **nv=0**.

sweep	Sweep Out Array Summaries	**sweep**

USAGE:

sweep(a, margin, stats, fun, arg ...)

ARGUMENTS:

a: array. Missing values (NAs) are allowed.

margin: vector describing the dimensions of **a** that correspond to **stats**.

stats: vector giving a summary statistic of array **a** that is to be swept out. Missing values (NAs) are allowed.

fun: character string name of function to be used in sweep operation. Default "−" (subtraction).

arg: optional arguments to **fun**.

VALUE:

an array like **a**, but with marginal statistics swept out as defined by the other arguments. In the most common cases, the function is "−" or "/" to subtract or divide by statistics that result from using the **apply** function on the array. For example: **colmean←apply(z,2,"mean")** computes column means of array **z**. **zcenter←sweep(z,2,colmean)** removes the column means.

More generally, based on **margin**, there are one or more values of **a** that would be used by **apply** to create **stats**. **sweep** creates an array like **a** where the corresponding value of **stats** is used in place of each value of **a** that would have been used to create **stats**. The function **fun** is then used to operate element-by-element on each value in **a** and in the constructed array.

EXAMPLES:

a ← sweep(a,2,apply(a,2,"mean")) # remove col means
a ← sweep(a,1,apply(a,1,"mean")) # row means
    # a simple two-way analysis

symbols	Draw Symbols on a Plot	**symbols**

USAGE:

symbols(x, y, circles=, squares=, rectangles=, stars=, add, inches)

ARGUMENTS:

x,y: coordinates of centers of symbols. A structure containing components named **x** and **y** can also be given.

circles=: the radii of the circles.

squares=: the lengths of the sides of the squares.

rectangles=: matrix whose two columns give widths and heights of rectangles.

stars=: a matrix with **n** columns, where **n** is the number of points to a star. The matrix should be scaled from 0 to 1.

add: logical flag. If FALSE (default), a new plot is set up. If TRUE, symbols are added to the existing plot.

inches: if FALSE, the symbol parameters are interpreted as being in units of **x** and **y**. If TRUE (default), the symbols are scaled so that the largest symbol is one inch in size. If a number is given, the parameters are scaled so that **inches** is the size of the largest symbol in inches.

Graphical parameters may also be given as arguments to this function (see **par**).

Only one of the arguments **circles=**, **squares=**, **rectangles=**, or **stars** can be given.

EXAMPLES:

symbols(x,y,circle=radii,inches=5)
    # largest circle has radius 5 inches

p ← ?col(parm,range) #find min/max of each column of parm
pscale ← scale(p,center=p[1,],scale=p[2,]-p[1,]) # scale to 0-1
symbols(x,y,stars=pscale) # plot stars

syntax	S Expressions	syntax

All user input to S is composed of expressions. The simplest expressions are

NUMBER   1, 5.3, 1e17

CHARACTER STRING   "abc", 'def ghi'

NAME   iris, city23

More complicated expressions can be made by combining expressions using the following rules:

EXPR op EXPR   where op is one of
$$+ \ - \ * \ / \ \hat{} \ ** $$
$$== \ != \ <= \ >= \ < \ > \ \& \ : \ \%\%$$

(EXPR)   parens specify order of evaluation

op EXPR   unary operators $+ - !$

NAME ← EXPR   assignment

EXPR $ NAME   selection of component

NAME [EXPR, EXPR, ...] subscripting

NAME (NAME1 = EXPR1, NAME2 = EXPR2, ...)   function call

NAME $ [EXPR]   component by number

for (NAME **in** EXPR) EXPR   loop

if (EXPR) EXPR1 **else** EXPR2 conditional expression

{ EXPR1 ; EXPR2 ; ... }   compound expression

Notice that expressions can be arbitrarily complicated by repeated uses of the above rules.

A typed expression may be continued on further lines by ending a line at a place where the line is obviously incomplete with a trailing comma, operator, or with more left parens than right parens (implying more rights will follow). Ordinarily the prompt character signifying that more input is necessary is "> ", but when continuation is expected the prompt is "+ ".

The NAME for subscripting may also be a parenthesized EXPR, function call, or component selection.

Functions may be called without arguments, may have missing arguments of the form NAME= or ,, and the keywords NAME= are optional.

The value of the input expression is automatically printed if it is not saved, producing a "desk calculator" mode. The expression **A** ← **1+2** is not printed because the result is saved in **A**. Certain functions (**save**) do not automatically print their results.

A line consisting of only a name is first interpreted as a function with no arguments, and if the function is not found, the line is treated as a dataset name and automatically printed.

If the top-most EXPR is a function call, it is not necessary to use parens to delimit its argument list, e.g. **plot x,y** is identical to **plot(x,y)**. All other functions must use parens, however.

A line whose first character is "!" is executed as a system command with no changes.

The value of a compound expression is the value of the last executed subexpression.

SEE ALSO:

precedence.

---

sys	Execute System Commands	sys

USAGE:

    sys(cmd)

ARGUMENTS:

cmd: character vector, each element of which is passed to the operating system and executed as a system command.

EXAMPLES:

    sys("cat myfile") # same as !cat myfile

---

t	Transpose a Matrix	t

USAGE:

    t(x)

ARGUMENTS:

x: matrix. Missing values (NAs) are allowed.

VALUE:

    transpose of **x** (rows of **x** are columns of result).

---

table	Create Contingency Table from Categories	table

USAGE:

    table(arg1, arg2, ...)

ARGUMENTS:

arg: categories, as variables for the contingency table. All **arg**s must be of equal length. Missing values (NAs) are allowed.

VALUE:

    contingency table structure, i.e., a multi-way array with an additional component **Label**. This is a structure with one component for each arg; the component corresponding to **arg1** is named "arg1" and is a character vector containing the values from **arg1$Label**.

The table contains counts of the number of times that each particular combination of values of **arg1**, **arg2**, ..., occurred. A combination is not counted if missing values are present in any **arg**s. If there is no missing data, **sum(table(a,b,...))** equals **len(a)**.

SEE ALSO:

Functions **cut** and **code** create categories; **tprint** prints tables, and **tapply** can be used for applying functions to observations in table cells.

EXAMPLES:

tprint(table(age,sex,race,income)) # print contingency table

---

tapply	Apply a Function to a Ragged Array	**tapply**

USAGE:

tapply(x, indices, fun, arg1, ...)

ARGUMENTS:

x: vector of data to be grouped by **indices**. Missing values (NAs) are allowed if **fun** accepts them.

indices: data structure containing a vector named **Data** and a labelling structure named **Label**. **Data** is a vector of the same length as **x** describing the position in a multi-way array corresponding to each **x** observation. **Label** is a label structure that contains one character vector component for each dimension of the multi-way array. This argument is generally the output from the **index** function. Missing values (NAs) are allowed.

fun: character string giving the name of the function to be applied to each group.

arg1:   optional arguments to be given to each invocation of **fun**.

VALUE:

if **fun** returns a single value for each group (e.g. "mean", "var", etc.) then **tapply** returns a multi-way table containing a multi-way array and a component named **Label**. If **fun** returns a vector of the same length as each group (e.g. "sort"), then **tapply** returns a vector result like **x**. **Tapply** generates an error if the **fun** returns a vector of any other length (although it should eventually handle any constant-length result, e.g. "range".)

EXAMPLES:

```
# generate mean republican votes for regions of the U.S.
# category that gives the region for each observation
region ← state.region[row(votes.repub)]
election ← code(votes.year)[col(votes.repub)]
pos ← index(region,election)
mn ← tapply(votes.repub,pos,"mean")
tprint(round(mn,1))  # table of mean vote by region and election
resid ← votes.repub - mn[pos]
```

---

tbl	Print a Table Using the UNIX tbl Command	**tbl**

USAGE:

tbl(x, file, head, rowlab, collab, options)

ARGUMENTS:

x: matrix to be printed as a table. May be any mode and may contain NA's.

file: name of the file on which the **tbl** output will be printed. Default is "tbl.out".

head: optional character string to appear as the heading of the table.

rowlab: optional character vector to be used as labels for the rows of the table.

collab: optional character vector to be used as labels for the columns of the table.

options: optional vector of options to the **tbl** preprocessor. Default is "center,box;".

The output file must be used with the tbl preprocessor to the UNIX nroff/troff commands. The **tbl** function in S is currently dumb about things like long lines or long fields. Either edit the output file or (better) break the matrix up into reasonable subsets of columns.

EXAMPLES:

       tbl(x) #simple table

       tbl(x>0, r=encode("Row",seq(10)), head="Is x positive?")

       # table of numeric (x,y) and character data (name, position)
       tbl(cbind(encode(x),name,encode(y),position))

T	t-Probability Distribution	**T**

USAGE:

       pt(q, par1)
       qt(p, par1)
       rt(n, par1)

ARGUMENTS:

q: vector of quantiles. Missing values (NAs) are allowed.
p: vector of probabilities. Missing values (NAs) are allowed.
n: sample size.
par1:   vector of degrees of freedom.

VALUE:

       probability (quantile) vector corresponding to the given quantile
       (probability) vector, or random sample (**rt**), from Students t-
       distribution on **par1** degrees of freedom.

---

tek10	see **tek12**	**tek10**

tek12	Tektronix Storage Scope Devices	**tek12**

USAGE:

      tek10(ask)
      tek12(ask)
      tek14(ask)
      tek14q(ask)

ARGUMENTS:

ask: logical, should device driver print the message "GO?" to ask permission to clear the screen? Default TRUE. If **ask**=−1 is given, a hardcopy will be made of each frame prior to clearing the screen, and no user interaction is necessary.

       tek10, **tek12**, **tek14**, and **tek14q** identify Tektronix storage scope graphics devices. **tek10** is designed for upper-case only models (4006, 4010), **tek12** for upper/lower case models such as 4012, while **tek14** and **tek14q** are designed specifically to support 4014 models. **tek14q** is somewhat quicker in its plotting than **tek14**, at the cost of slightly decreased resolution.

       If **ask** is TRUE, whenever a new plot is about to be produced the message "GO?" appears in the lower left corner of the screen. This allows further viewing, making a hard copy, etc., before the screen is erased. When ready for plotting to proceed, simply hit carriage return. The function **hardcopy** can be used to automatically generate a copy of the information on the screen. To produce unattended plots, use the **hardcopy** function with the **ask**=FALSE to these device functions.

       When graphic input (**identify**, **rdpen**) is requested from these terminals, the cursor lines appear horizontally and vertically across the entire screen. The thumb wheels should be adjusted to position the intersection of the cursors at the desired point, and then any single character should be typed (may need carriage return on some terminals). A carriage return alone will terminate graphic input without transmitting the point.

       Only **tek14** and **tek14q** change character size and line type. Character sizes are .9, 1, 1.5, and 1.6. There are 5 line types (1 through 5) giving solid, dotted, dot dash, short dash, and long dash lines. Characters can not be rotated on these devices.

A device must be specified before any graphics functions can be used.

---

tek14	see **tek12**	**tek14**

tek14q	see **tek12**	**tek14q**

tek4112	Tektronix 4112 Raster Scope	**tek4112**

USAGE:

> **tek4112(ask)**
> **!S TEK4112**   # UNIX command to set up function keys

ARGUMENTS:

ask: logical, should device driver print the message "GO?" to ask permission to clear the screen? Default TRUE.

On **tek4112** scopes, graphics is drawn in a "graphics area" while interaction with the computer is carried out in the "dialog area". The "Dialog" key on the terminal is a switch which controls visibility of the dialog area. The **S TEK4112** utility routine sets up function keys to provide control over the graphics/dialog areas. Function keys 1 through 4 are set up as "View Graph", "No Graph", "Small Dialog (2 lines)", and "Large Dialog (full screen)". In addition, the shifted version of function key 1 is "Erase Graph".

Graphics drawn on the screen are stored in a "segment" in local memory in the 4112. When the page button is pressed, the segment is re-drawn. Use can also be made of the pan and zoom keys on the terminal to view parts of the graphics in finer detail. Although the terminal only displays 640 by 480 rasters of information at any time, the terminal receives data to a 4096 by 3072 resolution, and the pan and zoom operations cause finer detail to be made available.

Refer to the 4112 Operators Manual for a description of how to execute the following "setup commands" which set non-default options: DAENABLE yes; FLAGGING input; BYPASSCANCEL null.

If **ask** is TRUE, whenever a new plot is about to be produced the message "GO?" appears in the lower left corner of the screen. This allows further viewing, making a hard copy, etc., before the screen is erased. When ready for plotting to proceed, simply hit carriage return. The function **hardcopy** can be used to generate a copy of the information on the screen. To produce unattended plots, use the **hardcopy** function with the **ask=FALSE** to this device function.

When graphic input (**identify**, **rdpen**) is requested from this terminal, the cursor lines appear horizontally and vertically across the entire screen. The thumb wheels should be adjusted to position the intersection of the cursors at the desired point, and then any single character should be typed. A carriage return alone will terminate graphic input without transmitting the point.

Character size, rotation, and line type may be changed. Characters may appear in any size or rotation angle. There are 5 line types (1 through 8) giving solid, dotted, dot dash, short dash, long dash, dash 3-dot, long dash dot, and dash space.

A device must be specified before any graphics functions can be used.

TEK4112	see **tek4112**	**TEK4112**

tek46	Tektronix 4662 Pen Plotter	**tek46**

USAGE:

> **tek46(width, height, ask, color, speed)**
> **tek46h**
> **tek46v**

ARGUMENTS:

width: width of plotted surface in inches. Default is 8 for **tek46v**, 10 for **tek46h**, 15 otherwise.

height: height of plotted surface in inches. Default is 8 for **tek46h**, 10 otherwise.

ask: logical, should user be prompted by "GO?" prior to advancing to new frame? Default TRUE.

color: integer reflecting the degree of color-plotting provided by the device. 0=color changes ignored, 1=prompt user when color changes (to allow manual pen changes, etc), 2= (default) device has automatic color changing capability (4662 Option 31).

speed: speed in mm/sec used in drawing long (>2.6 inch) vectors. Valid speeds are from 6 to 570 (default) mm/sec. Slow speeds cause more consistent line quality, and may be useful for high-quality work or for plotting with special pens, paper, or film. The **speed** argument is only valid if **color=2** was specified, since speed changing is only available with Option 31.

> These commands identify Tektronix 4662 series plotters. The plotter may be used with standard 8.5 by 11 inch paper with the long dimension horizontally (**tek46h**) or vertically (**tek46v**), or with 11 by 17 paper (**tek46**). The arguments listed after **tek46** can also be given with **tek46h** or **tek46v**.

> Whenever a new plot is about to be produced, the message "GO?" appears on the terminal. At this time, new paper may be loaded, pens changed, etc. When ready for plotting to proceed, simply hit carriage return.

> When graphic input (**identify, rdpen**) is requested, the plotter will beep and light its "prompt" light. The pen should be positioned by means of the joystick. When the desired pen position is reached, depress the "call" button momentarily. Hold the "call" button down for about 2 seconds (until it beeps once) to terminate the input. The coordinates of the terminating point are

not transmitted.

Character sizes may be changed to any desired value. Pens can be changed manually to provide different colors or automatically if the device has that capability. If argument **color** is 1, the user is prompted with the message "Load pen x" when graphical parameter **col** changes to **x**. The initial pen is assumed to be color 1. Characters can be rotated to any orientation. Only solid line style is available.

At 300 baud, the switch settings on the back of the device should be set to "0231" (no parity) or "02B1" (even parity). At 1200 baud, the no parity setting is "0233".

A device must be specified before any graphics functions can be used.

---

text	Plot Text	**text**

USAGE:

> **text(x, y, labels, cex, col)**

ARGUMENTS:

x,y: vector of position(s) at which to plot labels. A time-series or structure containing **x** and **y** may also be given for **x**. Missing values (NAs) are allowed.

labels: optional, controls labelling. The interpretation depends on the mode of **labels**. If it is a character vector, **labels[i]** is plotted at each point **(x[i],y[i])**. If it is a logical vector (of the same length as **x** and **y**), the value **i** is plotted at each **(x[i],y[i])** for which **labels[i]** is TRUE. If **labels** is integer or real, **labels[i]** is encoded and plotted at **(x[i],y[i])**. Missing values (NAs) are allowed. Default is **seq(x)**.

cex: character expansion parameter (see **par**) for each text string. May be a vector.

col: color parameter (see **par**) for each text string. May be a vector.

> If the lengths of **x, y, labels, cex,** and **col** are not identical, the shorter vectors are reused cyclically.

The **text** function can be used after **plot(x,y,type="n")** which draws axes and surrounding box without plotting the data.

EXAMPLES:

plot(x,y,type="n")
text(x,y) #plot i at (x[i],y[i])

---

time	Create Time Vector	**time**

USAGE:

time(x)
cycle(x)

ARGUMENTS:

x: time-series. Missing values (NAs) are allowed.

VALUE:

time-series with same **start**, **end** and **nper** as **x**. **time** returns the time value at each point, e.g., the value at January, 1973 is 1973.0, and the value at July 1973 is 1973.5. **cycle** returns the period associated with each observation, e.g., 1 for each January, etc.

EXAMPLES:

cgrowth ← code(cycle(growth), 1:12, month.name)
tapply(growth, cgrowth, "mean")   # mean growth for each month

---

title	Plot Titling Information and/or Axes	**title**

USAGE:

title(main, sub, xlab, ylab, axes)
axes(main, sub, xlab, ylab, axes)

ARGUMENTS:

main: optional character string for the main title, plotted on top in enlarged (cex=1.5) characters.

sub: optional character string for the sub-title, plotted on the bottom.

xlab: optional character string to label the x-axis.
ylab: optional character string to label the y-axis.
axes: logical, should axes be drawn? Default TRUE for function **axes**,
             FALSE for **title**.

             Graphical parameters may also be supplied as arguments to this
             function (see **par**).

             The two functions differ only in that **axes** also plots the x and y
             axes. All the character strings are empty by default.

EXAMPLES:

             title("residuals from final fit") # main title only
             axes(xlab="concentration",ylab="temperature")

---

tprint	Print Multi-way Tables	**tprint**

USAGE:

             tprint(table1, table2, ..., Label=, maxcol=, mincol=)

ARGUMENTS:
table: table or multi-way array. Contingency tables may be constructed
             (complete with **Label** component) from category data by the **table**
             function. If more than one table is given, values from the tables
             are printed above one-another, and the values are labelled by the
             table name, or by the **name** used if **table** is given in **name=value**
             form. All tables must have the same dimensionality and number
             of levels on each category (i.e., dimension of the tables). Missing
             values (NAs) are allowed.
Label=:   optional structure containing as many components as there are
             categories to the tables. The name of each component is the
             name used for the corresponding category, and each component
             should be a character vector with as many values as the number
             of levels of the category. If any of the tables already contain such
             a component named **Label**, this argument is not needed.
maxcol=:   optional, maximum category to be laid out across the page. **tprint**
             displays the first category of the tables down the page and the
             second across the page. If there is enough room, categories 3, 4, ...
             will also be laid out across the page. If **maxcol** is given this will

be the last category laid out this way, even if more would fit.

mincol=: optional, minimum category to be laid out across the page. Even if the category will not fit, **tprint** will lay it out across the page, breaking the table into blocks if necessary. By default, no category will be used across the page unless all levels fit.

EXAMPLES:

```
mydata ← table(age,sex,height,weight,eye.color)
tprint(mydata)    # print the table with labels for the levels
fitted ← loglin(mydata,model)    # fit log-linear model to table
    # three tables: data, fit, residuals for each cell
tprint(data=mydata,fit=fitted,residuals=mydata−fitted)
```

```
# the example plot is produced by:
repair.1978 ← code( auto.stats[,3],
    1:5, c("Poor","Fair","Average","Good","Excellent")
    )
improved ← cut( auto.stats[,3]−auto.stats[,4],
    c(−100,−.5,.5,100),c("Worse","Same","Better")
    )
price ← cut( auto.stats[,1],
    c(0,5000,8000,50000),c("Cheap","Mid","Expensive")
    )
tprint(table(repair.1978,improved,price))
```

```
price: Cheap
improved:        Worse     Same     Better
repair.1978:
    Poor          1         1         0
    Fair          0         3         0
    Average       0        12         3
    Good          0         2         2
    Excellent     0         2         3


price: Mid
improved:        Worse     Same     Better
repair.1978:
    Poor          0         0         0
    Fair          1         3         0
    Average       2         3         1
    Good          0         6         4
    Excellent     0         3         1


price: Expensive
improved:        Worse     Same     Better
```

```
repair.1978:
   Poor         0       0       0
   Fair         0       1       0
   Average      3       4       0
   Good         0       3       0
   Excellent    0       0       2
```

trunc	see **ceiling**	**trunc**

ts	Create Time-Series	**ts**

USAGE:

> **ts(data, start, nper, end, tsp)**

ARGUMENTS:

data: vector giving the data values for the time-series.

start: starting date for the series, in years, e.g., February, 1970 would be 1970+(1/12) or 1970.083. If **start** is a vector with at least two data values, the first is interpreted as the year, and the second as the number of npers into the year; e.g., February, 1970 could be c(1970,2).

nper: periodicity of the series, in observations per year, e.g. monthly data has **nper=12**, yearly has **nper=1**.

end: ending date for the series.

tsp: real vector giving the values for **start**, **end** and **nper**. **tsp** may be used in place of these 3 arguments.

VALUE:

> time-series with the given **data** and **start**, **end**, **nper**.

ts checks that the number of observations is compatible with the specified dates. Default values for **start**, **end** and **nper** are 1, **len(data)**, and 1. If only two of **start**, **end**, and **nper** are given, the other parameter will be computed based on the given values and **len(data)**.

Any periodic data can be fit into the time-series framework.

Hourly data might use days for **start** and **end** with **nper=24**; daily data could use weeks with **nper=7**.

tslines	see **tsplot**	**tslines**

tsmatrix	Create Matrix with Time-Series as Columns	**tsmatrix**

USAGE:

tsmatrix(arg1, arg2, ...)

ARGUMENTS:

argi: time-series to form a column of the resulting matrix. The time window for the matrix will be the intersection of time windows for the arguments (i.e., the maximum of the start dates and the minimum of the end dates). All series must have the same periodicity. Missing values (NAs) are allowed.

VALUE:

a matrix with one column for each argument and a component named **Tsp** with the start date, end date, and number of observations per year.

Note the distinction from the related function **cbind**. This allows vectors or matrices as arguments as well, but makes no effort to make time parameters consistent. Frequently, the two functions can be used together (see examples).

EXAMPLES:

tsmatrix(x,lag(x),diff(x)) #x, lag−1 of x and first diffs
cbind(tsmatrix(x,lag(x)),1) #two series, column of 1s

yx←tsmatrix(employment,gnp,lag(gnp), ...)
regress(yx[,1],yx[,−1]) # regression with x and y times aligned

---

| **tsplot** | Plot Multiple Time-Series | **tsplot** |

USAGE:

   tsplot(ts1, ts2, ..., type=, lty=, pch=, col=)

ARGUMENTS:

tsi: time-series to be plotted. The x-axis is set up as the union of the times spanned by the series. The y-axis includes the range of all the series. Missing values (NAs) are allowed.

type=:   an optional character string, telling which type of plot (points, lines, both, none or high-density) should be done for each plot. The first character of type defines the first plot, the second character the second, etc. Elements of type are used cyclically; e.g., "pl" alternately plots points and lines. Default is "l" (lines) for **tsplot** and **tslines** and "p" for **tspoints**.

lty=:   optional vector of line types. The first element is the line type for the first line, etc. Line types will be used cyclically until all plots are drawn. Default is **1:5**.

pch=:   optional character vector for plotting characters The first character is the plotting character for the first plot of type "p", the second for the second, etc. Default is the digits (1 through 9, 0) then the letters.

col=:   optional vector of colors. Default is **1:4**. Colors are also cycled if necessary.

   Graphical parameters may also be supplied as arguments to this function (see **par**).

   tsplot generates a new plot; **tspoints** and **tslines** add to the current plot.

EXAMPLES:

   tsplot(gnp,smooth(gnp),type="pl")
   tspoints(x,y,pch="*") # points with "*" for both series
   # the example plot is produced by:
   tsplot(hstart,smooth(hstart),type="pl")
   title(main="Housing Starts",sub="Data and Smoothed Values")

Housing Starts

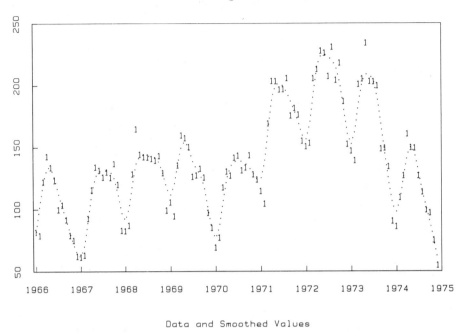

Data and Smoothed Values

| **tspoints** | see **tsplot** | **tspoints** |

| **twoway** | Analysis of Two-way Table | **twoway** |

USAGE:

twoway(x, trim, iter, eps, print)

ARGUMENTS:

x: matrix to be analyzed. Missing values (NAs) are allowed.

trim: optional trimming fraction for carrying out analysis. Default .5
             trimmed mean (median). **trim=0** will cause analysis by means,
             .25 by midmeans, etc.

iter: maximum number of full (row and column) sweeps. Default 6.

eps: error tolerance. If **eps** is given, the algorithm will iterate until
             the maximum change in row or column effects is < **eps**. Default

is to iterate until the specified number of iterations or until converged to the accuracy of the machine arithmetic. It is not always
possible to converge to a unique answer.

print: logical flag, should convergence printing be done? Default,
FALSE. If TRUE, the maximum change in row/column effects in
the last iteration is printed.

VALUE:

structure consisting of 4 components, **resid**, **row**, **col**, and **grand**,
such that

$x[i,j]$ equals **grand** + **row**[i] + **col**[j] + **resid**[i,j]

grand: overall location estimate of the data.
row: vector of row effects.
col: vector of column effects.
resid: matrix of residuals from the fit.

SEE ALSO:

Function **plotfit** produces a graphical display of the fit generated
by **twoway**.

EXAMPLES:

twoway(temperature,trim=.25) # analysis by midmeans

---

unif	Uniform Distribution.	**unif**

USAGE:

punif(q, par1, par2)
qunif(p, par2, par2)
runif(n, par1, par2)

ARGUMENTS:
q: vector of quantiles. Missing values (NAs) are allowed.
p: vector of probabilities. Missing values (NAs) are allowed.
n: sample size.
par1:   vector of lower limits. Default is 0.
par2:   vector of upper limits. Default is 1.

VALUE:

probability (quantile) vector corresponding to the given quantile (probability) vector, or random sample (**runif**), for the uniform distribution on the range **par1** to **par2**.

EXAMPLES:

runif(100,−1,1) #100 numbers uniform on −1 to 1

?union	Union of Several Lists	**?union**

USAGE:

?union(x1, x2, ...)

ARGUMENTS:

xi: vector containing a list.

VALUE:

vector of the (unique) values appearing in at least one of the lists.

---

uniq	see **unique**	**uniq**

---

unique	Unique Values in a Vector	**unique**

USAGE:

unique(x)
uniq(x)

ARGUMENTS:

x: vector.

VALUE:

a vector like **x** but with no duplicate values. The values will be in the same order as **x** except that repeated values will be deleted.

EXAMPLES:

sort(unique(names)) #sorted list of names with no duplicates

---

unixplot	Produce Graphics on UNIX Devices	**unixplot**

USAGE:

> **unixplot**
>
> This function produces a file of commands in a form compatible with UNIX graphics devices. The file, named "unixplot.out", can be given as input to any of the UNIX device driver filters (See **plot(1)** or **tplot(1)** in the UNIX Users Manual).
>
> When graphic input (**identify, rdpen**) is requested, the prompt **x,y:** appears on the users terminal, and the user should type the (x,y) coordinate pair of the point to be input. To terminate the input, hit carriage return. Graphic input is not allowed in batch mode.
>
> A device must be specified before any graphics functions can be used.

EXAMPLES:

> **unixplot**
> **plot(hstart)**
> **q**   # exit from S
> **tplot −Tver <unixplot.out**   # UNIX command for Versatec plotter
> **prtx unixplot.out|lpr**   # plotting on Printronix printer

---

usa	Plot United States Coast-line and States	**usa**

USAGE:

> **usa(states, coast, add, xlim, ylim, fifty)**

ARGUMENTS:

states: logical flag to control whether state boundaries are plotted. Default TRUE.

coast: logical flag to control whether coast-line is plotted. Default TRUE.

add: logical flag. If TRUE, plot is superimposed on existing plot. Otherwise, a new plot is generated. Default FALSE.

xlim: optional limits for the x-axis (longitude). Default is **c(65,135)**.
ylim: optional limits for the y-axis (latitude). Default is **c(24,50)**.
fifty: logical flag. If TRUE, boxes are drawn in the Pacific Ocean to represent Alaska and Hawaii. Default FALSE.

Graphical parameters may also be supplied as arguments to this function (see **par**).

The plot is done in correct physical proportion. The coordinate system set up for the plot uses negative longitude, so that x-values increase from left to right on the plot.

EXAMPLES:

usa(states=F) # the u.s. without state lines
usa(xlim=c(65,85),ylim=c(35,50)) #plot the north-east
the example plot is produced by:
usa
text(state.center,state.abb)

---

var	Variance, Covariance, and Correlation	**var**

USAGE:

> **var(x, y, trim)**
> **cor(x, y, trim)**

ARGUMENTS:

x: matrix or vector.

y: matrix or vector. If omitted, same as **x**.

trim: the proportion trimmed in the internal calculations. Default 0.

VALUE:

> correlations or variances (covariances), optionally with trimming. If **x** or **y** is a matrix, the result is a correlation or covariance matrix; the **[i,j]** element corresponds to the **i**th column of **x** and the **j**th column of **y**.

> Trimmed correlations are computed by the standardized sums and differences method. Untrimmed variances are sample variances: **sum((x−mean(x))ˆ2)/(len(x)−1)**.

> Trimmed covariances are not yet implemented.

EXAMPLES:

> **cor(cbind(longley.x,longley.y))**   # correlation matrix for longley data

> **std.dev ← sqrt(var(x))**   # standard deviation of a vector

---

?vecs2mat	Make Matrix from a Set of Vectors	**?vecs2mat**

USAGE:

> **?vecs2mat(name, arg1, arg2, ...)**

ARGUMENTS:

name: name to be used for output.

argi: data vectors to be made into the columns of constructed matrix. Missing values (NAs) are allowed.

EFFECT:

> The matrix produced will be called **name.data**, and the character vector containing column names will be called **name.collab**.

EXAMPLES:

> **?vecs2mat(mymat,height,weight,age)** #create 3-column matrix
> # mymat.data and vector of column labels mymat.collab

---

vu	Create Vu-graphs (Slides)	**vu**

USAGE:

> **vu(text, indent, width, height, line, csize, font)**

ARGUMENTS:

text: character string vector, the elements of which are either lines of text to plot, or else commands to the **vu** function (see below). The text will be plotted, with character size as large as possible to fill the page.

> For many purposes, the function **quickvu** will generate a dataset easily which **vu** can plot. In any case, the function **edit** is useful for editing a dataset for use with **vu**.

indent: number of spaces to indent lists. Default 3.

width: width of the page in inches. Default 7.

height: height of the paper in inches. Default 7.

line: thickness of lines in inches (needed for bold face and for drawing bullets - see below). Default is .015 inches (the thickness of the thick pens for Hewlett-Packard plotters). Thin pens on the hp7221 are .01 inches.

csize: vector of allowable character sizes for the plot, i.e., the values of graphical parameter **cex** that the device supports. This argument is important only for devices with a limited set of hardware-drawn character sizes when **font** is omitted. If **csize** is given as a scalar, any integer multiple of that size is allowed, otherwise **csize** is the vector of allowable sizes. Default is continuously variable sizes.

font: optional character name of a font to be used in plotting the vu-graph. If font is not specified, the hardware generated font of the plotting device will be used. Fonts can also be specified within

text by the .F command.

Abbreviation	Font Name
sr	Simplex Roman
cr	Complex Roman
dr	Duplex Roman
tr	Triplex Roman
ro	Roman (Constant Width)
ci	Complex Italic
ti	Triplex Italic
ss	Simplex Script
cs	Complex Script
sg	Simplex Greek
cg	Complex Greek
ge	Gothic English
gg	Gothic German
gi	Gothic Italian
cc	Complex Cyrillic
sp	Special Symbols

SOURCE:

Norman M. Wolcott and Joseph Hilsenrath, *A Contribution to Computer Typesetting Techniques*, "Tables of Coordinates for Hershey's Repertory of Occidental Type Fonts and Graphics Symbols", National Bureau of Standards Special Publication 424, April 1976.

The following are the commands recognized by **vu**. They are similar to commands in several **troff** macro systems, and they begin with a period (.) and one or two capital letters:

.C   Choose a color. This command takes an argument which is the color number, i.e., ".C 3" for color 3, etc.

.F   Choose a font. This command takes an argument which is the two-letter font name to be used for subsequent lines of text. If no font name is given, the font specified by the **font** argument is used. (To force use of the hardware font, give a non-existent font name.) Note that fonts currently cannot be changed in the middle of a line. See also argument **font**.

.L   Make the text larger. Each time this command appears, the size is increased by 25%. This command may also take an argument, which is the desired character size relative to initial size 1, thus ".L 2" would be double the initial size, and ".L 1" would return to initial size. The argument to this command is *not* the same as the graphical parameter **cex**: **vu** always adjusts the final value of **cex** so that all of the lines will fit on the final display. An argument to **.L** simply gives the size of the following text relative to the size of preceding text.

.S   Make the text size smaller, by the same amount that ".L" makes it larger. Thus, any number of ".S" commands cancel the same number of ".L" commands. This command may take an argument, in which case it behaves exactly as ".L".

.B   Embolden the text (done by replotting one line width to the right). The argument tells how many overstrikes to use, default 1.

.H   Highlight the text by plotting first in a highlighting color (given by the argument) and then plotting the text in the current color. With no argument, highlighting is turned off.

.R   Remove the emboldening.

.CE   The argument gives the number of following lines of text which are to be centered, default 1.

            The next commands all create lists of items. The items in the list can be preceded by bullets, diamonds, numbers, or any character string (for example, "--"). Lists may be nested, but general or numbered lists may only appear once.

.BL   Start a bullet list.

.DL   Start a diamond list.

.GL   Start a general list with each item preceded by the character string given after the command. For long strings, the value of argument **indent** may need to be increased.

.NL   Start a numbered list, each item preceded by a sequential number and trailing dash.

.LI   List item. This command must precede **each** item in the list.

.LE   End of the list (causes the indenting of the list to be cancelled).

The following text illustrates four of the fonts described above.

EXAMPLES:
```
.CE 4
.C 1
.F ro
.B
Slides Made With VU are
.R
.C 2
.S 1.5
.F ge
VERY SEXY
.S 1
.F cs
.C 3
And Also Simple To Make
.C 4
.F tr
QUICK and INEXPENSIVE
```

Slides Made With VU are

𝔙𝔈ℜ𝔜 𝔖𝔈𝔛𝔜

And Also Simple To Make

QUICK and INEXPENSIVE

warning	see **fatal**	warning

?which	Find Which Observations Satisfy a Condition	**?which**

USAGE:

   ?which(condition)

ARGUMENTS:

condition: logical expression. Missing values (NAs) are allowed.

VALUE:

   vector of subscripts that tell which observations satisfy **condition**.

EXAMPLES:

   # observation numbers for tall people under 30

?which(age<30 & height>72)

| **window** | Window a Time-Series | **window** |

USAGE:

window(x, start, end)

ARGUMENTS:

x: time-series. Missing values (NAs) are allowed.

start: optional new starting date for the series. If earlier than start date of **x**, no change is made.

end: optional new ending date for the series; if later than end date of **x**, no change is made.

VALUE:

a time-series like **x** giving the data between the new **start** and **end** dates.

EXAMPLES:

sgnp ← window(gnp,c(1970,1),c(1975,12))
    #subseries of gnp from Jan 1970 through Dec 1975

| **write** | Write Data to File | **write** |

USAGE:

write(data, file, ncol)

ARGUMENTS:

data: vector, missing values (NAs) are allowed.

file: optional character string naming the file on which to write the data. The name of **data** will be used by default; if **data** has no name, the file will be called "data".

ncol: optional number of data items to put on each line of **file**. Default is 5 per line for numeric data, 1 per line for character data.

Because of the way matrix data is stored, matrices are written in column by column order.

See also functions **dput**, and **dump**.

EXAMPLES:

write(x,"filex")

write(t(x),"byrows")   # write matrix row by row

xor	Exclusive Or	xor

USAGE:

xor(arg1, arg2)

ARGUMENTS:

argi: logical vectors to be combined by means of an exclusive-or operation. Missing values (NAs) are allowed.

VALUE:

logical result of exclusive-or applied to **arg1** and **arg2**.

Appendix 2
System Datasets

Where more than one name appears under a heading (e.g., **number** and **payoff** under **lottery**), the actual dataset names will be composed of the heading, a period, and the name (for example, **lottery.payoff**).

| **akima** | Waveform Distortion Data for Bivariate Interpolation | **akima** |

x,y,z: represents a smooth surface of **z** values at selected points irregularly distributed in the **x-y** plane.

SOURCE:

Hiroshi Akima, "A Method of Bivariate Interpolation and Smooth Surface Fitting for Irregularly Distributed Data Points", *ACM Transactions on Mathematical Software*, Vol 4, No 2, June 1978, pp 148-159. Copyright 1978, Association for Computing Machinery, Inc., reprinted by permission.

The data was taken from a study of waveform distortion in electronic circuits, described in: Hiroshi Akima, "A method of Bivariate Interpolation and Smooth Surface Fitting Based on Local Procedures", *CACM*, Vol 17, No 1, January 1974, pp 18-20.

| **author** | Character Counts for Books by Various Authors | **author** |

count: matrix of 26 columns corresponding to letters of the alphabet. Each row contains data for one work, giving the counts of each letter. There are different total counts for each work, and proper nouns were removed prior to counting.

collab: column labels, the letters of the alphabet.

rowlab: row labels, book titles and author names.

SOURCE:

W. A. Larsen and R. McGill, unpublished data collected in 1973.

auto	Statistics of Automobile Models	**auto**

collab: names of the columns (variables). These include price in dollars, mileage in miles per gallon, repair records for 1977 and 1978 (coded on a 5-point scale, 5 is best, 1 is worst), headroom in inches, rear seat clearance (distance from front seat back to rear seat back) in inches, trunk space in cubic feet, weight in pounds, length in inches, turning diameter (clearance required to make a U turn) in feet, displacement in cubic inches, and gear ratio for high gear.

rowlab: names of the automobile models summarized.

stats: actual statistics corresponding to the variables and models. The matrix gives data on 74 models by 12 statistics.

The data give statistics for automobiles of the 1979 model year as sold in the United States.

SOURCE:

Fuel consumption figures from United States Government EPA statistics. All other data from "CU Judges the 1979 Cars", *Consumer Reports*, April 1979. Copyright 1979 by Consumers Union of United States, Inc., Mount Vernon, NY 10550. Reprinted by permission from *CONSUMER REPORTS*, April 1979.

bicoal	Bituminous Coal Production in USA	**bicoal**

tons: production in millions of net tons per year, 1920-1968.

SOURCE:

U. S. Bureau of Mines, also in J. W. Tukey, *Exploratory Data Analysis*, 1977, Addison-Wesley Publishing Co., Massachusetts.

bonds	Daily Yields of 6 AT&T Bonds	bonds

yield: matrix of daily bond yields; rows represent 192 days from April 1975 to December 1975. Columns are one of 6 bonds, characterized by their coupon rate.

coupon: vector of 6 different coupon rates corresponding to columns of **yield**.

collab,label: Column and overall label.

car	Fuel Consumption Data	car

time: number of days since initial car purchase.

miles: number of miles driven between this fill-up and previous fill-up.

gals: number of gallons required to fill tank.

This data pertains to a new, 1974 model automobile.

SOURCE:

R. A. Becker, personal data.

cereal	Consumer Attitudes Towards Breakfast Cereals	cereal

attitude: matrix giving percentage of people agreeing with 11 statements (rows) about 8 brands of cereals (columns).

rowlab: statements about cereals, i.e., "Reasonably Priced", etc.

collab: brand of cereal.

SOURCE:

T. K. Chakrapani and A. S. C. Ehrenberg, "An Alternative to Factor Analysis in Marketing Research--Part 2: Between Group Analysis", *PMRS Journal*, Vol. 1, Issue 2, October 1981, pp 32-38. Republished by permission of the Professional Marketing Research Society.

---

chernoff2   Mineral Contents Data (used by Chernoff)   **chernoff2**

The data is a 53 by 12 matrix representing mineral analysis of a 4500 foot core drilled from a Colorado mountainside. Twelve variables (columns) represent assays of seven mineral contents by one method and repeated assays of five of these by a second method. Fifty-three equally spaced specimens (rows) along the core were assayed. Specimen ID numbers were 200 to 252.

SOURCE:

H. Chernoff, "The use of Faces to Represent Points in k-dimensional Space Graphically", *Journal of the American Statistical Association*, Vol. 68, No. 342, 1973, pp 361-368. Republished by permission of the American Statistical Association.

---

city   Names and Location of Selected US Cities   **city**

---

name: character vector of city names.
state: character vector giving state for each city.
x,y: location of city: **x** is negative longitude to correspond to coordinate system set up by function **usa**. **y** is latitude (both measured in degrees).

These cities cover the geographical regions of the US, and were not chosen by population.

---

co2   Mauna Loa Carbon Dioxide Concentration   **co2**

This data represents monthly CO2 concentrations in parts per million (ppm) from January 1958 to December 1975.

SOURCE:

Data collected by Charles D. Keeling, Scripps Institute of Oceanography, La Jolla, California.

An updated version of the data may be found in *Carbon Dioxide*

Review 1982, Oxford University Press, New York, p 378.

---

corn	Corn Yields and Rainfall	**corn**

yield,rain: yearly corn yield in bushels per acre, and rainfall measurements in inches, in six Corn Belt states (Iowa, Illinois, Nebraska, Missouri, Indiana, and Ohio) from 1890 to 1927.

SOURCE:

M. Ezekiel and K. A. Fox, *Methods of Correlation and Regression Analysis*, p 212. Copyright 1959, John Wiley and Sons, Inc., New York.

Data originally from E. G. Misner, "Studies of the Relationship of Weather to the Production and Price of Farm Products, I. Corn", mimeographed publication, Cornell University, March 1928.

---

evap	Soil Evaporation Data	**evap**

collab: column labels for **x**.

x: matrix of independent variables: maximum daily soil temperature, minimum daily soil temperature average soil temperature (integrated area under daily soil temperature curve), maximum, minimum, and average air temperature, maximum, minimum, and average relative humidity, and total wind (miles per day).

y: daily amount of evaporation from the soil.

"It is desired to estimate the daily amount of evaporation from the soil as a function of air temperature, relative humidity, and wind. Since these factors vary considerably throughout the day it is not clear what function or aspect of these variables is most important. For this reason the following ten variables relating to these factors are recorded." (from Freund).

Observations represent 46 consecutive days from June 6 through July 21.

SOURCE:

> R. J. Freund, "Multicollinearity etc., Some "new" Examples", *American Statistical Association Proceedings of Statistical Computing Section*, 1979, pp 111-112. Republished by permission of the American Statistical Association.

---

font	Vector Drawn Fonts	**font**

The font datasets describe vector drawn characters which make up sixteen different fonts. Each font is known by a 2-character name, e.g. "sr" for simplex roman. The corresponding font dataset is named "font.sr".

Abbreviation	Font Name
cc	Complex Cyrillic
cg	Complex Greek
ci	Complex Italic
cr	Complex Roman
cs	Complex Script
dr	Duplex Roman
ge	Gothic English
gg	Gothic German
gi	Gothic Italian
ro	Roman (Constant Width)
sg	Simplex Greek
sp	Special Symbols
sr	Simplex Roman
ss	Simplex Script
ti	Triplex Italic
tr	Triplex Roman

These datasets are normally used by graphics functions which have vector font capability. The following information is not needed for the use of the fonts, but may be helpful for constructing new fonts.

Fonts are digitized on a coordinate system centered at (0,0) and ranging to at most 50 in all directions. An em is 32 units. Since the range of each coordinate is restricted, a single x or y coordinate can be represented in one (ASCII) character. Given an ASCII character, the coordinate value is found by taking the integer character number and subtracting 77 ("A" is 65 and thus represents coordinate −12).

Fonts contain information for drawing the 95 ASCII characters from space (32) to tilde (126). Each font is a data structure with the following components:

left: Character vector containing a single string of 95 characters giving the leftmost coordinates of the characters in the font.

right: Character vector giving rightmost coordinates.

data: Character vector with one string for each of the 95 vector drawn characters in the font. Each string gives coordinate pairs used in drawing the character. A tab character appearing in the string indicates that the pen should be lifted at that point.

SOURCE:

Norman M. Wolcott and Joseph Hilsenrath, *A Contribution to Computer Typesetting Techniques*, "Tables of Coordinates for Hershey's Repertory of Occidental Type Fonts and Graphics Symbols", National Bureau of Standards Special Publication 424, April 1976.

---

freeny	Revenue Data	**freeny**

y: quarterly revenue, 39 observations from (1962,2Q) to (1971,4Q).

x: matrix of independent variables, Columns are **y** lagged 1 quarter, price index, income level, and market potential.

SOURCE:

A. E. Freeny, "A Portable Linear Regression Package with Test Programs", Bell Laboratories memorandum, 1977.

| **hstart** | US Housing Starts | **hstart** |

U. S. Housing Starts, monthly, January 1966 to December 1974.

SOURCE:

U. S. Bureau of the Census, Construction Reports.

| **iris** | Fisher's Iris Data | **iris** |

Array giving 4 measurements on 50 flowers from each of 3 species of iris. Sepal length and width, and petal length and width are measured in centimeters. Species are Setosa, Virginica, and Versicolor.

SOURCE:

R. A. Fisher, "The Use of Multiple Measurements in Taxonomic Problems", *Annals of Eugenics,* Vol 7, Part II, 1936, pp 179-188. Republished by permission of Cambridge University Press.

The data were collected by Edgar Anderson, "The irises of the Gaspe Peninsula", *Bulletin of the American Iris Society,* Vol. 59, 1935, pp 2-5.

| **liver** | Carcinogeneity Studies of Rat Livers | **liver** |

cells: number of cells injected into each animal.
exper: category for the three experiments (A, B, C) in the study.
gt: matrix (52 by 4) for the 4 lobes (ARL, PRL, PPC, AC), with counts for each (observation,lobe) pair of the GT(+) colonies.
section: category for the section (replication) for each observation. Successive observations are pairs of sections for the same specimen.

SOURCE:

Data were collected by Brian Laishes, University of Wisconsin, Madison, Wisconsin.

The data were used in: B. A. Laishes and P. B. Rolfe, "Quantitative Assessment of Liver Colony Formation and Hepatocellular Carcinoma Incidence in Rats Receiving Intravenous Injections of Isogeneic Liver Cells Isolated during Hepatocarcinogenesis", *Cancer Research*, Vol 40, pp 4133-4143, November 1980. Republished by permission of Cancer Research Journal.

The data appear in: "Detection of Damaged Animals in Carcinogeneity Studies with Laboratory Animals", Camil Fuchs, 1980, University of Wisconsin Statistics Laboratory Report 80/3.

---

longley	Longley's Regression Data	**longley**

y: number of people employed, yearly from 1947 to 1962.

x: matrix with 6 columns, giving GNP implicit price deflator (1954=100), GNP, unemployed, armed forces, non-institutionalized population 14 years of age and over, and year.

This regression is known to be highly collinear.

SOURCE:

J. W. Longley, "An appraisal of least-squares programs from the point of view of the user", *Journal of the American Statistical Association*, Vol. 62, 1967, pp 819-841. Republished by permission of the American Statistical Association.

---

lottery	New Jersey Pick-it Lottery Data	**lottery**

number: winning 3-digit number (from 000 to 999) for drawings from May 22, 1975 to Mar 16, 1976. (This was the beginning of the Pick-it lottery).

payoff: payoff corresponding to each **number** (dollars).

Republished by permission of the New Jersey State Lottery Commission.

lottery2	New Jersey Pick-it Lottery Data (Second Set)	**lottery2**

number: winning 3-digit number (from 000 to 999) for drawings from Nov 10, 1976 to Sep 6, 1977.

payoff: payoff corresponding to each **number** (dollars).

lottery3	New Jersey Pick-it Lottery Data (Third Set)	**lottery3**

number: winning 3-digit number (from 000 to 999) for drawings from Dec 1, 1980 to Sep 22, 1981.

payoff: payoff corresponding to each **number** (dollars).

lynx	Canadian Lynx Trappings	**lynx**

The data give annual number of lynx trappings in the Mackenzie River District of North-West Canada for the period 1821 to 1934.

SOURCE:

Elton and Nicholson, "The ten-year cycle in numbers of lynx in Canada", *Journal of Animal Ecology*, Vol. 11, 1942, pp 215-244. Republished by permission of Blackwell Scientific Publications Ltd., Oxford, U. K., and the British Ecological Society, Reading, U. K.

Analyzed in: M. J. Campbell and A. M. Walker, "A Survey of Statistical Work on the Mackenzie River Series of Annual Canadian Lynx Trappings for the Years 1821-1934 and a New Analysis", *Journal of the Royal Statistical Society A*, Vol. 140, Part 4, 1977, pp 411-431. Republished by permission of the Royal Statistical Society, London, U. K..

month	Month Names and Abbreviations	**month**

name,abb: character name and abbreviations for months of the year.

ozone	Ozone Concentrations in North-East US	**ozone**

median,quartile: median and upper quartile of daily maxima ozone concentration for June-August, 1974. Concentrations in parts per billion (ppb).

city: character name of ozone monitoring site.

xy: structure containing components named **x** and **y**, that give the negative longitude and latitude of the monitoring sites (in the coordinate system used by function **usa**).

SOURCE:

W. S. Cleveland, B. Kleiner, J. E. McRae, J. L. Warner, and R. E. Pasceri, "The Analysis of Ground-Level Ozone Data from New Jersey, New York, Connecticut, and Massachusetts: Data Quality Assessment and Temporal and Geographical Properties", Bell Laboratories Memorandum, July 17, 1975.

Original data collected by New Jersey Department of Environmental Protection, New York State Department of Environmental Protection, Boyce Thompson Institute (Yonkers, NY data), Connecticut Department of Environmental Protection, and Massachusetts Department of Public Health.

prim	Particle Physics Data	**prim**

prim4 is a matrix of 500 experimental observations, with 4 parameters describing each. **prim9** is a similar matrix, but with 9 parameters describing each observation.

SOURCE:

Examples are believed to have been produced from work on the Prim-9 graphics system at Stanford Linear Accelerator. Probably related to the data cited in J. H. Friedman and J. W. Tukey, *IEEE Transactions on Computing*, Vol. C-23, 1974, pp 881-889. If so, these are particle-scattering experiments, and the columns of the matrices represent some chosen set of parameters (energy, momentum, etc.) to describe the nuclear reactions.

rain	New York City Precipitation	**rain**

nyc1:   Yearly total New York City precipitation, in inches, 1869-1957; currently listed in World Weather Records.

nyc2:   Yearly total New York City precipitation, in inches, 1869-1957; formerly listed in World Weather Records and found in some almanacs.

SOURCE:

U. S. Department of Commerce, Weather Bureau (**nyc1**). Formerly published by Smithsonian Institution (**nyc2**).

Random.seed	Seeds for Random Number Generators	**Random.seed**

A vector of starting values for the various random number generators used in S. The generators (for all distributions) are organized so that successive random numbers are equivalent to a long sample from the underlying uniform distribution. This allows the long-term properties of the generator to be maintained. The first time a user generates some random numbers, the dataset **Random.seed** is found on the system database. The values contained are the standard initial values of the underlying generator algorithm. After each random sample, the dataset **Random.seed** is assigned on the users working database, to maintain consistent values on subsequent calls.

There is a useful technique for reproducing random samples in later work. Just copy **Random.seed** before generating the sample for the first time, and then restore it when the sample is to be reproduced.

EXAMPLES:

> **oldseed** ← **Random.seed** #save it
> **y** ← **rnorm(1000)** # get sample, analyze, etc.
>
> ...
>
> **Random.seed** ← **oldseed** #restore seed
> **yagain** ← **rnorm(1000)** # will be the same as y

Source: Adapted from G. Marsaglia, et al. *Random Number Package: "Super-Duper"*, School of Computer Science, McGill University, 1973. (Our generator, **uni** is a portable version of their generator.)

---

saving	Savings Rates for Countries	**saving**

x: matrix with 50 rows (countries) and 5 columns (variables)
rowlab: country names corresponding to rows of **x**.
collab: names of the variables (columns) in **x**.

SOURCE:

David A. Belsley, Edwin Kuh, Roy E. Welsch, *Regression Diagnostics: Identifying Influential Data and Sources of Collinearity*, Wiley, 1980, pp 39-42. Copyright 1980 by John Wiley and Sons, Inc., New York.

Originally from unpublished data of Arlie Sterling, these are averages over 1960-1970 (to remove business cycle or other short-term fluctuations). Income is per-capita disposable income in U.S. dollars; growth is the per cent rate of change in per-capita disposable income; savings rate is aggregate personal saving divided by disposable income.

ship	Manufacturing Shipments	**ship**

Value of shipments, in millions of dollars, monthly from January, 1967 to December, 1974. This represents manufacturers' receipts, billings, or the value of products shipped, less discounts, and allowances, and excluding freight charges and excise taxes. Shipments by foreign subsidiaries are excluded, but shipments to a foreign subsidiary by a domestic firm are included.

SOURCE:

U. S. Bureau of the Census, Manufacturer's Shipments, Inventories and Orders.

stack	Stack-loss Data	**stack**

loss: percent of ammonia lost (times 10).
x: matrix with 21 rows and 3 columns representing air flow to the plant, cooling water inlet temperature, and acid concentration as a percentage (coded by subtracting 50 and then multiplying by 10).
collab: character labels for columns of **x**.

The data is from operation of a plant for the oxidation of ammonia to nitric acid, measured on 21 consecutive days.

SOURCE:

K. A. Brownlee, *Statistical Theory and Methodology in Science and Engineering*, Wiley 1965, p 454. Copyright 1965 by John Wiley & Sons, Inc., New York. Also in Draper and Smith, *Applied Regression Analysis*, Wiley, 1966, Ch 6, and Daniel and Wood, *Fitting Equations to Data*, Wiley, 1971, p 61.

state	States of the US	**state**

name,abb: character names and abbreviations for 50 US states, sorted alphabetically by name.

center: structure with components named **x** and **y** giving the approximate geographic center of each state in negative longitude and latitude (as used by function **usa**). Alaska and Hawaii are placed just off the West Coast.

region: category giving region (Northeast, South, North Central, West) that each state belongs to.

division: category giving state divisions (New England, Middle Atlantic, South Atlantic, East South Central, West South Central East North Central, West North Central, Mountain and Pacific).

x77:   matrix giving statistics for the states. Columns are Population estimate as of July 1, 1975; Per capita Income (1974); Illiteracy (1970, percent of population); Life Expectancy in years (1969-71); Murder and non-negligent manslaughter rate per 100,000 population (1976); Percent High-school Graduates (1970); Mean Number of days with min temperature < 32 degrees (1931-1960) in capital or large city; and Land Area in square miles.

col77:   character vector of column labels for x77 matrix.

SOURCE:

Statistical Abstract of the United States, 1977, and *County and City Data Book,* 1977, U.S. Department of Commerce, Bureau of the Census.

steam	Steam Usage Data	**steam**

y: pounds of steam used monthly.

x: matrix with 9 columns, giving pounds of real fatty acid in storage, pounds of crude glycerine made, average wind velocity (mph), calendar days per month, operating days per month, days below 32 degrees Fahrenheit, average atmospheric temperature, average wind velocity (squared), number of startups.

collab: character vector giving short column labels for **x**.

SOURCE:

Norman Draper and Harry Smith, *Applied Regression Analysis,* Wiley, 1966, pp 351-364. Copyright 1966 by John Wiley & Sons, Inc., New York.

| **sunspots** | Monthly Mean Relative Sunspot Numbers | **sunspots** |

Monthly means of daily relative sunspot numbers, which are based upon counts of spots and groups of spots.

SOURCE:

D. F. Andrews and A. M. Herzberg, ~~~~*Data*, 1980. Republished by permission of the authors.

The data was previously collected from different observers by the Swiss Federal Observatory, Zurich, Prof. M. Waldmeier, Director. M. Waldmeier, *The Sunspot Activity in the Years 1610-1960*, Schulthess, Zurich, 1961.

The data is currently collected by the Tokyo Astronomical Observatory, Tokyo, Japan.

| **swiss** | Fertility Data for Switzerland in 1888 | **swiss** |

fertility: standardized fertility measure I[g] for each of 47 French-speaking provinces of Switzerland at about 1888.

x: matrix whose columns give 5 socioeconomic indicators for the provinces: 1) percent of population involved in agriculture as an occupation. 2) percent of "draftees" receiving highest mark on army examination. 3) percent of population whose education is beyond primary school. 4) percent of population who are Catholic. 5) percent of live births who live less than 1 year: infant mortality.

collab: short column labels for **x**.

SOURCE:

Mosteller and Tukey, *Data Analysis and Regression*, Addison Wesley, 1977, pp 549-551.

Unpublished data used by permission of Francine van de Walle, Population Study Center, University of Pennsylvania, Philadelphia, PA.

switzerland	Heights of Switzerland on 12 by 12 Grid	**switzerland**

Height in 1000's of feet to nearest 1000 feet. Accuracy of the data is questionable.

SOURCE:

Beat Kleiner, personal communication.

telsam	Interviewer Response Data	**telsam**

response: table giving counts of number of answers poor, fair, good, and excellent (the columns) for a number of different interviewers (the rows).

collab: character label for columns of **response**.

rowlab: interviewer identification number.

tone	Bricker's Tone-Ringer Preference Data	**tone**

appeal: matrix giving standardized ratings of 100 tones (rows) by 43 subjects (columns). Standardized so that each subject has mean 0 and biased variance 1.

rowlab: character identifiers for tones.

SOURCE:

P. J. Bricker, "Listener Evaluation of Simulated Telephone Calling Signals", *Bell System Technical Journal*, Vol. 50, No. 5, 1971, pp 1559-1578.

util	Earnings and Market/Book Ratio for Utilities	**util**

earn: earnings of 45 utilities.
mktbook: market price to book value ratio for the utilities.

votes	Votes for Republican Candidate in Presidential Elections	**votes**

repub: percent of votes given to republican candidate in presidential elections from 1856 to 1976. Rows represent the 50 states (See dataset **state.**), and columns the 31 elections.

year: year of each election (corresponds to columns of **x**).

Contains missing values (NAs) for years prior to statehood.

SOURCE:

S. Peterson, *A Statistical History of the American Presidential Elections,* Frederick Ungar Publishing Co., New York, 1973. Republished by permission.

Data from 1964 to 1976 is from R. M. Scammon, *American Votes 12,* Congressional Quarterly. Republished by permission of Richard M. Scammon, Editor, Elections Research Center.

Appendix 3
Index to S Functions

This appendix is meant to aid you in finding S functions that perform specific tasks. Here, S functions, macros, and utilities are organized into groups by the kind of operations that they perform.

The first section gives a brief description of the groups and what is included in them. The second section lists functions, macros, and utilities that apply to each group.

Groups of S Functions

Built-in Operators/Syntax

Operators; grammar; subscripting.

Categorical Data   Create, tabulate, and analyze discrete data.

Data Attributes   Information about datasets aside from their data values, e.g., length, mode, components.

Data Directories   Create datasets; manipulate data directories and the directory search list.

Data Manipulation   Manipulate vectors by sorting, reversing, replication, combination, selection, mode changes.

Data Structures   Create   and   modify   data   structures, including hierarchical structures.

Documentation
Keep track of analysis; find documentation, and aid in documenting datasets and macros.

Error Handling
Recover from or correct errors in expressions or macros.

Graphics — Add to Existing Plot
Modify an existing plot; display or modify graphical parameters.

Graphics — Computations Related to Plotting
Create data structures that can be used to augment or build up plots. Also includes graphical functions that return special data structures.

Graphics — Interacting with Plots
Utilize the interactive capabilities of graphical devices.

Graphics — High-Level Plots
Create a complete plot, including appropriate coordinate systems and axes.

Graphics — Specific Devices
Hardware (terminals and plotters) for graphics; also, functions specific to certain devices.

Hierarchical Clustering
Compute and display clustering; operate on cluster trees and distance structures.

Input/Output — Files
Read or write UNIX system files.

Linear Algebra
Includes numerical analysis and optimization.

Logical Operations
Operate on logical (TRUE/FALSE) data.

Looping and Iteration
Perform iteration, e.g., **apply**.

Macros
The S macro facility and associated functions.

Mathematical Operations
Carry out basic mathematical operations on vectors; trigonometric functions, square root, sum, etc.

Matrices
Create, modify, or analyze matrix data; see also **Linear Algebra**.

Multi-Dimensional Arrays
Operate on higher-order arrays (3-way, 4-way, etc). See also **Matrices**.

Multivariate Statistical Techniques
Statistical techniques that operate on multivariate data, not including **hierarchical clustering** or **regression** which are listed separately.

Probability Distributions and Random Numbers
Compute probabilities (begin with "p"), quantiles (begin with "q"), and generate random numbers (begin with "r") for many distributions. Also includes probability plots and sampling.

Regression
Linear regression, including robust regression and non-linear smoothing.

Reports
Print or display data; includes presentation graphics.

Robust/Resistant Techniques
Operations which provide resistance to outliers.

Session Control
Keep track of an S session, set options, interface with the UNIX system.

Statistical Operations
Univariate statistical functions, descriptive statistics: mean, var, min, max, etc. See also **Probability Distributinos and Random Numbers**.

Time-Series
Operate on time-series: seasonal adjustment, smoothing, etc.

Utilities
Operate on external data files; execute S non-interactively; Computations which are carried out by programs running under the UNIX operating system rather than by S functions.

Index to S Functions by Group

In the remainder of this appendix, the left column gives the name of a function, macro, or utility; the right hand side gives a brief description. For details on a particular routines, see Appendix 1 under the routine name. (Note: macro names begin with a question mark "?". Utility routines generally have names that are all upper-case, for example **HELP**, **REPORT**, and are invoked from the UNIX shell as

    S HELP

etc.)

Built-in Operators/Syntax

+ – / * ^ ** %% %/	
	Arithmetic Operators
< <= > >= == !=	
	Comparison Operators
← → _ **assign**	Assignment
& | !	Logical Operations; And, Or, Not
$ **select**	Component Selection
[ ]	Subsets of Data
: seq	Sequences
macros	The S Macro Facility
precedence	Order of Expression Evaluation
print	Print Data
Syntax	S Expressions

Categorical Data

?bartable	Produce Bar Plot of a Category
code	Create Category from Discrete Data
cut	Create Category by Cutting Continuous Data
index	Compute Position in Array
loglin	Contingency Table Analysis
split	Split Data by Groups
table	Create Contingency Table from Categories
tapply	Apply a Function to a Ragged Array
tbl	Print a Table Using the UNIX tbl Command
tprint	Print Multi-way Tables

Data Attributes

Data Directories

Data Manipulation

c	Combine Values
cbind rbind	Form Matrix from Columns or Rows
coerce %m	Change Mode
cstr	Create Structure from Components
edit again	Edit Dumped Expressions or Character Vectors
encode	Encode Text and Numeric Data
ifelse	Conditional Data Selection
?intersect	Intersection of Two Lists
?mat2vecs	Change Columns of Matrix to Individual Vectors
?matedit	Text Editing of Matrices
mstr	Modify Structure
rep	Replicate Data Values
rev	Reverse the Order of Elements in a Vector
split	Split Data by Groups
?union	Union of Several Lists
unique uniq	Unique Values in a Vector
?vecs2mat	Make Matrix from a Set of Vectors

Data Structures

$ select	Component Selection
array	Create an Array
compname	Component Names
cstr	Create Structure from Components
matrix	Create Matrix
mstr	Modify Structure
ncomp	Number of Components
sapply	Apply a Function to Components of a Structure
split	Split Data by Groups
ts	Create Time-Series
tsmatrix	Create Matrix with Time-Series as Columns

Documentation

?comment	Add a Comment to a Dataset
diary	Keep Diary of S Commands
help call	On-line Documentation
HELP	Print S Function Documentation
?listfun	List Names of all S Functions
mprint	Print Macros
mprompt	Produce Shell of Documentation for Macros

Graphics - Computations Related to Plotting

approx	Approximate Function from Discrete Values
?barycentric	Compute Barycentric Coordinates for Mixtures
boxplot	Box Plots
chull	Convex Hull of a Planar Set of Points
defer	Control Deferred Graphics
density	Estimate Probability Density Function
hist	Plot a Histogram
identify	Identify Points on Plot
interp	Bivariate Interpolation for Irregular Data
lowess	Scatter Plot Smoothing
mstree	Minimal Spanning Tree and Multivariate Planing
par	Graphical Parameters
pardump	Dump of Graphical Parameters
plclust	Plot Trees From Hierarchical Clustering
pretty	Return Vector of Prettied Values
qqplot qqnorm	Quantile-Quantile Plots
quickvu	Make Slides with Simple Lists
range	Range of Data (minimum, maximum)
smooth	Non-linear Smoothing Using Running Medians

Graphics - High-Level Plots

barplot	Bar Graph
?bartable	Produce Bar Plot of a Category
boxplot	Box Plots
bxp	Boxplots From Processed Data
contour	Contour Plotting
?Cp	All Subsets Regression and Cp Plot
faces	Plot Symbolic Faces
hist	Plot a Histogram
?idplot	Scatter Plot with Identifiers for Each Point
?mdsplot	Plot the Results of Multidimensional Scaling
?mixplot	Barycentric Plot of Mixture
monthplot	Seasonal Subseries Plot
mulbar	Multiple Bar Plot
?pairs	All Pair-wise Scatter Plots
pairs	Plot Pair-wise Scatters of Multivariate Data
par	Graphical Parameters
persp	3-Dimensional Perspective Plots

Graphics - Interacting with Plots

Graphics - Specific Devices

pic	Produce Graphics on Phototypesetter via pic
printer	Line Printer Graphics
ram6211	Ramtek 6211 Color Graphics Terminal
show	Show Current Printer Plot
tek10 tek12 tek14 tek14q	
	Tektronix Storage Scope Devices
tek4112	Tektronix 4112 Raster Scope
tek46	Tektronix 4662 Pen Plotter
unixplot	Produce Graphics on UNIX Devices

Hierarchical Clustering

clorder	Re-Order Leaves of a Cluster Tree
cutree	Create Groups from Hierarchical Clustering
dist	Distance Matrix Calculation
?dist2full	Distance Structure to Full Symmetric Matrix
?full2dist	Full Symmetric Matrix to Distance Structure
hclust	Hierarchical Clustering
labclust	Label a Cluster Plot
plclust	Plot Trees From Hierarchical Clustering
subtree	Extract Part of a Cluster Tree

Input/Output - Files

dget	Retrieve a General Structure from File
dput	Save a Dataset on a File
dump	Dump Datasets to a File
?extract	Extract Columns of Data from File
extract	Extract Columns or Fields as S Datasets
read	Read Data from a File
restore	Restore Datasets from Dump File
scandata	Scan fixed format data file
sink	Send S Output to the Diary or to a File
source	Execute S Expressions from a File
write	Write Data to File

Linear Algebra

%*	Matrix Multiplication Operator
apply	Apply a Function to Sections of an Array
backsolve	Backsolve Upper- or Lower-triangular Equations

chol	Triangular Decomposition of Symmetric Matrix
crossprod %c	Matrix Cross Product Operator
diag	Diagonal Matrices
eigen	Eigen Analysis of Symmetric Matrix
?fmin	Macro to minimize an S expression
gs	Gram-Schmidt Decomposition
?kronecker	Form Kronecker Product of Matrices
napsack	Solve Knapsack Problems
outer %o	Generalized Outer Products
prcomp	Principal Component Analysis
scale	Scale Columns of a Matrix
solve	Solve Linear Equations and Invert Matrices
svd	Singular Value Decomposition
t	Transpose a Matrix

Logical Operations

< <= > >= == !=	
	Comparison Operators
& | !	Logical Operations; And, Or, Not
all any	Logical Sum and Product
ifelse	Conditional Data Selection
?missing	Find Observations Containing NAs
na NA	Pick Out Missing Values
?which	Find Which Observations Satisfy a Condition
xor	Exclusive Or

Looping and Iteration

apply	Apply a Function to Sections of an Array
?apply	Apply a Macro to a Number of Arguments
?col	Apply a Function to the Columns of a Matrix
?csweep	Sweep Out Column Effects from Matrix
?row	Apply a Function to the Rows of a Matrix
?rsweep	Sweep Out Row Effects from Matrix
sapply	Apply a Function to Components of a Structure
sweep	Sweep Out Array Summaries
tapply	Apply a Function to a Ragged Array

Macros

?apply	Apply a Macro to a Number of Arguments
?cleanup	Selectively Remove Datasets
define	Define a Macro
?define	Define a Macro at Execution Time
diary	Keep Diary of S Commands
macros	The S Macro Facility
?medit	Edit a Macro
medit	Edit a Macro
menu	Menu Interaction Function
?mlist	List Names of Macros
mprint	Print Macros
mprompt	Produce Shell of Documentation for Macros
?prompt	Create Documentation for Macros or Datasets
source	Execute S Expressions from a File

Mathematical Operations

+ − / * ^ ** %% %/	
	Arithmetic Operators
< <= > >= == !=	
	Comparison Operators
abs	Absolute Value
acos asin	Inverse Trigonometric Functions
approx	Approximate Function from Discrete Values
atan	Inverse Tangent
ceiling floor trunc	Integer Values
cos sin	Trigonometric Functions
cumsum	Cumulative Sums
diff	Create a Differenced Series
exp log log10 sqrt	Math Functions
gamma lgamma	Gamma Function (and its Natural Logarithm)
max min	Extremes
mean	Mean Value
median	Median
order	Ordering to Create Sorted Data
pmax pmin	Parallel Maximum or Minimum
prod sum	Products and Sums
range	Range of Data (minimum, maximum)
rank	Ranks of Data

reg	Regression
regress regprt	Regression Printing
regsum	Regression Summaries
?row	Apply a Function to the Rows of a Matrix
rreg	Robust Regression
?rsweep	Sweep Out Row Effects from Matrix
scale	Scale Columns of a Matrix
smatrix	Print a Symbolic Matrix for Multivariate Data
solve	Solve Linear Equations and Invert Matrices
subtree	Extract Part of a Cluster Tree
svd	Singular Value Decomposition
t	Transpose a Matrix
tsmatrix	Create Matrix with Time-Series as Columns
twoway	Analysis of Two-way Table
?vecs2mat	Make Matrix from a Set of Vectors

Multi-Dimensional Arrays

aperm	Array Permutations
apply	Apply a Function to Sections of an Array
array	Create an Array
index	Compute Position in Array
?kronecker	Form Kronecker Product of Matrices
sweep	Sweep Out Array Summaries
tapply	Apply a Function to a Ragged Array

Multivariate Statistical Techniques

cancor	Canonical Correlation Analysis
cmdscale	Classical Metric Multi-dimensional Scaling
contour	Contour Plotting
cutree	Create Groups from Hierarchical Clustering
discr	Discriminant Analysis
?discr	Discriminant Analysis with Grouping Vector
dist	Distance Matrix Calculation
?dist2full	Distance Structure to Full Symmetric Matrix
faces	Plot Symbolic Faces
?full2dist	Full Symmetric Matrix to Distance Structure
hclust	Hierarchical Clustering
labclust	Label a Cluster Plot
loglin	Contingency Table Analysis

Probability Distributions and Random Numbers

?sample	Random Sample

Regression

?Cp	All Subsets Regression and Cp Plot
hat	Hat Matrix Regression Diagnostic
l1fit	Minimum Absolute Residual (L1) regression
leaps	All-subset Regressions by Leaps and Bounds
lowess	Scatter Plot Smoothing
?missing	Find Observations Containing NAs
rbiwt	Robust Simple Regression by Biweight
reg	Regression
regress regprt	Regression Printing
regsum	Regression Summaries
rreg	Robust Regression
smooth	Non-linear Smoothing Using Running Medians

Reports

barplot	Bar Graph
?bartable	Produce Bar Plot of a Category
legend	Put a Legend on a Plot
pie	Pie Charts
print	Print Data
quickvu	Make Slides with Simple Lists
REPORT	Report writing facility
table	Create Contingency Table from Categories
tbl	Print a Table Using the UNIX tbl Command
tprint	Print Multi-way Tables
vu	Create Vu-graphs (Slides)

Robust/Resistant Techniques

lowess	Scatter Plot Smoothing
mean	Mean Value
median	Median
?ranktest	Two-sample Rank Test
rbiwt	Robust Simple Regression by Biweight
rreg	Robust Regression
sabl	Seasonal Decomposition
smooth	Non-linear Smoothing Using Running Medians

twoway	Analysis of Two-way Table

Session Control

defer	Control Deferred Graphics
diary	Keep Diary of S Commands
help call	On-line Documentation
mail	Suggestions, Questions and Reports
menu	Menu Interaction Function
options option	Set or Print Options
pardump	Dump of Graphical Parameters
q	Terminating Execution
stamp	Time Stamp Output, Graph, and Diary
sys	Execute System Commands

Statistical Operations

density	Estimate Probability Density Function
hist	Plot a Histogram
max min	Extremes
mean	Mean Value
median	Median
prod sum	Products and Sums
range	Range of Data (minimum, maximum)
?ranktest	Two-sample Rank Test
stem	Stem and Leaf Display
var cor	Variance, Covariance, and Correlation

Time-Series

diff	Create a Differenced Series
end nper start	Time-Series Parameters
lag	Create a Lagged Time-Series
monthplot	Seasonal Subseries Plot
sabl	Seasonal Decomposition
?sablplot	Sabl Decomposition - Data and Components Plot
smooth	Non-linear Smoothing Using Running Medians
time cycle	Create Time Vector
ts	Create Time-Series
tsmatrix	Create Matrix with Time-Series as Columns
tsplot tslines tspoints	

	Plot Multiple Time-Series
window	Window a Time-Series

Utilities

BATCH	Batch (non-interactive) execution of S
?extract	Extract Columns of Data from File
extract	Extract Columns or Fields as S Datasets
HELP	Print S Function Documentation
REPORT	Report writing facility
scandata	Scan fixed format data file

Index